SOLID MECHANICS

SOLID MECHANICS AND ITS APPLICATIONS
Volume 15

Series Editor:

G.M.L. GLADWELL
Solid Mechanics Division, Faculty of Engineering
University of Waterloo
Waterloo, Ontario, Canada N2L 3G1

Aims and Scope of the Series

The fundamental questions arising in mechanics are: *Why?*, *How?*, and *How much?* The aim of this series is to provide lucid accounts written by authoritative researchers giving vision and insight in answering these questions on the subject of mechanics as it relates to solids.

The scope of the series covers the entire spectrum of solid mechanics. Thus it includes the foundation of mechanics; variational formulations; computational mechanics; statics, kinematics and dynamics of rigid and elastic bodies; vibrations of solids and structures; dynamical systems and chaos; the theories of elasticity, plasticity and viscoelasticity; composite materials; rods, beams, shells and membranes; structural control and stability; soils, rocks and geomechanics; fracture; tribology; experimental mechanics; biomechanics and machine design.

The median level of presentation is the first year graduate student. Some texts are monographs defining the current state of the field; others are accessible to final year undergraduates; but essentially the emphasis is on readability and clarity.

For a list of related mechanics titles, see final pages.

Solid Mechanics

An Introduction

by

J. P. WARD
Loughborough University of Technology, U.K.

Kluwer Academic Publishers
Dordrecht / Boston / London

ISBN 0-7923-1949-4

Published by Kluwer Academic Publishers,
P.O. Box 17, 3300 AA Dordrecht, The Netherlands.

Kluwer Academic Publishers incorporates
the publishing programmes of
D. Reidel, Martinus Nijhoff, Dr W. Junk and MTP Press.

Sold and distributed in the U.S.A. and Canada
by Kluwer Academic Publishers,
101 Philip Drive, Norwell, MA 02061, U.S.A.

In all other countries, sold and distributed
by Kluwer Academic Publishers Group,
P.O. Box 322, 3300 AH Dordrecht, The Netherlands.

Printed on acid-free paper

Printed in the Netherlands

To the memory of my stepfather and friend Ciprijans Oleks

Preface

This book is intended as an introductory text on Solid Mechanics suitable for engineers, scientists and applied mathematicians. Solid mechanics is treated as a subset of mathematical engineering and courses on this topic which include theoretical, numerical and experimental aspects (as this text does) can be amongst the most interesting and accessible that an undergraduate science student can take. I have concentrated entirely on linear elasticity being, to the beginner, the most amenable and accessible aspect of solid mechanics.

It is a subject with a long history, though its development in relatively recent times can be traced back to Hooke (circa 1670). Partly because of its long history solid mechanics has an 'old fashioned' feel to it which is reflected in numerous texts written on the subject. This is particularly so in the classic text by Love (*A Treatise on the Mathematical Theory of Elasticity* 4th ed., Cambridge, Univ. Press, 1927). Although there is a wealth of information in that text it is not in a form which is easily accessible to the average lecturer let alone the average engineering student. This classic style avoiding the use of vectors or tensors has been mirrored in many other more 'modern' texts.

This present book is a relatively short text covering the foundations of stress, strain and elasticity theory and consistently employs the use of vectors and (particularly) Cartesian tensor notation. Vectors and Cartesian tensors are introduced and discussed in detail before being applied. Of course there is an initial difficulty to be faced with the use of Cartesian tensors – this would be true with the introduction of almost any mathematical topic – but I feel that the advantages of using the Cartesian tensor notation, in a consistent manner, by far outweighs any initial discomfort. For the mathematical engineer a knowledge of Cartesian tensors is a positive advantage to their understanding of other engineering disciplines.

As far as possible I have attempted to keep the mathematical content to the minimum required for a proper understanding of the development of the basic equations of linear elasticity and for an appreciation of some of the basic results which are deduced. At all stages numerous worked examples are included to ease understanding.

The layout of the book is straightforward. The first chapter is a (substantial) refresher course on vectors with particular emphasis being paid to applications which arise in later chapters. In this chapter I cover the index notation for vectors which helps to ease the transition from vectors to tensors. Chapter 2 introduces Cartesian tensors and describes some of their important applications. In particular, finite and infinitessimal rotations are examined as are isotropic tensors and second order symmetric tensors. The last topic of this chapter includes a full discussion on eigenvalues and eigenvectors.

There are separate introductions, in Chapters 3 and 4, to stress and strain and to their practical measurement using, respectively, photoelastic methods and strain gauges.

In Chapter 5 the concepts of stress and strain are brought together and, in conjunction with Newton's equilibrium equations, used to deduce the basic equations of linear elasticity theory. These fundamental equations are then examined and analysed by obtaining simple exact solutions, including solutions which describe twisting, bending and stretching of beams.

Chapter 6 introduces the fundamental concept of strain energy and uses this concept to derive the Kirchoff uniqueness theorem, Rayleigh's reciprocal theorem and the important Castigliano relations. The chapter concludes with a thorough treatment of the theorem of minimum potential energy and examines some of its applications.

The final three chapters examine the application of the fundamental equations to the theory of torsion, to structural analysis and to the treatment of two dimensional elastostatics by analytical and approximate (finite element) methods.

My aim has been to produce a text which is reasonably concise and readable. It is intended to be largely self-contained and a genuine introduction – not a research text, which does not cover too much material. Only time will tell whether or not my aim has been accurate.

J.P. Ward
Loughborough, July 1992

CONTENTS

6 Energy

7 The General Torsion Problem

8 The Matrix Analysis of Structures

9 Two Dimensional Elastostatics

Appendix 1 The Variational Calculus

Chapter 1

VECTORS

1.1 Introduction to Vector Algebra

This chapter is intended as a relatively brief 'refresher' course on vectors. Few of the results will be derived rigorously. If a more detailed description is required the reader is referred to the authors earlier text 'Vector Analysis for Engineers and Scientists' (Addison-Wesley 1989).

The use of vectors and of tensors (described in Chapter 2) in solid mechanics is undoubted. The very 'building blocks' of the subject – stresses and strains – are tensors whilst the more fundamental quantatities – forces and displacements – are vectors. (We shall see later that vectors are examples of so-called first order tensors).

A number of 'older' texts and surprisingly many so-called 'modern' texts on solid mechanics avoid the use of vectors and tensors. Whilst it is certainly possible to discuss most scientific subjects without the use of either vectors or tensors, in the case of the subject of this text, it is inadvisable. Only through the use of tensors can one fully grasp the basic underlying principles of elasticity theory. As well as being fundamental for the mathematical description of 'objects' which arise naturally in the theory their use has the added bonus of providing a great notational advantage over the classical scalar approach.

However, we shall not use 'general' tensors in our treatment of elasticity; rather we shall only make use of Cartesian tensors which are far more suitable in an introductory text of this kind.

We begin the description of vectors using standard notation, which is familiar to most students. After the basic theory has been described and some of the fundamental results of vector analysis have been obtained, we introduce the reader to the index notation for vectors. The reader is encouraged to use the classical notation and the index notation interchangeably. An understanding of, and a facility with, the index notation will ease the transition from vectors to tensors considerably.

A vector is defined as a quantity which is characterised by both a magnitude and a direction and is such that two vectors combine ('add-up') according to a rule known as the parallelogram law. The archetypal example of a vector is the directed line segment. A directed line segment (or displacement) from a point A to a point B (see Figure 1.1a) is represented by a straight line segment with an arrow to indicate the direction. We shall refer to it as $\overrightarrow{AB}$.

It is an obvious conclusion that if we move from point A to point B along the directed line segment $\overrightarrow{AB}$ and then from point B to point C along the directed line segment $\overrightarrow{BC}$ then the same displacement is achieved if we move directly from point A to point C along the directed line segment $\overrightarrow{AC}$ (see Figure 1.1b).

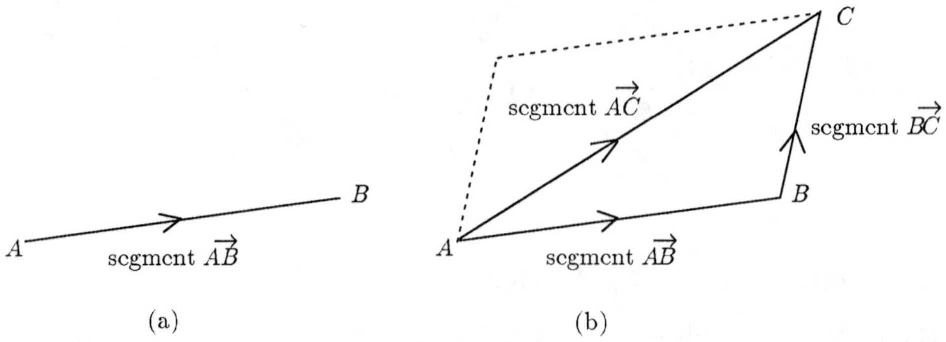

Figure 1.1

The 'resultant' displacement $\overrightarrow{AC}$ is the diagonal of the parallelogram with sides parallel to $\overrightarrow{AB}$ and $\overrightarrow{BC}$ and this way of producing a new directed line segment $\overrightarrow{AC}$ from given segments $\overrightarrow{AB}$ and $\overrightarrow{BC}$ is known as the parallelogram law.

The directed line segment is thus a **vector** as it obeys the parallelogram law of combination and is characterised by both a magnitude (the length of the line) and (obviously) a direction in space.

Notationally we shall often use a single underlined letter, $\underline{a}$ say, as an alternative to the artifice of using an arrow above two letters to denote a vector.

- **The Magnitude of a Vector**

We remember that if s is a scalar (either positive or negative) then the magnitude (or modulus) of s is written $|s|$ and is defined by

$$|s| = \begin{cases} s & \text{if} \quad s > 0 \\ -s & \text{if} \quad s < 0 \end{cases} \tag{1.1}$$

We use a similar notation to denote the magnitude of a vector. If $\overrightarrow{AB}$ (or $\underline{a}$) is a vector then its magnitude is denoted by $|\overrightarrow{AB}|$ (or $|\underline{a}|$) and is often called the **modulus** of the vector. As far as the directed line segment is concerned the modulus of this vector is simply its length. The modulus of a vector is a scalar (a number)

A **null** or **zero** vector is any vector with zero modulus. (The 'direction' in this case is a meaningless concept).

- **Equality of vectors**

We will say that two vectors $\underline{a}$ and $\underline{b}$ are **equal** if and only if both vectors have the same magnitude and both are pointing in the same direction: when this is the case we shall write

$$\underline{a} = \underline{b}$$

Note that we do not demand that both vectors are in exactly the same position in space. Hence all of the vectors shown in Figure 1.2 are equal

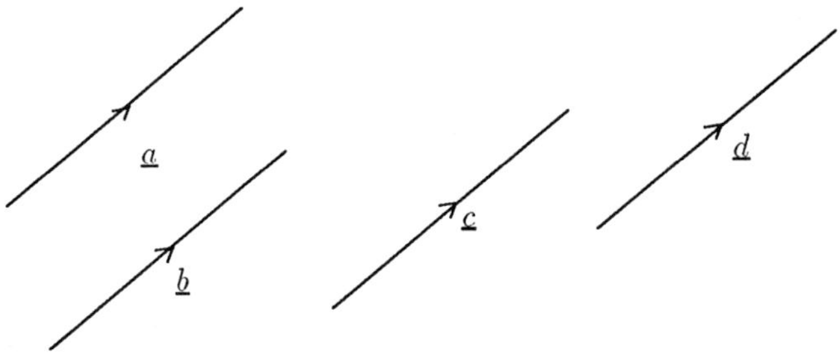

Figure 1.2

This definition is strictly only valid for what is called a 'free vector'. We shall soon find out that not all vectors – satisfying the three basic characteristics – are of the same kind. As well as free vectors we shall also encounter bound vectors (such as forces and vector fields), sliding vectors (like torques) and position vectors.

- **Addition of Vectors**

The sum of two vectors $\underline{a}$ and $\underline{b}$ will be written $\underline{a}+\underline{b}$ and is equal to a vector $\underline{c}$:

$$\underline{a}+\underline{b}=\underline{c}$$

such that $\underline{c}$ is the diagonal of a parallelogram with sides represented by $\underline{a}$ and $\underline{b}$ (Figure 1.3a).

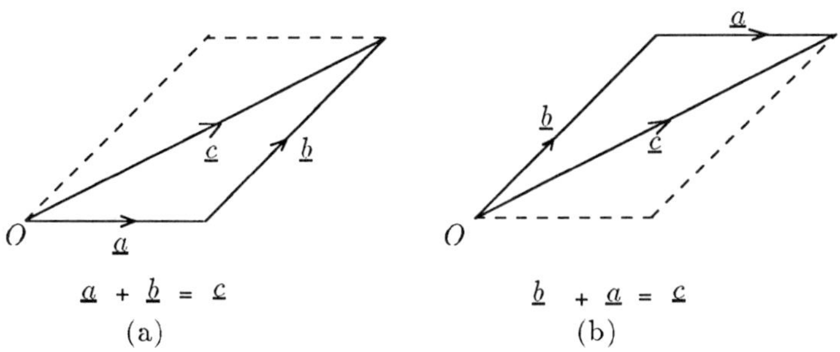

Figure 1.3

However, for free vectors we can slide $\underline{b}$ (keeping the direction fixed) so that it originates at O and similarly slide $\underline{a}$ so that it extends out from the end point of $\underline{b}$. The result is clearly the same (Figure 1.3b) i.e.

$$\underline{a}+\underline{b}=\underline{c}=\underline{b}+\underline{a} \tag{1.2}$$

Vector addition is thus commutative (order is not important).

A simple extension of this construction will further show that if we have three vectors $\underline{a}$, $\underline{b}$, $\underline{c}$ then

$$\underline{a}+(\underline{b}+\underline{c})=(\underline{a}+\underline{b})+\underline{c}$$

(The reader is invited to obtain this result). Therefore vector addition is associative (grouping is not important) and we can write without ambiguity,

$$\underline{a}+\underline{b}+\underline{c}$$

If we reverse the direction of $\underline{c}$ in Figure 1.3 so that all three vectors $\underline{a}$, $\underline{b}$ and $\underline{c}$ point the same way round the edges of the triangle then, using the parallelogram rule $\underline{a}+\underline{b}+\underline{c}=0$. This simple result can be generalised:

If $\underline{a}$, $\underline{b}, \ldots, \underline{f}$ are vectors representing the sides of a closed polygon (all vectors pointing the same way around the polygon's edge) then it is a simple deduction (by segmenting the polygon into triangles) that

$$\underline{a}+\underline{b}+\ldots+\underline{f}=0 \tag{1.3}$$

- **Multiplication by a Scalar**

If $\underline{a}$ is a vector and $s>0$ is a scalar (a positive number) then we define the product of s with $\underline{a}$, written $s\underline{a}$, to be the vector in the same direction as $\underline{a}$ but with magnitude $s|\underline{a}|$. If $s<0$ then we define the product $s\underline{a}$ to be a vector in the opposite direction to $\underline{a}$ and of magnitude $|s|\ |\underline{a}|$ (See Figure 1.4)

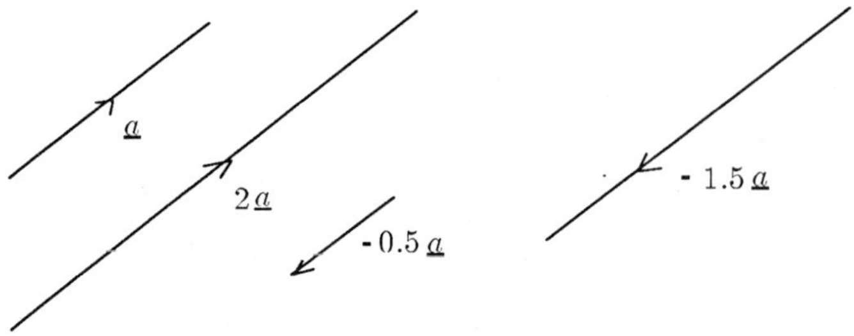

Figure 1.4

- **Unit Vectors**

A unit vector will **always** be denoted by a letter with a circumflex, e.g. $\hat{a}$, and is such that it has a magnitude of 1 i.e.

$$|\hat{a}|=1$$

If a vector $\underline{a}$ and a unit vector $\hat{a}$ are in the same direction then clearly we may write

$$\underline{a}=|\underline{a}|\hat{a} \tag{1.4}$$

since the vectors $\underline{a}$ and $|\underline{a}|\hat{a}$ are in the same direction and have the same magnitude. Alternatively if we wish to find a unit vector in the direction of $\underline{a}$ we simply write

$$\hat{a}=\frac{\underline{a}}{|\underline{a}|} \quad (=\frac{1}{|\underline{a}|}\underline{a} \qquad \text{i.e. multiplication by a scalar}) \tag{1.5}$$

- **Cartesian Coordinates and the Unit Vectors $\hat{\imath}$, $\hat{\jmath}$, $\hat{k}$**

Before delving further into vectors we shall remind the reader of the Cartesian description of three dimensional space. We first choose the position of an origin O.

From O we draw two perpendicular lines (or axes) the x- and y-axes. The origin splits the axes into positive and negative parts; the positive side being indicated by an arrow (Figure 1.5a)

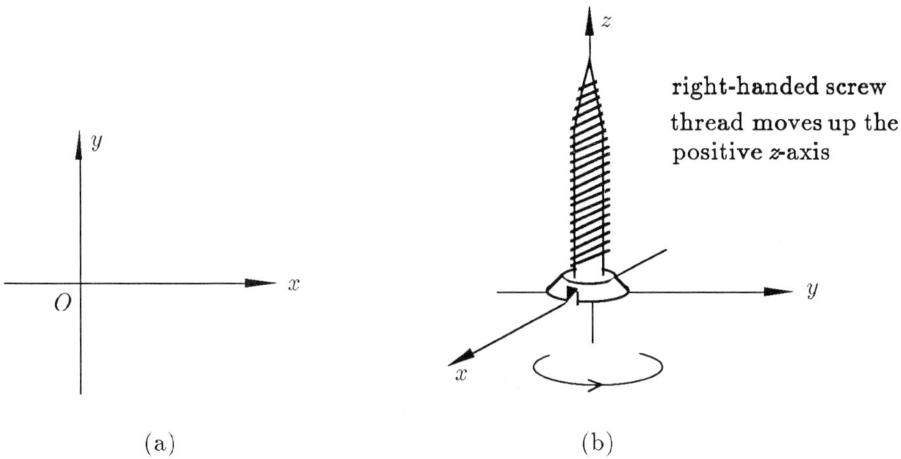

Figure 1.5

We choose a third axis – the z-axis – through O perpendicular to Ox and Oy. The positive direction Oz is chosen in such a way that if a screw with a right-handed thread (most screws are of this type!) was placed at O with its threaded end pointing along Oz then on rotating the screw from Ox to Oy through 90° the screw would move in the positive direction (Figure 1.5b).

This choice of coordinate axes allows any point P in space to be uniquely defined by three numbers x_p, y_p, z_p (often called coordinates). See Figure 1.6.

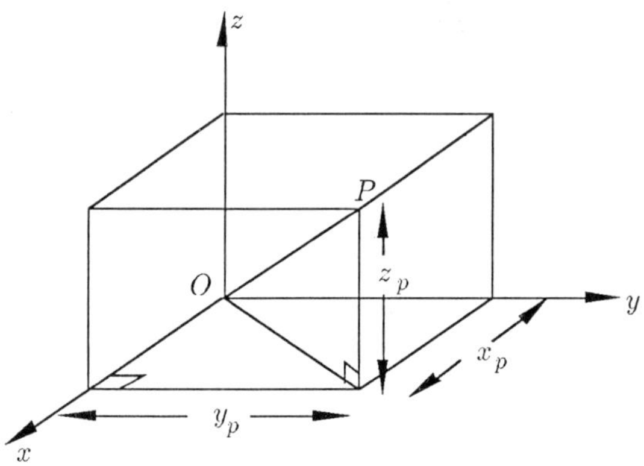

Figure 1.6

We now introduce three mutually perpendicular unit vectors $\hat{\imath}$, $\hat{\jmath}$, $\hat{k}$ pointing along the positive Ox, Oy and Oz axes respectively (Figure 1.7).

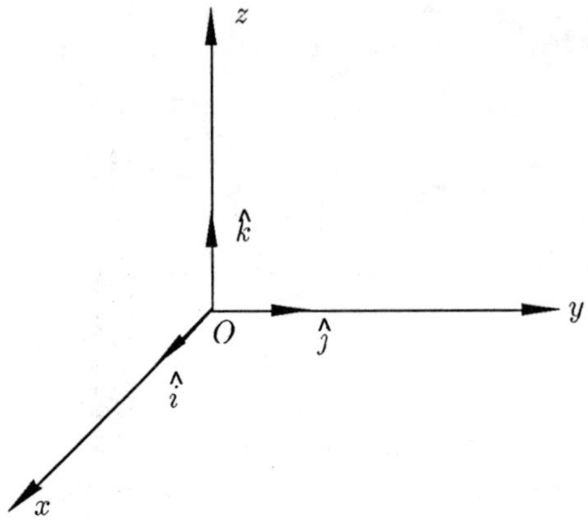

Figure 1.7

These vectors are called **basis vectors** since, as we shall now show, any arbitrary vector $\underline{a}$ may be expressed uniquely in terms of them.

Consider a vector $\underline{a}$. We can always slide the vector $\underline{a}$ so that it points directly out of the origin (Figure 1.8)

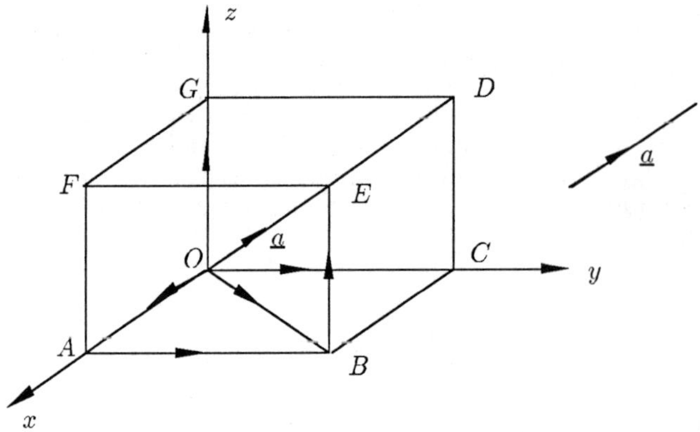

Figure 1.8

Using the properties of vector addition

$$\overrightarrow{OB} + \overrightarrow{BE} = \underline{a}$$

But

$$\overrightarrow{OB} = \overrightarrow{OA} + \overrightarrow{AB} \qquad \text{and} \qquad \overrightarrow{BE} = \overrightarrow{OG} \qquad \text{while} \qquad \overrightarrow{AB} = \overrightarrow{OC}$$

Therefore, for any vector

$$\underline{a} = \overrightarrow{OA} + \overrightarrow{OC} + \overrightarrow{OG}$$

However, using the unit vectors $\hat{\imath}$, $\hat{\jmath}$, $\hat{k}$ we have

$$\overrightarrow{OA} = |\overrightarrow{OA}|\ \hat{\imath}$$

since $\overrightarrow{OA}$ is in the same direction as $\hat{\imath}$ (along the x-axis) and has magnitude $|\overrightarrow{OA}|$. Similarly

$$\overrightarrow{OC} = |\overrightarrow{OC}|\ \hat{\jmath} \quad \text{and} \quad \overrightarrow{OG} = |\overrightarrow{OG}|\ \hat{k}$$

Hence *any* vector $\underline{a}$ may be represented in the form

$$\underline{a} = |\overrightarrow{OA}|\ \hat{\imath} + |\overrightarrow{OC}|\ \hat{\jmath} + |\overrightarrow{OG}|\ \hat{k} \tag{1.6}$$

Remembering, from our earlier discussion, that the numbers $(|\overrightarrow{OA}|,\ |\overrightarrow{OC}|,\ |\overrightarrow{OG}|)$ are the coordinates of the end point of the vector $\underline{a}$ (with its other end point at the origin) and allows us an alternative expression for the vector namely:

$$\underline{a} = (|\overrightarrow{OA}|,\ |\overrightarrow{OC}|,\ |\overrightarrow{OG}|) \tag{1.7}$$

We must realise that the right side of (1.7) is a short-hand notation for the expansion in (1.6). The expression in (1.6) is often called the component form of $\underline{a}$ whilst (1.7) is the **coordinate** form.

We should also note, again from Figure 1.8 that

$$|\underline{a}| = \sqrt{|\overrightarrow{OA}|^2 + |\overrightarrow{OC}|^2 + |\overrightarrow{OG}|^2} \tag{1.8}$$

In summary: any vector $\underline{a}$ may be written in the Cartesian form

$$\underline{a} = a_x\hat{\imath} + a_y\hat{\jmath} + a_z\hat{k} \qquad (a_x = |\overrightarrow{OA}|,\quad a_y = |\overrightarrow{OC}|,\quad a_z = |\overrightarrow{OG}|) \tag{1.9}$$

where $a_x,\ a_y,\ a_z$ are called the **components** of $\underline{a}$ and, with this representation,

$$|\underline{a}| = \sqrt{a_x^2 + a_y^2 + a_z^2} \tag{1.10}$$

A unit vector in the direction of $\underline{a}$ is

$$\hat{a} = \frac{\underline{a}}{|\underline{a}|} = \frac{a_x\hat{\imath} + a_y\hat{\jmath} + a_z\hat{k}}{\sqrt{a_x^2 + a_y^2 + a_z^2}} \tag{1.11}$$

If two vectors are written in $\hat{\imath},\ \hat{\jmath},\ \hat{k}$ form:

$$\underline{a} = a_x\hat{\imath} + a_y\hat{\jmath} + a_z\hat{k} \qquad\qquad \underline{b} = b_x\hat{\imath} + b_y\hat{\jmath} + b_z\hat{k}$$

then addition (and subtraction) follows directly:

$$\begin{aligned} \underline{a} + \underline{b} &= a_x\hat{\imath} + a_y\hat{\jmath} + a_z\hat{k} + b_x\hat{\imath} + b_y\hat{\jmath} + b_z\hat{k} \\ &= (a_x + b_x)\hat{\imath} + (a_y + b_y)\hat{\jmath} + (a_z + b_z)\hat{k} \end{aligned} \tag{1.12a}$$

$$\begin{aligned} \underline{a} - \underline{b} &= a_x\hat{\imath} + a_y\hat{\jmath} + a_z\hat{k} - b_x\hat{\imath} - b_y\hat{\jmath} - b_z\hat{k} \\ &= (a_x - b_x)\hat{\imath} + (a_y - b_y)\hat{\jmath} + (a_z - b_z)\hat{k} \end{aligned} \tag{1.12b}$$

that is, we add (or subtract) corresponding components.

Multiplication by a scalar is easily defined:

$$\text{If}\quad \underline{a} = a_x\hat{i} + a_y\hat{j} + a_z\hat{k} \quad\text{then}\quad 3\underline{a} = 3a_x\hat{i} + 3a_y\hat{j} + 3a_z\hat{k} \quad\text{etc.}$$

The test for the equality of two vectors is now clear. Two vectors $\underline{a}$ and $\underline{b}$ will be equal if and only if each of their corresponding components are equal:

$$\underline{a} = \underline{b} \quad\text{if and only if}\quad a_x = b_x \qquad a_y = b_y \quad\text{and}\quad a_z = b_z \tag{1.13}$$

The use of $\hat{i}\hat{j}\hat{k}$ or coordinate notation is essentially the computational part of vector algebra. It allows for the quick evaluation of numerical results concerning vectors whereas the underlined symbolism (or coordinate free notation) allows a more general discussion of vector properties to be carried out.

EXERCISES

1.1 A vector $\underline{a}$ is parallel to the line through $A : (1, 2, 3)$ and $B : (-1, 0, 2)$ whilst a vector $\underline{b}$ is parallel to the line through $C : (0, 0, 1)$ and $D : (-1, -2, -3)$. Both vectors have magnitude 4. Find $\underline{a} + \underline{b}$ and $\underline{a} - \underline{b}$ in component form.

1.2 $ABCD$ is a quadrilateral with $\overrightarrow{AB} = \overrightarrow{DC}$. Prove that $ABCD$ is a parallelogram.

1.2 The Scalar Product

The **scalar product** of two vectors $\underline{a}$ and $\underline{b}$ is written $\underline{a}.\underline{b}$ (often called the dot product) and is defined as the scalar quantity

$$\underline{a}.\underline{b} = |\underline{a}|\ |\underline{b}| \cos\theta \tag{1.14}$$

Here θ is the smaller angle between the vectors $\underline{a}$ and $\underline{b}$ when they are directed away from a common point and so $0 \leq \theta \leq \pi$ (Figure 1.9).

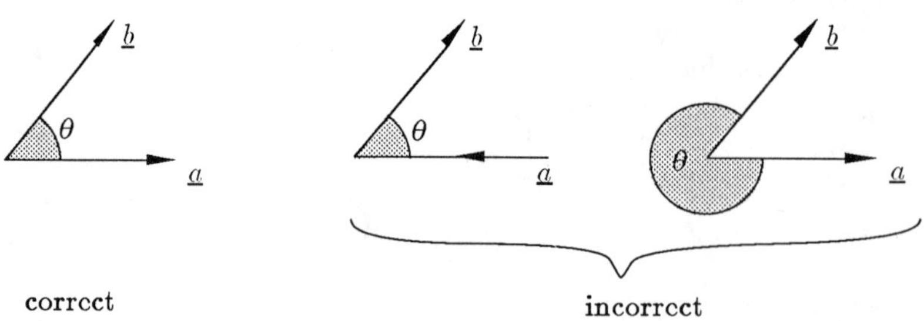

Figure 1.9

It is clear from the definition that

$$\underline{b}.\underline{a} = |\underline{b}|\ |\underline{a}| \cos\theta = \underline{a}.\underline{b} \tag{1.15}$$

that is, the scalar product is commutative.

It is easily shown that the distributive law for scalar multiplication:

$$s(t+u) = st + su \tag{1.16}$$

also holds for the scalar product. This is done by realising, from the definition (1.14), that $\underline{a}.\underline{b}$ can be regarded as the product of the orthogonal projection of $\underline{b}$ onto $\underline{a}$ (namely $|\underline{b}|\cos\theta$) times $|\underline{a}|$. Now consider any three vectors $\underline{a}$, $\underline{b}$ and $\underline{c}$. They can always be arranged (by sliding if necessary) as in Figure 1.10.

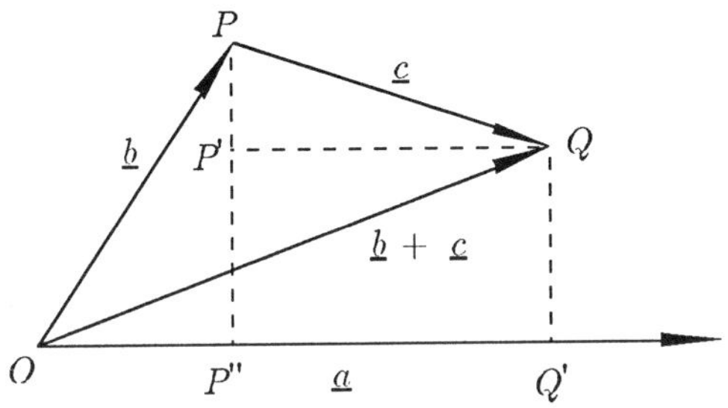

Figure 1.10

From the figure it is clear that the sum of the orthogonal projections of $\underline{b}$ and $\underline{c}$ onto $\underline{a}$; that is, $OP'' + P'Q$ is equal to $OP'' + P''Q'$ which is the orthogonal projection of $(\underline{b}+\underline{c})$ onto $\underline{a}$: Therefore

$$\underline{a}.(\underline{b}+\underline{c}) = \underline{a}.\underline{b} + \underline{a}.\underline{c} \tag{1.17}$$

By repeatedly applying this basic result it is clear that the scalar product of two vector sums may be expanded as in scalar algebra. Thus

$$(\underline{a}+\underline{b}+\ldots).(\underline{p}+\underline{q}\ldots) = \underline{a}.\underline{p}+\underline{a}.\underline{q}+\ldots+\underline{b}.\underline{p}+\underline{b}.\underline{q}+\ldots \tag{1.18}$$

- **Component Form of the Scalar Product**

Using the definition of the scalar product we have immediately

$$\hat{\imath}.\hat{\imath} = 1 \quad \hat{\imath}.\hat{\jmath} = 0 \quad \hat{\imath}.\hat{k} = 0$$

$$\hat{\jmath}.\hat{\imath} = 0 \quad \hat{\jmath}.\hat{\jmath} = 1 \quad \hat{\jmath}.\hat{k} = 0$$

$$\hat{k}.\hat{\imath} = 0 \quad \hat{k}.\hat{\jmath} = 0 \quad \hat{k}.\hat{k} = 1$$

Hence, using the distributive property (1.18), if $\underline{a} = a_x\hat{\imath} + a_y\hat{\jmath} + a_z\hat{k}$, and $\underline{b} = b_x\hat{\imath} + b_y\hat{\jmath} + b_z\hat{k}$ then $\underline{a}.\underline{b} = (a_x\hat{\imath} + a_y\hat{\jmath} + a_z\hat{k}).(b_x\hat{\imath} + b_y\hat{\jmath} + b_z\hat{k})$. That is,

$$\underline{a}.\underline{b} = a_x b_x + a_y b_y + a_z b_z \tag{1.19}$$

which is the component form of the scalar product.

The result (1.19), in conjunction with (1.14), provides a convenient method for determining the angle θ between two vectors

$$\underline{a}.\underline{b} = |\underline{a}|\ |\underline{b}| \cos\theta = a_x b_x + a_y b_y + a_z b_z$$

therefore

$$\theta = \cos^{-1}\left[\frac{a_x b_x + a_y b_y + a_z b_z}{|\underline{a}|\ |\underline{b}|}\right]$$

The dot product can be used to express the component form of a vector (1.19) in an alternative way which shall prove to be useful when we discuss tensors. Any vector $\underline{a}$ may always be written in the form:

$$\underline{a} = (\underline{a}.\hat{\imath})\hat{\imath} + (\underline{a}.\hat{\jmath})\hat{\jmath} + (\underline{a}.\hat{k})\hat{k} \qquad \textbf{(1.20)}$$

(which is easily verified by taking the scalar product of $\underline{a}$ with $\hat{\imath}$, $\hat{\jmath}$ and $\hat{k}$ in turn)

EXERCISES

1.3 Three forces, one of magnitude $8N$ in the direction of $\hat{\imath}$, a second of magnitude $10N$ in the direction of $3\hat{\imath} + 4\hat{\jmath}$ and a third of magnitude $13N$ in the direction of $5\hat{\imath} + 12\hat{k}$ displace a particle from $A(3, 2, -6)$m to $B(5, -1, 0)$m. Find the work done on the particle. (NB When a force $\underline{F}$ displaces a particle through a displacement represented by the vector $\underline{d}$ then the work done by the force is defined to be $|\underline{F}||\underline{d}| \cos\theta$ where θ is the angle between $\underline{F}$ and $\underline{d}$. That is work done $=\underline{F}.\underline{d}$).

1.4 If $\underline{a} = \cos\theta\hat{\imath} + \sin\theta\hat{\jmath}$ and $\underline{b} = \cos\phi\hat{\imath} + \sin\phi\hat{\jmath}$ show, using the scalar product that

$$\cos(\theta - \phi) = \cos\theta\cos\phi + \sin\theta\sin\phi$$

1.5 A particle moves under the action of a constant force $\underline{F}$ around a *closed* polygon. Show that the work done on the particle is zero.

1.3 The Vector Product

The **vector product** of two vectors $\underline{a}$ and $\underline{b}$ is written $\underline{a} \wedge \underline{b}$ (often called the cross product and alternatively written as $\underline{a} \times \underline{b}$). It is defined as the vector quantity

$$\underline{a} \wedge \underline{b} = |\underline{a}|\ |\underline{b}| \sin\theta\hat{n} \qquad \textbf{(1.21)}$$

where θ is the smaller angle between $\underline{a}$ and $\underline{b}$ (as used in the scalar product) and $\hat{n}$ is a unit vector perpendicular to both $\underline{a}$ and $\underline{b}$. The sense of $\hat{n}$ is such that $\underline{a}$, $\underline{b}$, $\hat{n}$ taken **in this order** form a right-handed system of vectors, i.e. as we rotate from $\underline{a}$ to $\underline{b}$ through the angle θ then a screw with a right-handed thread would move in the direction of $\hat{n}$. See Figure 1.11.

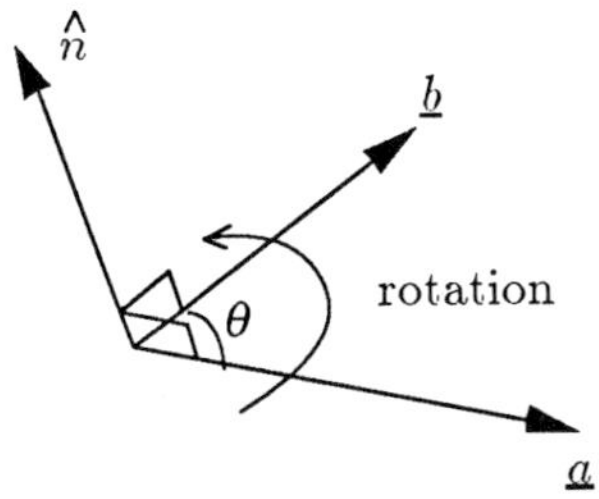

Figure 1.11

If we reverse the order of $\underline{a}$ and $\underline{b}$ then according to the definition

$$\underline{b} \wedge \underline{a} = |\underline{a}|\ |\underline{b}| \sin\theta \hat{m}$$

where $\hat{m}$ is perpendicular to both $\underline{b}$ and $\underline{a}$ and such that $\underline{b}$, $\underline{a}$, $\hat{m}$ taken in this order from a right-handed system. But clearly, as we rotate from $\underline{b}$ to $\underline{a}$ a screw with a right-handed thread would move in the direction of $-\hat{n}$ i.e. $\hat{m} = -\hat{n}$. Hence

$$\underline{b} \wedge \underline{a} = -|\underline{b}|\ |\underline{a}| \sin\theta \hat{n} = -\underline{a} \wedge \underline{b} \tag{1.22}$$

i.e. the vector product is **not** commutative and order **is** important.

We state without proof that the distributive law also holds for the vector product:

$$\underline{a} \wedge (\underline{b} + \underline{c}) = \underline{a} \wedge \underline{b} + \underline{a} \wedge \underline{c} \tag{1.23}$$

• Component Form of the Vector Product

From the above properties we can quickly deduce:

$$\begin{array}{lll} \hat{\imath} \wedge \hat{\imath} = 0 & \hat{\imath} \wedge \hat{\jmath} = \hat{k} & \hat{\imath} \wedge \hat{k} = -\hat{\jmath} \\ \hat{\jmath} \wedge \hat{\imath} = -\hat{k} & \hat{\jmath} \wedge \hat{\jmath} = 0 & \hat{\jmath} \wedge \hat{k} = \hat{\imath} \\ \hat{k} \wedge \hat{\imath} = \hat{\jmath} & \hat{k} \wedge \hat{\jmath} = -\hat{\imath} & \hat{k} \wedge \hat{k} = 0 \end{array} \tag{1.24}$$

An easy way to remember these results is to write the vectors $\hat{\imath}, \hat{\jmath}, \hat{k}$ in a list $\hat{\imath}, \hat{\jmath}, \hat{k}, \hat{\imath}, \hat{\jmath}, \hat{k}$. Now the vector product of any two consecutive vectors (reading from left to right) in the list is equal to the immediately following vector. If instead, we read from right to left then the vector product of any two consecutive vectors in the list is equal to minus the immediately preceding vector

We are now in a position to derive the component form of the vector product. If

$$\underline{a} = a_x\hat{\imath} + a_y\hat{\jmath} + a_z\hat{k} \qquad \text{and} \qquad \underline{b} = b_x\hat{\imath} + b_y\hat{\jmath} + b_z\hat{k}$$

so

$$\underline{a} \wedge \underline{b} = (a_x\hat{\imath} + a_y\hat{\jmath} + a_z\hat{k}) \wedge (b_x\hat{\imath} + b_y\hat{\jmath} + b_z\hat{k})$$

and by the distributive law of the vector product we find, using the results of (1.24)

$$\underline{a} \wedge \underline{b} = (a_yb_z - a_zb_y)\hat{\imath} + (a_zb_x - a_xb_z)\hat{\jmath} + (a_xb_y - a_yb_x)\hat{k} \tag{1.25a}$$

For those who have experience of determinants this may be more conveniently expressed as

$$\underline{a} \wedge \underline{b} = \begin{vmatrix} \hat{\imath} & \hat{\jmath} & \hat{k} \\ a_x & a_y & a_z \\ b_x & b_y & b_z \end{vmatrix} \qquad \textbf{(1.25b)}$$

where expansion about the top row gives us back (1.25a).

- **Vector Area**

At this stage it is pertinent to introduce the idea that area may be regarded as a vector quantity and with the definition of the vector product, above, this idea can now be formalised. A parallelogram with sides represented by the vectors $\underline{a}$ and $\underline{b}$ (Figure 1.12) is **defined** to have the vector area $\underline{A}$, given by the expression

$$\underline{A} = \underline{a} \wedge \underline{b} = |\underline{a}|\ |\underline{b}| \sin\theta \hat{n} \qquad \textbf{(1.26)}$$

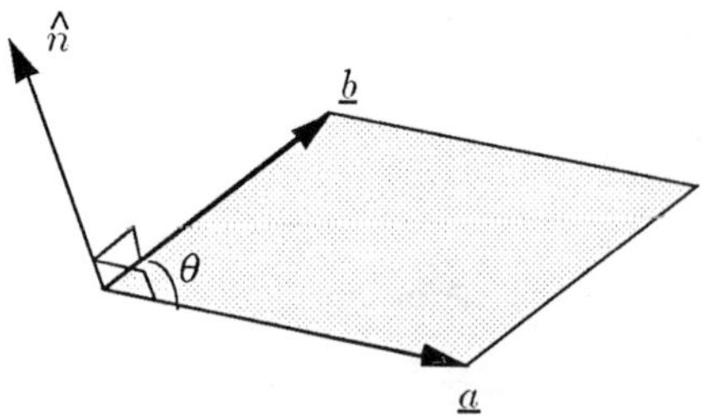

Figure 1.12

Clearly $|\underline{A}|$ is the correct numerical value of the area $(= |\underline{a}|\ |\underline{b}| \sin\theta)$ but now we are giving a direction to the area viz. the direction of $\underline{a} \wedge \underline{b}$, which is normal to the plane of the parallelogram.

How are the components of this area to be interpreted? If $\underline{A}$ represents a directed line segment:

$$\underline{A} = A_x\hat{\imath} + A_y\hat{\jmath} + A_z\hat{k}$$

then the components A_x, A_y, A_z have an obvious interpretation (noted in (1.9)) in terms of the resolution of $\underline{A}$ along the three coordinate axes. However, if $\underline{A}$ represents an area then the interpretation of the three components is different. The x-component $A_x = \underline{A}.\hat{\imath}$ is interpreted as the numerical value of the area $\underline{A}$ obtained by projection of this area onto the zy plane (with normal $\hat{\imath}$). Similarly A_y $(= \underline{A}.\hat{\jmath})$ and A_z $(= \underline{A}.\hat{k})$ are interpreted as the areas of projection of $\underline{A}$ onto the zx and the xy planes respectively. To illustrate, we consider a parallelogram with sides $\underline{a}$, $\underline{b}$. If

$$\underline{a} = a_x\hat{\imath} + a_y\hat{\jmath} + a_z\hat{k}, \qquad \underline{b} = b_x\hat{\imath} + b_y\hat{\jmath} + b_z\hat{k}$$

then, using (1.26) and (1.25b), the area of this parallelogram is the vector $\underline{A}$,

$$\underline{A} = \underline{a} \wedge \underline{b} = \begin{vmatrix} \hat{i} & \hat{j} & \hat{k} \\ a_x & a_y & a_z \\ b_x & b_y & b_z \end{vmatrix}$$

i.e. the components of $\underline{A}$ are

$$A_x = a_y b_z - a_z b_y \qquad A_y = a_z b_x - a_x b_z \qquad A_z = a_x b_y - a_y b_x$$

If this parallelogram is projected onto the xy plane (see Figure 1.13a) then it will form a parallelogram with sides $\underline{a}'$, $\underline{b}'$ such that $\underline{a}' = a_x\hat{i} + a_y\hat{j}$ and $\underline{b}' = b_x\hat{i} + b_y\hat{j}$

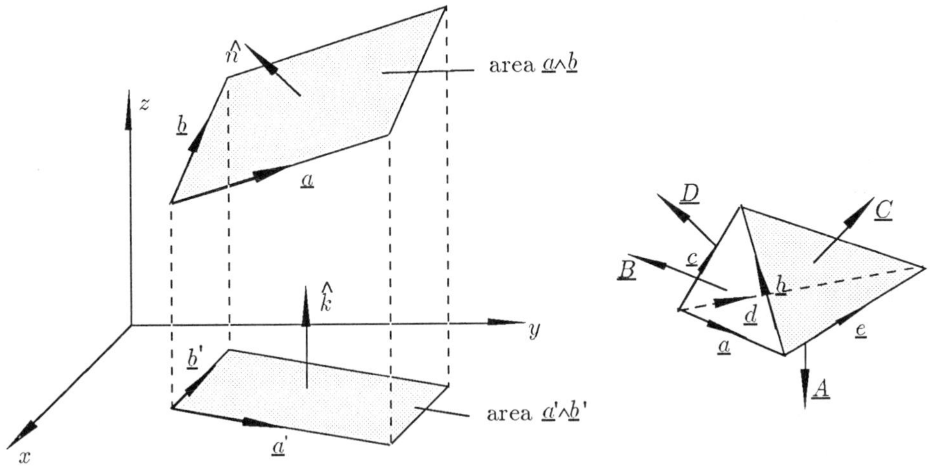

(a) (b)

Figure 1.13

The area of this second parallelogram is

$$\underline{a}' \wedge \underline{b}' = \hat{k}\,(a_x b_y - a_y b_x) = \hat{k}\; A_z$$

which concurs with our interpretation of A_z outlined above. We can obtain a similar verification of our interpretation of A_x, A_y if $\underline{A}$ is projected onto the other two coordinate planes.

A result corresponding to (1.3) may be deduced for area vectors. If we consider the tetrahedron of Figure 1.13b then expressions for the vector areas of each face may be obtained in terms of the edge vectors $\underline{a}$, $\underline{b}$, $\underline{c}$, $\underline{d}$ and $\underline{e}$. If we organise matters so that the area vectors point *out* of the closed region then

$$\underline{A} = \frac{1}{2}\underline{d} \wedge \underline{a} \qquad \underline{B} = \frac{1}{2}\underline{a} \wedge \underline{c} \qquad \underline{C} = \frac{1}{2}\underline{e} \wedge \underline{b} \qquad \underline{D} = \frac{1}{2}\underline{c} \wedge \underline{d}$$

Hence

$$\begin{aligned}\underline{A}+\underline{B}+\underline{C}+\underline{D} &= \frac{1}{2}\left[\underline{d}\wedge\underline{a}+\underline{a}\wedge\underline{c}+\underline{e}\wedge\underline{b}+\underline{c}\wedge\underline{d}\right]\\ &= \frac{1}{2}\left[\underline{d}\wedge\underline{a}+\underline{a}\wedge\underline{c}+(\underline{d}-\underline{a})\wedge(\underline{c}-\underline{a})+\underline{c}\wedge\underline{d}\right]\\ &= \frac{1}{2}\left[\underline{d}\wedge\underline{a}+\underline{a}\wedge\underline{c}+\underline{d}\wedge\underline{c}-\underline{a}\wedge\underline{c}-\underline{d}\wedge\underline{a}+\underline{c}\wedge\underline{d}\right]\\ &= 0\end{aligned}$$

This result can be generalised. If $\underline{A}$, $\underline{B}, \ldots, \underline{K}$ represent the vector areas of each of the faces of a prism constructed from planar elements then it may be shown (by segmenting the prism into tetrahedra) that

$$\underline{A}+\underline{B}+\ldots+\underline{K}=0 \tag{1.27}$$

Here it is assumed that each vector representing the areas all point out of (or all point in through) the prism.

EXERCISES

1.6 Find the area of a triangle whose vertices are at $(1,1,-1)$, $(2,1,0)$ and $(0,0,0)$.

1.7 A triangle has vertices at $(0,1,3)$, $(3,4,5)$ and $(2,0,1)$. Determine its vector area. Find the magnitude of the area of projection onto the xy plane and also onto a plane with normal $2\hat{\imath}-3\hat{\jmath}+4\hat{k}$.

1.8 Show that the perpendicular distance between a corner of a unit cube and a diagonal which does not pass through it is 0.816.

1.4 Applications of Vectors to Forces

The effect of a force acting on a body is dependent not only on the magnitude and direction of the force but also, generally, on the point of application. However in some physical systems (such as rigid bodies) only the line of action of the force, as opposed to its actual point of application, is of importance. Two simple examples will illustrate the point.

Consider a block of rubber held fixed in some way (Figure 1.14a). If we apply a force $\underline{F}_1$ at point A in the direction of AB then the deformation of the solid would be similar to that indicated in Figure 1.14b.

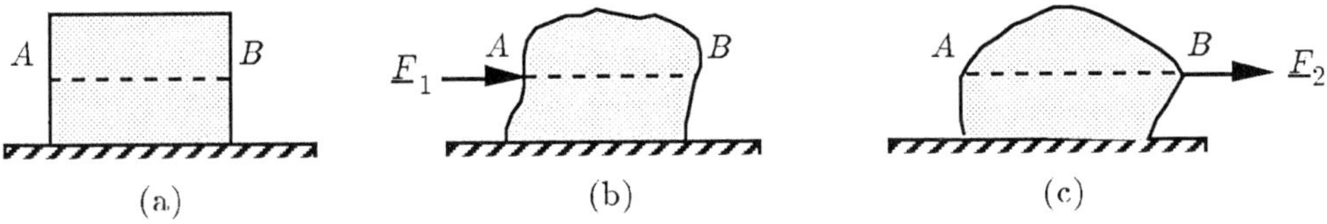

Figure 1.14

However if a force $\underline{F}_2$ of the same magnitude and direction as $\underline{F}_1$ was applied at point B (i.e. along the same line of action) then the deformed shape would be similar to that shown in Figure 1.14c, that is, a completely distinct deformation has resulted simply by changing the point of application of the force.

As a second example consider the mechanical configuration of a cantilever shown in Figure 1.15

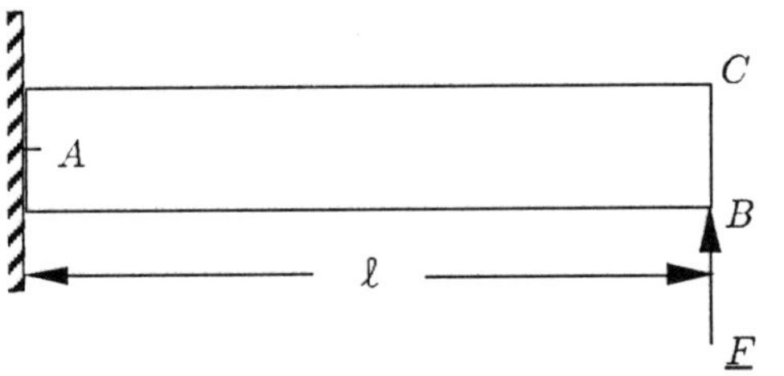

Figure 1.15

If we are interested in the moment developed at point A due to the force $\underline{F}$ then it is irrelevant whether the point of application of $\underline{F}$ is at B or at C or indeed at any point on the line BC (the line of action of $\underline{F}$) because exactly the same moment (which is $|\underline{F}|\ \ell$) is obtained in all cases.

It is unfortunate, but nonetheless a fact, that forces cannot generally be considered as 'free' vectors. However, in the special case where forces are concurrent, i.e. all the forces have a common point of application, then they may be regarded as 'free' vectors. This special application is not quite as artificial as it seems since we shall discover later in this chapter that we can often arrange matters so that a complicated system of forces is replaced by an equivalent system of concurrent forces (plus, to be precise, a torque).

- **The Moment of a Force**

When a force $\underline{F}$ (with a given line of action) acts on a body then we say that the force has a **moment** about a point O equal to the product of the magnitude of $\underline{F}$ with the perpendicular distance from the line of action of $\underline{F}$ to O. (See Figure 1.16).

If the body is pivoted at O then the physical effect of the moment is to induce a 'turning' about an axis through O perpendicular to the plane of $\underline{r}$ and $\underline{F}$ where $\underline{r}$ is the position vector of **any** point on the line of action of $\underline{F}$. This suggests the definition of the vector moment $\underline{M}$ (or **torque**) of $\underline{F}$ about O as

$$\underline{M} = \underline{r} \wedge \underline{F} \qquad \textbf{(1.28)}$$

This concise vector equation gives us the magnitude of the moment (this being $|\underline{M}| = |\underline{r}|\ |\underline{F}| \sin\theta = |\underline{F}|\ d$) as well as the direction of the turning axis referred to above, this being the direction of $\underline{r} \wedge \underline{F}$.

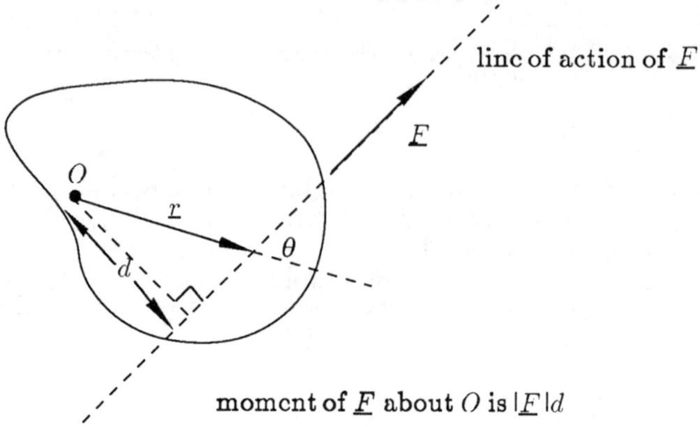

Figure 1.16

- **Couples**

If we have a pair of forces, equal in magnitude but opposite in direction and with distinct lines of action then this system of forces constitutes a **couple** (Figure 1.17).

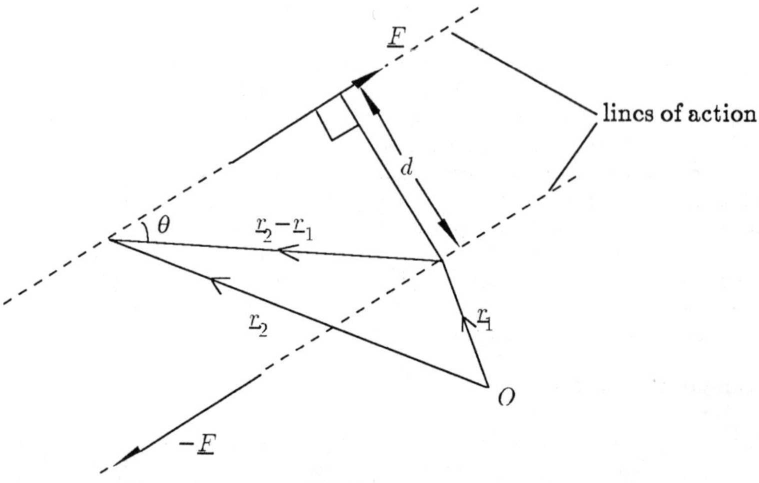

Figure 1.17

Let $\underline{r}_1$ and $\underline{r}_2$ be the position vectors of **any** two points on the lines of action of the forces with respect to an origin at O. The resultant torque of the forces about O is defined simply as the vector sum of the two vectors $\underline{r}_1 \wedge (-\underline{F})$ and $\underline{r}_2 \wedge (\underline{F})$, these being the separate torques of the forces about O.

The resultant torque is $\underline{r}_1 \wedge (-\underline{F}) + \underline{r}_2 \wedge (\underline{F}) = (\underline{r}_2 - \underline{r}_1) \wedge \underline{F}$ (using the distributive law for vector products). But $\underline{r}_2 - \underline{r}_1$ is **any** vector joining the lines of action of the two forces and so this resultant is independent of the choice of origin. The direction of the resultant torque is perpendicular to the plane containing the two lines of action. From the definition of the vector product we see that the magnitude of the torque is $|\underline{r}_2 - \underline{r}_1|\,|\underline{F}|\sin\theta$. But, from Figure 1.17, $d = |\underline{r}_2 - \underline{r}_1|\sin\theta$ where d is the perpendicular distance between the parallel lines of action. Hence the torque has the simple magnitude $|\underline{F}|d$.

• The Resolution of Forces

We are now in a position to prove that, suitably treated, forces may be regarded as free vectors. We show that **any** system of forces may be reduced to a single force acting through a convenient point O and a single torque about O.

Let $\underline{F}_1, \underline{F}_2, \ldots, \underline{F}_n$ be the forces and $\underline{r}_1, \underline{r}_2, \ldots, \underline{r}_n$ be the position vectors from an origin O to any point on the lines of action of the forces (Figure 1.18).

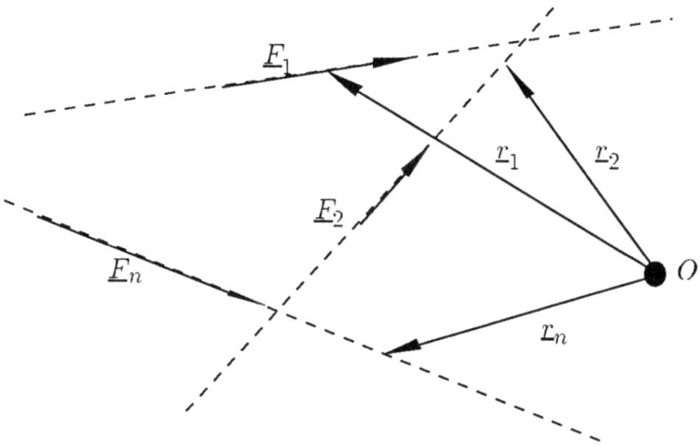

Figure 1.18

If we consider a particular force $\underline{F}_i$ then we can add and subtract a force parallel to $\underline{F}_i$ but acting through O (Figure 1.19). (In other words we have simply added the zero vector to the system). Now this collection of three forces is equivalent to a single force $\underline{F}_i$ acting through O together with a single couple about O with torque $\underline{r}_i \wedge \underline{F}_i$. Repeating this construction for each force in the system, we see that we can reduce the original system to

(i) a single **resultant** force $\sum_{i=1}^{n} \underline{F}_i$ through O

(ii) a single **resultant** couple with torque $\sum_{i=1}^{n} \underline{r}_i \wedge \underline{F}_i$ about O.

A body acted on by a system of forces is said to be in equilibrium if **both** the resultant force **and** the resultant couple is zero. What happens if we choose a different origin, at O' say? (See Figure 1.20).

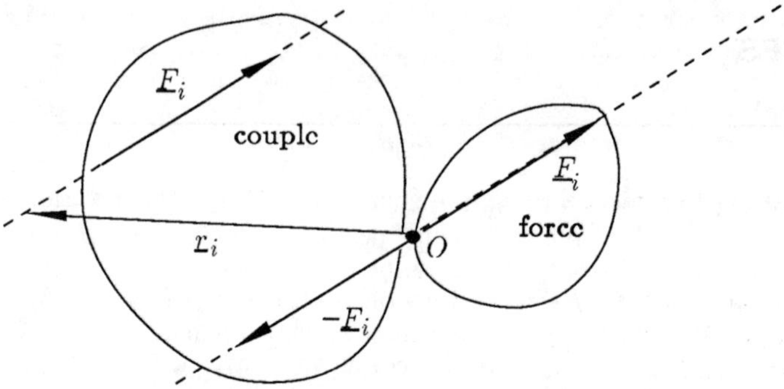

Figure 1.19

We have already seen that about O, the system of forces is equivalent to a single force $\sum_{i=1}^{n} \underline{F}_i$ and a single couple $\sum_{i=1}^{n} \underline{r}_i \wedge \underline{F}_i$. Had we considered O' instead, we would conclude, using exactly the same construction as that described above that the system of forces is equivalent to a single force $\sum_{i=1}^{n} \underline{F}_i$ (as before) and a single couple:

$$\sum_{i=1}^{n} \underline{r}_i' \wedge \underline{F}_i \equiv \sum_{i=1}^{n} (\underline{r}_i - \underline{s}) \wedge \underline{F}_i \equiv -\underline{s} \wedge \sum_{i=1}^{n} \underline{F}_i + \sum_{i=1}^{n} \underline{r}_i \wedge \underline{F}_i$$

which is clearly dependent on the position of O' (the value of $\underline{s}$) unless the resultant force is zero.

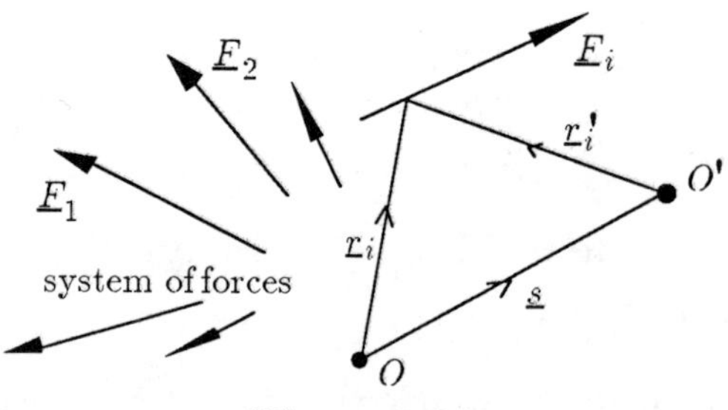

Figure 1.20

We conclude that when the resultant force in a system is zero, the resultant couple (which is now independent of position) may be evaluated about any convenient point. This result will be used quite often in later chapters.

EXERCISES

1.9 A force of 12 N acts through the point $A(3,4,5)$m in the direction of the vector $3\hat{\imath}+4\hat{\jmath}-5\hat{k}$. What is the moment of the force about $B(2,1,0)$m?

1.10 Verify that the force $\underline{F}_1 = 3\hat{\imath}-\hat{k}$ acting at $(1,2,0)$ and $\underline{F}_2 = 2\hat{\imath}+3\hat{\jmath}$ acting at $(0,-1,3)$ reduce to a resultant force $5\hat{\imath}+3\hat{\jmath}-\hat{k}$ at the origin together with a couple $-11\hat{\imath}+7\hat{\jmath}-4\hat{k}$. Show that the system of forces cannot be reduced to a single resultant force.

1.11 Three forces of magnitudes F, $2F$ and $3F$ act along the sides AB, BC and CA respectively of an equilateral triangle ABC. Determine the resultant of these forces and show that the line of action of the resultant cuts BC in a point P such that $2BP = 3PC$.

1.5 Triple Products

There are two types of triple products i.e. products involving three vectors: the scalar triple product $\underline{a}.(\underline{b}\wedge\underline{c})$ and the vector triple product $\underline{a}\wedge(\underline{b}\wedge\underline{c})$. These products occur sufficiently often in practice to warrant a detailed examination.

- **Scalar Triple Product**

We have already seen that $\underline{b}\wedge\underline{c}$ represents the area of a parallelogram with sides $\underline{b}$ and $\underline{c}$. Hence $\underline{a}.(\underline{b}\wedge\underline{c})$ is simply the volume of a prism (parallelepiped) with sides $\underline{a}$, $\underline{b}$ and $\underline{c}$. Since the volume can also be calculated using the base $\underline{a}\wedge\underline{b}$ or $\underline{c}\wedge\underline{a}$ we can conclude, from symmetry considerations that

$$\underline{a}.(\underline{b}\wedge\underline{c}) = \underline{b}.(\underline{c}\wedge\underline{a}) = \underline{c}.(\underline{a}\wedge\underline{b}) \qquad \textbf{(1.29)}$$

A useful way of expressing the scalar triple product is in determinant form. Since

$$\underline{b}\wedge\underline{c} = \begin{vmatrix} \hat{\imath} & \hat{\jmath} & \hat{k} \\ b_x & b_y & b_z \\ c_x & c_y & c_z \end{vmatrix}$$

it readily follows from (1.19) that

$$\underline{a}.(\underline{b}\wedge\underline{c}) = \begin{vmatrix} a_x & a_y & a_z \\ b_x & b_y & b_z \\ c_x & c_y & c_z \end{vmatrix} \qquad \textbf{(1.30)}$$

The equalities of (1.29) then follow from the properties of determinants.

- **Vector Triple Product**

From the properties of the cross-product it follows that the vector $\underline{a} \wedge (\underline{b} \wedge \underline{c})$ is perpendicular both to $\underline{a}$ and to $(\underline{b} \wedge \underline{c})$. Since $(\underline{b} \wedge \underline{c})$ is perpendicular to both $\underline{b}$ and $\underline{c}$ it follows that $\underline{a} \wedge (\underline{b} \wedge \underline{c})$ must lie in the plane containing both $\underline{b}$ and $\underline{c}$ (See Figure 1.21)

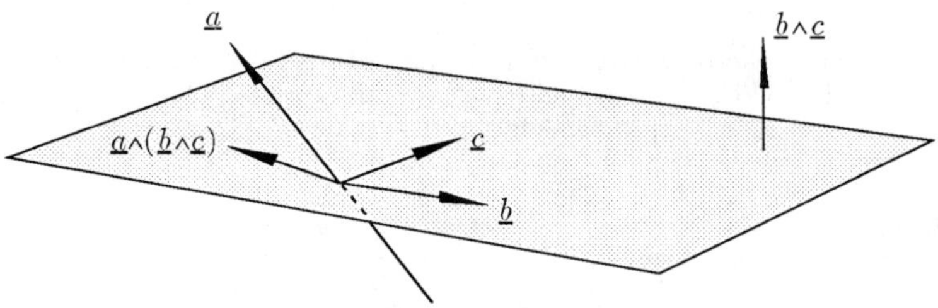

Figure 1.21

Hence we may write

$$\underline{a} \wedge (\underline{b} \wedge \underline{c}) = \alpha \underline{b} + \beta \underline{c} \tag{1.31}$$

where α and β are yet to be determined.

We shall evaluate the vector product on the left hand side and use the right hand side in order to determine α and β. Now if $\underline{a} = a_x \hat{i} + a_y \hat{j} + a_z \hat{k}$ and similarly for $\underline{b}$ and $\underline{c}$ then

$$\underline{b} \wedge \underline{c} = \hat{i}(b_y c_z - b_z c_y) - \hat{j}(b_x c_z - c_x b_z) + \hat{k}(b_x c_y - c_x b_y)$$

and so, using (1.25b)

$$\begin{aligned}\underline{a} \wedge (\underline{b} \wedge \underline{c}) &= \begin{vmatrix} \hat{i} & \hat{j} & \hat{k} \\ a_x & a_y & a_z \\ b_y c_z - b_z c_y & c_x b_z - b_x c_z & b_x c_y - c_x b_y \end{vmatrix} \\ &= \hat{i}\,\{a_y(b_x c_y - c_x b_y) - a_z(c_x b_z - b_x c_z)\} \\ &\quad - \hat{j}\,\{a_x(b_x c_y - c_x b_y) - a_z(b_y c_z - b_z c_y)\} \\ &\quad + \hat{k}\,\{a_x(c_x b_z - b_x c_z) - a_y(b_y c_z - b_z c_y)\}\end{aligned}$$

Now regrouping as far as possible according to the form of the right hand side of (1.31) we find

$$\begin{aligned}\underline{a} \wedge (\underline{b} \wedge \underline{c}) =& \hat{i} b_x(a_y c_y + a_z c_z) + \hat{j} b_y(a_x c_x + a_z c_z) + \hat{k} b_z(a_x c_x + a_y c_y) \\ & - \hat{i} c_x(a_y b_y + a_z b_z) - \hat{j} c_y(a_x b_x + a_z b_z) - \hat{k} c_z(a_x b_x + a_y b_y)\end{aligned} \tag{1.32}$$

Adding and subtracting $\hat{\imath}(b_x a_x c_x) + \hat{\jmath}(b_y a_y c_y) + \hat{k}(b_z a_z c_z)$ to the right hand side of (1.32) gives finally

$$\underline{a} \wedge (\underline{b} \wedge \underline{c}) = (\underline{a}.\underline{c})\underline{b} - (\underline{a}.\underline{b})\underline{c} \tag{1.33}$$

By similar reasoning, the vector triple product $(\underline{a} \wedge \underline{b}) \wedge \underline{c}$ is a vector which lies in the plane containing $\underline{a}$ and $\underline{b}$. A detaialed calculation shows that

$$(\underline{a} \wedge \underline{b}) \wedge \underline{c} = (\underline{a}.\underline{c})\underline{b} - (\underline{b}.\underline{c})\underline{a}$$

Clearly $\underline{a} \wedge (\underline{b} \wedge \underline{c})$ is a vector with a different magnitude and direction from the vector $(\underline{a} \wedge \underline{b}) \wedge \underline{c}$. In other words the associative law fails for the vector triple product.

EXERCISES

1.12 If $\underline{a} = \hat{\imath} + \hat{\jmath}$, $\underline{b} = \hat{\imath} + 2\hat{\jmath} + \hat{k}$, and $\underline{c} = \hat{\imath} - 3\hat{\jmath}$ evaluate:
(1) $\underline{a}.[(\underline{a} - \underline{b}) \wedge \underline{c}]$
(ii) $(\underline{a} \wedge \underline{c}).(\underline{b} \wedge \underline{c})$
(iii) $\underline{a} \wedge (\underline{b} \wedge \underline{c})$
(iv) $(\underline{a}.\underline{c})\underline{b} - (\underline{a}.\underline{b})\underline{c}$

1.13 Using properties of the vector triple product and the scalar triple product prove that

$$(\underline{a} \wedge \underline{b}).(\underline{c} \wedge \underline{d}) = (\underline{a}.\underline{c})(\underline{b}.\underline{d}) - (\underline{b}.\underline{c})(\underline{a}.\underline{d})$$

1.14 (a) The volume of a tetrahedron is given by the product of one third perpendicular height with the area of the base. If $\underline{d}$, $\underline{a}$ and $\underline{f}$ are vectors representing the three sides of a tetrahedron which intersect at a corner show that the volume of the tetrahedron may be expressed in the form:

$$V = \tfrac{1}{6}|\underline{d}.(\underline{a} \wedge \underline{f})|$$

(b) If $\underline{a}$, $\underline{b}$, $\underline{c}$ and $\underline{d}$ are the position vectors of four points A, B, C and D respectively show that the condition for the four points to be coplanar is

$$\underline{a}.(\underline{b} \wedge \underline{c}) + \underline{a}.(\underline{c} \wedge \underline{d}) + \underline{a}.(\underline{d} \wedge \underline{b}) = \underline{b}.(\underline{c} \wedge \underline{d})$$

1.6 The Index Notation for Vectors

The notation we have adopted in this text in order to describe vectors is perhaps the most widely accepted. Certainly, in our view, it is many ways the 'best' notation with which to **introduce** vectors. However, there are a number of alternative notations and in this section we will record the main results in terms of one of the most useful of these alternatives – the index notation.

The index notation for the description of vectors enjoys a number of advantages. The first is **conciseness**: it allows standard results to be written in an immediately transparent though compact form. A second, less obvious, advantage is that the index approach often allows a surprisingly quick proof of many of the useful identities of vector analysis. There are many other advantages (and one or two disadvantages) that readers will ascertain for themselves. We would also point out that the index approach to the description of vectors is the most 'natural' notation if one wishes to consider a generalisation of the vector which has found many applications in Mathematics, Physics and Engineering - the **tensor**. Cartesian tensors are introduced in Chapter 2.

We shall find it convenient to re-label the coordinates (x, y, z) as (x_1, x_2, x_3) or simply x_i with the agreement that the index i can take any of the values $1, 2, 3$.

Using this index notation a vector $\underline{a}$ may be alternatively denoted by a_i. Thus the components of the vector a_i are (a_1, a_2, a_3).

- **Equality**

Two vectors a_i, b_i are equal if and only if

$$a_i = b_i \tag{1.34}$$

We shall take it for granted that if the index is unspecified then the equation in which it occurs is assumed to be valid for each of the three possible values that the index can take. Thus (1.34) is shorthand for the *three* equations

$$a_1 = b_1 \qquad a_2 = b_2 \qquad a_3 = b_3 \tag{1.35}$$

- **The Summation Convention**

If, in a term, an index is repeated then it is assumed that the term is summed as the repeated index takes the values 1,2,3: Thus

$$a_i b_i \qquad \text{means} \qquad \sum_{i=1}^{3} a_i b_i \equiv a_1 b_1 + a_2 b_2 + a_3 b_3 \tag{1.36}$$

Using this convention if a_i, b_i are two vectors (representing $\underline{a}$, $\underline{b}$) then the scalar product $\underline{a}.\underline{b}$ is represented by $a_i b_i$. (c.f. equation (1.19))

An index which is not repeated (in a term) is called a **free** index. A repeated index (in a term) is called a **dummy** index since the precise letter used is irrelevant. For example $a_i b_i$ has exactly the same meaning as $a_m b_m$ or $a_j b_j$ or $a_s b_s$ etc each meaning $a_1 b_1 + a_2 b_2 + a_3 b_3$.

In the index version of vector algebra there are two objects which occur again and again; these are the kronecker delta δ_{ij} and the permutation symbol η_{ijk}.

- **The Kronecker Delta**

We define a two-indexed quantity δ_{ij} called the Kronecker delta by

$$\delta_{ij} = \begin{cases} 1 & \text{if} \quad i = j \\ 0 & \quad \text{otherwise} \end{cases} \tag{1.37}$$

Thus $\delta_{11} = \delta_{22} = \delta_{33} = 1$ and $\delta_{12} = \delta_{21} = \delta_{13} = \delta_{31} = \delta_{23} = \delta_{32} = 0$. Another name for this quantity is the **substitution operator**, since it can be used to substitute one index for another. This follows from the easily verified result.

$$a_j = \delta_{ji} a_i \tag{1.38}$$

(The right hand side is $\delta_{j1}a_1 + \delta_{j2}a_2 + \delta_{j3}a_3$ and so if $j = 1$ in (1.38) $a_1 = \delta_{11}a_1 + \delta_{12}a_2 + \delta_{13}a_3 = a_1$ with similar results for $j = 2$ and $j = 3$). In matrix terms the Kronecker delta corresponds to the unit matrix.

Just as in ordinary algebra, an indexed equation will be formed from a collection of terms added together and equated to zero. Each term will be comprised of one or more indexed objects multiplied together. It is an inherent rule when using the index notation that the number and type of free indices in any term of a given equation must be the same. Thus

$$a_i c_j + g_{ij} e_k d_k - h_i c_j = 0 \qquad \text{is meaningful}$$

(i, j are free indices in each term, k is a dummy index) but

$$a_i c_k + d_{ij} a_t b_t = 0 \qquad \text{is not meaningful}$$

(i, k free in the first term, i, j free in the second term, t is a dummy index.)

EXERCISES

1.15 If a_i and b_i correspond to vectors $\underline{a}$ and $\underline{b}$ determine the index form of the vector equation

$$\underline{a} + (\underline{a}.\underline{b})\underline{b} = 0$$

1.16 Which of the following equations are meaningful indexed expressions?
(a) $a_i b_j + c_i d_j a_k a_k = 0$
(b) $a_i b_j + a_i = 0$
(c) $a_k b_k c_i + a_k d_i = 0$
(d) $\delta_{ii} - 3 = 0$
(e) $c_i + \varepsilon_{ijk} b_j d_k = 0$

1.17 Verify that
(a) $\delta_{ij}\delta_{jk} = \delta_{ik}$
(b) $\delta_{ij}\delta_{ij} = 3$

• The Permutation Symbol

This is defined by the relations:

$$\eta_{ijk} = \begin{cases} 1 & \text{if } i,j,k \text{ is a cyclic permutation of } 1,2,3 \\ -1 & \text{if } i,j,k \text{ is an anti} - \text{cyclic permutation of } 1,2,3 \\ 0 & \text{otherwise} \end{cases} \quad \mathbf{(1.39)}$$

Each subscript can only take on values 1,2 or 3 so η_{ijk} has 27 'components'. From the definition it follows that

$$\eta_{123} = \eta_{231} = \eta_{312} = +1 \qquad \eta_{132} = \eta_{321} = \eta_{213} = -1 \quad \mathbf{(1.40a)}$$

$$\text{all other} \quad \eta_{ijk} = 0 \qquad \text{e.g.} \qquad \eta_{233} = \eta_{223} = \eta_{311} = 0 \quad \mathbf{(1.40b)}$$

These results imply the following identities

$$\eta_{ijk} = \eta_{jki} = \eta_{kij} \quad \mathbf{(1.40c)}$$

Thus a cyclic permutation of the indices on the permutation symbol does not alter its value. Also

$$\eta_{ijk} = -\eta_{kji} \qquad \eta_{ijk} = -\eta_{jik} \qquad \eta_{ijk} = -\eta_{ikj} \quad \mathbf{(1.40d)}$$

That is, an interchange of any two indices on the permutation symbol introduces a minus sign. This coresponds to an anti-cyclic permutation of indices.

With the introduction of this symbol we can show that the ith component of the vector product of two vectors $\underline{a}, \underline{b}$ (or $a_i,\ b_i$) is:

$$(\underline{a} \wedge \underline{b})_i = \eta_{ijk} a_j b_k \quad \mathbf{(1.41)}$$

We can check that this is a valid representation of the vector product simply by allocating values to the free index 'i' and then expanding any summations inherent in the expression. The right-hand side of (1.41) involves two repeated indices and so involves a double summation. Firstly

$$\eta_{ijk} a_j b_k = \eta_{i1k} a_1 b_k + \eta_{i2k} a_2 b_k + \eta_{i3k} a_3 b_k \qquad \text{(summing over the index } j\text{)}$$

Now summing over the index k (and using the result implicit in (1.40b) that η_{ijk} is zero if it contains two identical indices)

$$\eta_{ijk} a_j b_k = \eta_{i12} a_1 b_2 + \eta_{i13} a_1 b_3 + \eta_{i21} a_2 b_1 + \eta_{i23} a_2 b_3 + \eta_{i31} a_3 b_1 + \eta_{i32} a_3 b_2$$

Therefore if index i takes the value 1

$$\eta_{1jk} a_j b_k = \eta_{123} a_2 b_3 + \eta_{132} a_3 b_2 = a_2 b_3 - a_3 b_2$$

Similarly if $i = 2$

$$\eta_{2jk} a_j b_k = \eta_{213} a_1 b_3 + \eta_{231} a_3 b_1 = a_3 b_1 - a_1 b_3$$

and if $i = 3$

$$\eta_{3jk} a_j b_k = \eta_{312} a_1 b_2 + \eta_{321} a_2 b_1 = a_1 b_2 - a_2 b_1$$

But these three terms are the components of $\underline{a} \wedge \underline{b}$ (see equation (1.25a)) and so (1.41) is verified.

- **The η-δ identity**

There exists a basic identity connecting the η-symbol and the δ-symbol. It is

$$\eta_{ijk}\eta_{ipr} = \delta_{jp}\delta_{kr} - \delta_{jr}\delta_{kp}$$

This identity is proved using the properties of determinants. The reader will find it a useful exercise to verify this result (simply by allocating values to the free indices) which has numerous applications when developing other vector identities. For example let us prove, using the index approach, the identity (1.33) viz.

$$\underline{a} \wedge (\underline{b} \wedge \underline{c}) = (\underline{a}.\underline{c})\underline{b} - (\underline{a}.\underline{b})\underline{c}$$

Now the ith component of the left hand side is, using (1.41) twice,

$$\eta_{ijk}a_j(\underline{b} \wedge \underline{c})_k = \eta_{ijk}\eta_{klm}a_jb_lc_m$$

After a cyclic permutation on the first η-symbol on the right hand side we find

$$\begin{aligned}\eta_{kij}\eta_{klm}a_jb_lc_m &= (\delta_{il}\delta_{jm} - \delta_{im}\delta_{jl})a_jb_lc_m \\ &= a_jb_ic_j - a_jb_jc_i \\ &= (\underline{a}.\underline{c})b_i - (\underline{a}.\underline{b})c_i \\ &= \{(\underline{a}.\underline{c})\underline{b} - (\underline{a}.\underline{b})\underline{c}\}_i\end{aligned}$$

which proves the result.

EXERCISES

1.18 Show that
(a) $\eta_{iik} = 0$
(b) $\eta_{ijk}a_ja_k = 0$
(c) $\eta_{ijk}\eta_{ijk} = 6$

1.19 Use the index notation to show that

$$\underline{a}.(\underline{b} \wedge \underline{c}) = (\underline{a} \wedge \underline{b}).\underline{c}$$

- **Symmetry and Skew-Symmetry**

As we have already explained a 'term' is comprised of one or more indexed objects multiplied together. Such a term (or group of terms) may be denoted by $T_{ij\ldots k}$ where $ij\ldots k$ refer to the free indices in the term. Thus we might, for convenience, denote the collection $\eta_{ijk}\eta_{ipr} - \delta_{jp}\delta_{kr} + \delta_{jr}\delta_{kp}$ by T_{jkpr} whilst η_{iik} might be denoted by T_k.

Consider now a general term $T_{ij\ldots k}$. If when two particular free indices are interchanged we obtain the result:

$$T_{ij\ldots p\ldots q\ldots k} = T_{ij\ldots q\ldots p\ldots k}$$

then we say the term is **symmetric** in the indices p and q. Alternatively if instead we obtain the result

$$T_{ij\ldots p\ldots q\ldots k} = -T_{ij\ldots q\ldots p\ldots k}$$

then we say the term is **skew-symmetric** in the indices p and q.

The typical examples of symmetric and skew-symmetric objects are the Kronecker delta and the permutation symbol respectively. Since $\delta_{ij} = \delta_{ji}$ (from (1.37)) for each value of i and j it follows that the Kronecker delta is symmetric in its indices (or simply symmetric). Also from (1.40d) it follows that the permutation symbol is skew-symmetric in *any* pair of indices. For this reason it is often referred to as the completely skew-symmetric object.

Not every indexed object (with two or more indices) is either symmetric or skew-symmetric though every indexed objects (with two or more indices) can be written (for each pair of indices) as the sum of a symmetric and a skew-symmetric object. For example if A_{ij} is a given two indexed object then by simple algebra

$$A_{ij} = \frac{1}{2}(A_{ij} + A_{ji}) + \frac{1}{2}(A_{ij} - A_{ji})$$

and it is easily verified that the object $C_{ij} \equiv \frac{1}{2}(A_{ij} + A_{ji})$ is symmetric whilst the object $D_{ij} \equiv \frac{1}{2}(A_{ij} - A_{ji})$ is skew-symmetric.

Example 1.1

Show that if $C_{ij\ldots p\ldots q\ldots k}$ is symmetric in p and q and $D_{lm\ldots p\ldots q\ldots n}$ is skew-symmetric in p and q then

$$C_{ij\ldots p\ldots q\ldots k}\ D_{lm\ldots p\ldots q\ldots n} = 0$$

This is a result which often arises.

Solution

Since both p and q are dummy indices we may use two different indices:

$$C_{ij\ldots p\ldots q\ldots k}\ D_{lm\ldots p\ldots q\ldots n} = C_{ij\ldots a\ldots b\ldots k}\ D_{lm\ldots a\ldots b\ldots n}$$

Now changing back from a, b to p, q but with $a \to q$ and $b \to p$ gives

$$C_{ij\ldots a\ldots b\ldots k}\ D_{lm\ldots a\ldots b\ldots n} = C_{ij\ldots q\ldots p\ldots k}\ D_{lm\ldots q\ldots p\ldots n}$$

But since C is symmetric in p and q

$$C_{ij\ldots q\ldots p\ldots k} = C_{ij\ldots p\ldots q\ldots k}$$

and since D is skew-symmetric in p and q

$$D_{ij\ldots q\ldots p\ldots k} = -D_{ij\ldots p\ldots q\ldots k}$$

We can write

$$C_{ij\ldots p\ldots q\ldots k}\ D_{lm\ldots p\ldots q\ldots n} = -C_{ij\ldots p\ldots q\ldots k}\ D_{lm\ldots p\ldots q\ldots n}$$

which is only possible if both sizes are zero (If $A = -A$ then $A = 0$). This proves the result.

- **Contraction**

If in a given equation:

$$C_{ij\ldots p\ldots q\ldots k} = 0$$

we make two of the free indices the same (implying summation)

$$C_{ij\ldots p\ldots p\ldots k} = 0$$

then we say we have **contracted** the equation over the indices p and q. The process of contraction reduces (or contracts) the number of free indices in an equation by 2. This may be done for any pair of free indices and the equation will remain valid. As the reader might well have perceived, contraction is a shorthand procedure for recording the effect of multiplying both sides of the original equation by the Kronceker delta δ_{pq}.

$$0 = \delta_{pq} C_{ij\ldots p\ldots q\ldots k} = C_{ij\ldots p\ldots p\ldots k}$$

Although a contracted equation is as valid as the original equation, the former contains less information since we have two fewer indices. For example it may be shown that

$$\eta_{ijk}\delta_{mn} - \eta_{mjk}\delta_{in} - \eta_{imk}\delta_{jn} - \eta_{ijm}\delta_{kn} = 0$$

(There are 243 equations here) However, if we contact k with m we find

$$\eta_{ijn} - \eta_{mjm}\delta_{in} - \eta_{imm}\delta_{jn} - \eta_{ijn} = 0$$

and using the properties of the permutation symbol $\eta_{mjm} = 0$ and $\eta_{imm} = 0$ then we get 0=0 which, although valid, contains no information whatsoever!

1.7 Vector Differential Calculus

Consider the situation in which we are given a vector function of a single indpendent variable, t, specifically a position vector:

$$\underline{r}(t) = x(t)\hat{i} + y(t)\hat{j} + z(t)\hat{k} \qquad \textbf{(1.42)}$$

As t changes the end-point of the position vector $\underline{r}$ traces out a curve in space. Obviously the direction and magnitude of $\underline{r}$ changes as the *parameter* t changes.

We can ask 'what is the rate of change of $\underline{r}$ with respect to t'? As with scalar functions of a single variable this leads directly to the concept of the derivative.

We define the derivative of $\underline{r}$ with respect to t to be

$$\frac{d\underline{r}}{dt} = \frac{dx}{dt}\hat{\imath} + \frac{dy}{dt}\hat{\jmath} + \frac{dx}{dt}\hat{k} \tag{1.43}$$

(The notation $\dot{\underline{r}}$ is sometimes used instead of $\dfrac{d\underline{r}}{dt}$)

If $y = f(x)$ (a curve in the xy plane) then dy/dx has the interpretation of being the gradient of the tangent line to the curve at the point of calculation. In a similar way one can show that the interpretation of $d\underline{r}/dt$ is as the **tangent vector** to the curve $\underline{r} = \underline{r}(t)$ at the point of calculation.

Although we have considered only the derivative of a position vector with respect to a parameter it is clear that **any** vector function of a single parameter t, say $\underline{a}(t)$, may be regarded as a position vector since any vector may be translated so as to emanate from the origin of coordinates. It follows that the derivative as defined above holds for any such vector function of a single parameter.

The familiar properties of the derivative of scalar functions $f(t)$, $g(t)$:

(i) $\quad \dfrac{d}{dt}(c) = 0 \qquad$ if c is a constant

(ii) $\quad \dfrac{d}{dt}(f(t) + g(t)) = \dfrac{df}{dt} + \dfrac{dg}{dt}$

(iii) $\quad \dfrac{d}{dt}(f(t)g(t)) = \dfrac{df}{dt}g + \dfrac{dg}{dt}f$

generalise in an obvious way to vector functions. For example, if $\underline{c}$ is a constant vector $\underline{c} = c_x\hat{\imath} + c_y\hat{\jmath} + c_z\hat{k}$ where c_x, c_y, c_z are constants (implying that the direction and magnitude of $\underline{c}$ remain fixed) then

$$\frac{d\underline{c}}{dt} = \frac{dc_x}{dt}\hat{\imath} + \frac{dc_y}{dt}\hat{\jmath} + \frac{dc_z}{dt}\hat{k} = 0$$

Also if $\underline{a}(t) = a_x(t)\hat{\imath} + a_y(t)\hat{\jmath} + a_z(t)\hat{k}$ and $\underline{b}(t) = b_x(t)\hat{\imath} + b_y(t)\hat{\jmath} + b_z(t)\hat{k}$ then

$$\begin{aligned}\frac{d}{dt}(\underline{a} + \underline{b}) &= \frac{d}{dt}\left[(a_x + b_x)\hat{\imath} + (a_y + b_y)\hat{\jmath} + (a_z + b_z)\hat{k}\right] \\ &= \frac{d\underline{a}}{dt} + \frac{d\underline{b}}{dt}\end{aligned}$$

Similarly

$$\frac{d}{dt}(f(t)\,\underline{a}(t)) = \frac{df}{dt}\,\underline{a} + f\frac{d\underline{a}}{dt} \qquad \text{a vector equation}$$

$$\frac{d}{dt}(\underline{a}.\underline{b}) = \frac{d\underline{a}}{dt}.\underline{b} + \underline{a}.\ \frac{d\underline{b}}{dt} \quad \text{a scalar equation}$$

$$\frac{d}{dt}(\underline{a} \wedge \underline{b}) = \frac{d\underline{a}}{dt} \wedge \underline{b} + \underline{a} \wedge \frac{d\underline{b}}{dt} \quad \text{a vector equation}$$

We should emphasize that in this last result (as in all formulae involving the vector product) the order of the vectors is important.

The vector derivative may be extended in two ways. Firstly we can obtain the second derivative of $\underline{a}(t)$ with respect to t; this is denoted by $\dfrac{d^2\underline{a}}{dt^2}$ and is defined in an obvious way:

$$\frac{d^2\underline{a}}{dt^2} = \frac{d}{dt}\left(\frac{d\underline{a}}{dt}\right)$$

Higher derivatives are similarly defined, for example $\dfrac{d^3\underline{a}}{dt^3} = \dfrac{d}{dt}\left(\dfrac{d^2\underline{a}}{dt^2}\right)$.

Secondly we can reverse the differentiation operation by the process of **vector integration**. If we have two vector functions $\underline{a}(t)$, $\underline{A}(t)$ related to each other by

$$\frac{d\underline{A}}{dt} = \underline{a} \tag{1.44}$$

then $\underline{A}$ is called the indefinite integral of $\underline{a}$ and we write

$$\underline{A} = \int \underline{a}\ dt \tag{1.45}$$

Now, since $\dfrac{d\underline{c}}{dt} = 0$ for any constant vector $\underline{c}$ the indefinite integral is only unique up to the addition of a constant vector. For example, if

$$\underline{a} = t\hat{i} + t^2\hat{j} + \cos t\hat{k} \qquad \text{then} \qquad \underline{A} = \frac{t^2}{2}\hat{i} + \frac{t^3}{3}\hat{j} + \sin t\hat{k} + \underline{c}$$

since $\dfrac{d\underline{A}}{dt} = \underline{a}$ for any choice of constant vector $\underline{c}$.

The most obvious application of the derivative of a vector occurs when the parameter t is the time and when $\underline{r}(t)$ describes the path of a moving particle in space. In this case $\dfrac{d\underline{r}}{dt}$ represents the **velocity** of the particle in both magnitude and direction and $\dfrac{d^2\underline{r}}{dt^2}$ represents the **acceleration** of the particle, again in magnitude and direction.

EXERCISES

1.20 Show that every point on the curve $\underline{r} = t\cos t\hat{\imath} + t\sin t\hat{\jmath} + t\hat{k}$ lies on a cone with vertex at the origin. Show that straight lines on the cone passing through the vertex are aligned at 45° to the z-axis.

1.21 Show, for any vector $\underline{r}(t)$, that $\dfrac{d}{dt}\left(\underline{r}\wedge\dfrac{d\underline{r}}{dt}\right) = \underline{r}\wedge\dfrac{d^2\underline{r}}{dt^2}$.

1.22 Determine the indefinite integrals of the following vectors:

(i) $\underline{a}(\theta) = 3\sin\theta\hat{\imath} + 4\sin\theta\cos\theta\hat{\jmath} + \theta^3\hat{k}$

(ii) $\underline{a}(t) = \frac{2}{3}t^{\frac{3}{2}}\hat{\imath} + t\hat{\jmath} + t\hat{k}$

1.23 A particle moves along the path (a helix) $\underline{r}(t) = \cos t\hat{\imath} + \sin t\hat{\jmath} + t\hat{k}$ where the parameter t is the time in seconds. At $t = 3$ seconds the particle flies off the path along a tangent. Determine the velocity and acceleration of the particle at the time of separation and locate its position 1 second later assuming that it continues along the tangent line at the same speed.

• The Gradient of a Scalar Field

Whereas a scalar is simply a number, a **scalar field** is a rule which assigns a number at each point of space; thus it is simply a function of the coordinates. If $G(x, y, z)$ is a scalar field then we define the gradient of G, written $\underline{\nabla}G$ or **grad**G by

$$\underline{\nabla}G = \hat{\imath}\frac{\partial G}{\partial x} + \hat{\jmath}\frac{\partial G}{\partial y} + \hat{k}\frac{\partial G}{\partial z} \qquad \textbf{(1.46)}$$

The gradient of a scalar field is a vector

It will sometimes be convenient to consider the vector operator $\underline{\nabla}$ alone:

$$\underline{\nabla} \equiv \hat{\imath}\frac{\partial}{\partial x} + \hat{\jmath}\frac{\partial}{\partial y} + \hat{k}\frac{\partial}{\partial z} \qquad \textbf{(1.47)}$$

so that the gradient of G, can be thought of as the operator $\underline{\nabla}$ applied to G to produce $\hat{\imath}\dfrac{\partial G}{\partial x} + \hat{\jmath}\dfrac{\partial G}{\partial y} + \hat{k}\dfrac{\partial G}{\partial z}$. The operator $\underline{\nabla}$ has properties similar to the ordinary derivative operator $\dfrac{d}{dt}$ in that if G_1, G_2 are any two scalar functions then

$$\text{(i)}\quad \underline{\nabla}(G_1 + G_2) = \underline{\nabla}G_1 + \underline{\nabla}G_2 \qquad \textbf{(1.48)}$$

and

(ii) $\underline{\nabla}(G_1G_2) = G_1\underline{\nabla}G_2 + G_2\underline{\nabla}G_1$ **(1.49)**

- **Vector Fields**

Formally a vector field is a rule which associates a vector with each point in a region of space. In mathematical terms a vector field is a vector whose components are functions of the variables x, y, z. (Here we are specifically considering 3-D space. In some cases we may want to consider 2-D space or even 4-D space in which case the components of the vector will be functions of two variables (x, y) or of four variables (x, y, z, t) respectively). Vector fields are often denoted by $\underline{A}(\underline{r})$ or $\underline{A}(x, y, z)$:

$$\underline{A}(x,y,z) = A_x(x,y,z)\hat{i} + A_y(x,y,z)\hat{j} + A_z(x,y,z)\hat{k} \quad \textbf{(1.50)}$$

- **The Divergence and Curl of a Vector Field**

The divergence of a vector field $\underline{F}$ is written $\underline{\nabla}.\underline{F}$ (or div $\underline{F}$) and is defined as

$$\underline{\nabla}.\underline{F} = \frac{\partial F_x}{\partial x} + \frac{\partial F_y}{\partial y} + \frac{\partial F_z}{\partial z} \quad \textbf{(1.51)}$$

The divergence of a vector field is a scalar field.

The curl of $\underline{F}$ written $\underline{\nabla} \wedge \underline{F}$ (often written **curl** $\underline{F}$ or $\underline{\nabla} \times \underline{F}$) is defined most conveniently in determinant form as:

$$\underline{\nabla} \wedge \underline{F} = \begin{vmatrix} \hat{i} & \hat{j} & \hat{k} \\ \frac{\partial}{\partial x} & \frac{\partial}{\partial y} & \frac{\partial}{\partial z} \\ F_x & F_y & F_z \end{vmatrix} \quad \textbf{(1.52a)}$$

or, when expanded

$$\underline{\nabla} \wedge \underline{F} = \hat{i}[\frac{\partial F_z}{\partial y} - \frac{\partial F_y}{\partial z}] + \hat{j}[\frac{\partial F_x}{\partial z} - \frac{\partial F_z}{\partial x}] + \hat{k}[\frac{\partial F_y}{\partial x} - \frac{\partial F_x}{\partial y}] \quad \textbf{(1.52b)}$$

The curl of a vector field is a vector field

- **Properties of Gradient, Divergence and Curl**

If $G(x, y, z)$ is any scalar field and $\underline{A}(x, y, z)$, $\underline{B}(x, y, z)$ are any two vector fields then the following results are valid:

(i) $\underline{\nabla}.(\underline{A} + \underline{B}) = \underline{\nabla}.\underline{A} + \underline{\nabla}.\underline{B}$

(ii) $\underline{\nabla} \wedge (\underline{A} + \underline{B}) = \underline{\nabla} \wedge \underline{A} + \underline{\nabla} \wedge \underline{B}$

(iii) $\underline{\nabla}.(G\underline{A}) = (\underline{\nabla}G).\underline{A} + G(\underline{\nabla}.\underline{A})$

(iv) $\underline{\nabla} \wedge (G\underline{A}) = (\underline{\nabla} G) \wedge \underline{A} + G(\underline{\nabla} \wedge \underline{A})$

These are the 'obvious' properties to be expected of any derivative operator. Note that the order of the vectors $(\underline{\nabla} G) \wedge \underline{A}$ is important in (iv). Other properties of note are the identities:

(v) $\underline{\nabla}.(\underline{\nabla} \wedge \underline{A}) = 0$ (div(**curl** $\underline{A}$) is zero for any vector field $\underline{A}$)

(vi) $\underline{\nabla} \wedge (\underline{\nabla} G) = 0$ (**curl**(**grad** G) is zero for any scalar field G)

The proofs of all of these properties (i) to (vi) are quite straightforward if one works directly from the definitions. We shall see below that the proofs are elementary if one uses indices.

The divergence of the vector field obtained by taking the gradient of a scalar field Φ is of considerable importance in applications. It is a scalar quantity called the **Laplacian** of Φ and is denoted by $\nabla^2\Phi$.

$$\underline{\nabla}.(\underline{\nabla}\Phi) \equiv \nabla^2\Phi = \text{div } (\textbf{grad } \Phi) \tag{1.53}$$

Written out explicitly we have

$$\begin{aligned} \nabla^2\Phi &= \underline{\nabla}.\left[\hat{i}\frac{\partial\Phi}{\partial x} + \hat{j}\frac{\partial\Phi}{\partial y} + \hat{k}\frac{\partial\Phi}{\partial z}\right] \\ &= \frac{\partial}{\partial x}\left(\frac{\partial\Phi}{\partial x}\right) + \frac{\partial}{\partial y}\left(\frac{\partial\Phi}{\partial y}\right) + \frac{\partial}{\partial z}\left(\frac{\partial\Phi}{\partial z}\right) \\ &= \frac{\partial^2\Phi}{\partial x^2} + \frac{\partial^2\Phi}{\partial y^2} + \frac{\partial^2\Phi}{\partial z^2} \end{aligned} \tag{1.54}$$

A scalar function is said to be **harmonic** in a region if its Laplacian vanishes in that region i.e. Φ is harmonic if

$$\nabla^2\Phi = 0$$

This partial differential equation is called Laplace's equation.

The gradient of a scalar field Φ is, as we have seen, a vector field $\underline{\nabla}\Phi$. We cannot therefore apply the gradient operator twice since after the first operation we would be trying to take the gradient of a vector field which is not defined. Similarly the divergence of a vector field $\underline{A}$ gives a scalar field and we could not again take this divergence. In other words neither $\underline{\nabla}(\underline{\nabla}\Phi)$ nor $\underline{\nabla}.(\underline{\nabla}.\underline{A})$ are defined.

On the other hand the curl of a vector field is itself a vector field and the curl operation could be repeated i.e. $\underline{\nabla} \wedge \underline{\nabla} \wedge \underline{A}$ is a meaningful quantity (it is another vector field). It is possible to show, by direct expansion in Cartesian coordinates, that the expansion of $\underline{\nabla} \wedge \underline{\nabla} \wedge \underline{A}$ is the same as that of the quantity

$$\underline{\nabla}(\underline{\nabla}.\underline{A}) - \nabla^2\underline{A}$$

where $\nabla^2\underline{A}$ is **defined** as a vector field whose components are $\nabla^2 A_x$, $\nabla^2 A_y$, $\nabla^2 A_z$. In other words we have an important vector identity:

$$\underline{\nabla} \wedge \underline{\nabla} \wedge \underline{A} \equiv \underline{\nabla}(\underline{\nabla}.\underline{A}) - \nabla^2\underline{A} \qquad \text{or} \qquad \textbf{curl curl } \underline{A} \equiv \textbf{ grad } \text{div } \underline{A} - \nabla^2\underline{A}$$

(We do not rewrite $\nabla^2\underline{A}$ as div **grad** $\underline{A}$.)

- **The Index Notation for Derivatives**

If G is a scalar function of the coordinates (x_1, x_2, x_3) then the partial derivative of G with respect to variable x_i : $\dfrac{\partial G}{\partial x_i}$ will be denoted by $G_{,i}$. For example, $G_{,1}$ means $\dfrac{\partial G}{\partial x_1}$ $\left(\text{or } \dfrac{\partial G}{\partial x}\right)$. Thus the object $G_{,i}$ stands for the collection of derivatives $\left(\dfrac{\partial G}{\partial x_1}, \dfrac{\partial G}{\partial x_2}, \dfrac{\partial G}{\partial x_3}\right)$: that is, it is the indexed version of the gradient $\underline{\nabla} G$. Similarly $G_{,ij}$ means $\dfrac{\partial^2 G}{\partial x_i \partial x_j}$.

The divergence can be treated similarly: a vector field $\underline{A}(x, y, z)$ will be denoted by A_i and so the divergence of $\underline{A}$ is:

$$\underline{\nabla}.\underline{A} = \frac{\partial A_1}{\partial x} + \frac{\partial A_2}{\partial y} + \frac{\partial A_3}{\partial z} = \frac{\partial A_1}{\partial x_1} + \frac{\partial A_2}{\partial x_2} + \frac{\partial A_3}{\partial x_3}$$

$$= A_{1,1} + A_{2,2} + A_{3,3}$$

which, if we employ the summation convention can be written in the form:

$$\underline{\nabla}.\underline{A} = A_{i,i}$$

We have seen above that if a scalar function G satisfies Laplace's equation $(\underline{\nabla}.(\underline{\nabla}G) = 0)$ then it is called harmonic. In terms of the index notation and again employing the summation convention G is harmonic if it satisfies $G_{,ii} = 0$

The curl of $\underline{A}$ is (notationally) the cross product of the operator $\underline{\nabla}$ with $\underline{A}$ and so the ith component of the curl is

$$(\underline{\nabla} \wedge \underline{A})_i = \eta_{ijk} \frac{\partial}{\partial x_j} A_k = \eta_{ijk} A_{k,j} \qquad \textbf{(1.55)}$$

For example $(\underline{\nabla} \wedge \underline{A})_1 = \eta_{1jk} \dfrac{\partial}{\partial x_j} A_k = \eta_{123} \dfrac{\partial A_3}{\partial x_2} + \eta_{132} \dfrac{\partial A_2}{\partial x_3}$ on using the summation convention and the properties of η_{ijk}. But $\eta_{132} = -1$ so $(\underline{\nabla} \wedge \underline{A})_1 = \dfrac{\partial A_3}{\partial x_2} - \dfrac{\partial A_2}{\partial x_3}$ as we know.

To illustrate the use of the index notation we will prove property (vi) above: $\underline{\nabla} \wedge (\underline{\nabla} G) = 0$. The ith component of $\underline{\nabla} \wedge (\underline{\nabla} G)$ is

$$\eta_{ijk} (\underline{\nabla} G)_{k,j} = \eta_{ijk} (G_{,k})_{,j} = \eta_{ijk} G_{,kj}$$

But

$$G_{,kj} \equiv \frac{\partial^2 G}{\partial x_k \partial x_j} \equiv \frac{\partial^2 G}{\partial x_j \partial x_k} \equiv G_{,jk}$$

i.e. it is symmetric in the indices k and j. But η_{ijk} is skew-symmetric and therefore, from the result of Example 1.1, $\eta_{ijk} G_{,jk} = 0$ which proves the identity.

Example 1.2

Using the index approach prove the identities
(a) $\underline{\nabla} \wedge (\underline{F} \wedge \underline{G}) = (\underline{G}.\underline{\nabla})\underline{F} - (\underline{F}.\underline{\nabla})\underline{G} + \underline{F}(\underline{\nabla}.\underline{G}) - \underline{G}(\underline{\nabla}.\underline{F})$
(b) $\underline{\nabla} \wedge (\underline{\nabla} \wedge \underline{A}) = \underline{\nabla}(\underline{\nabla}.\underline{A}) - \nabla^2 \underline{A}$

Solution

(a) Let F_i and G_i be the ith components of $\underline{F}$ and $\underline{G}$ respectively and let H_r denote the rth component of $\underline{F} \wedge \underline{G}$

$$H_r = (\underline{F} \wedge \underline{G})_r = \eta_{rjk} F_j G_k$$

The pth component of $\underline{\nabla} \wedge \underline{H}$ $(= \underline{\nabla} \wedge (\underline{F} \wedge \underline{G}))$ is, using (1.55)

$$(\underline{\nabla} \wedge \underline{H})_p = \eta_{pqr} H_{r,q}$$

But

$$H_{r,q} = (\eta_{rjk} F_j G_k)_{,q} = \eta_{rjk}(F_{j,q} G_k + F_j G_{k,q})$$

Therefore

$$(\underline{\nabla} \wedge \underline{H})_p = \eta_{pqr} \eta_{rjk}(F_{j,q} G_k + F_j G_{k,q}) \qquad \mathbf{(1.56)}$$

(simply taking the partial derivative of a product and realising that terms in η are constants). We now employ the basic η-δ described earlier:

$$\eta_{pqr}\eta_{rjk} = \eta_{rpq}\eta_{rjk} = \delta_{pj}\delta_{qk} - \delta_{pk}\delta_{qj}$$

therefore

$$\eta_{pqr}\eta_{rjk}(F_{j,q}G_k + F_j G_{k,q}) = (\delta_{pj}\delta_{qk} - \delta_{pk}\delta_{qj})(F_{j,q}G_k + F_j G_{k,q})$$

$$= F_{p,q}G_q + F_p G_{q,q} - F_{q,q}G_p - F_q G_{p,q}$$

Substituting into (1.56) the pth component of $\underline{\nabla} \wedge (\underline{F} \wedge \underline{G})$ may be expressed as

$$[(\underline{G}.\underline{\nabla})\underline{F}]_p + [(\underline{\nabla}.\underline{G})\underline{F}]_p - [(\underline{\nabla}.\underline{F})\underline{G}]_p - [(\underline{F}.\underline{\nabla})\underline{G}]_p$$

which proves the identity.
(b) This identity is verified using the techniques outlined in (a). Firstly the mth component of curl $\underline{A}$ is:

$$(\underline{\nabla} \wedge \underline{A})_m = \eta_{mrq} A_{q,r}$$

therefore

$$
\begin{aligned}
[\underline{\nabla} \wedge (\underline{\nabla} \wedge \underline{A})]_i &= \eta_{ipm}[\eta_{mrq}A_{q,r}]_{,p} \\
&= \eta_{ipm}\eta_{mrq}A_{q,rp} \\
&= [\delta_{ir}\delta_{pq} - \delta_{iq}\delta_{pr}]A_{q,rp} \\
&= A_{p,ip} - A_{i,rr}
\end{aligned}
$$

But $A_{p,ip} = A_{p,pi}$ (order of differentiation is not important) and also

$$A_{p,pi} = (A_{p,p})_{,i} = [\underline{\nabla}(\underline{\nabla}.\underline{A})]_i \qquad \text{and} \qquad A_{i,pp} = [\nabla^2 \underline{A}]_i$$

and so

$$\underline{\nabla} \wedge (\underline{\nabla} \wedge \underline{A}) = \underline{\nabla}(\underline{\nabla}.\underline{A}) - \nabla^2 \underline{A} \tag{1.57}$$

EXERCISES

1.24 Compute the gradient $\underline{\nabla}\Phi$ at the point (1,1,1) for the following scalar fields:
(i) $\Phi = xy + yz + xz$
(ii) $\Phi = xyz \cos \pi y$

1.25 If $\Phi = \dfrac{xy}{z}$ and $\underline{A} = \sin x \; \hat{i} + \cos x \; \hat{j} + xy \; \hat{k}$ verify the identities
curl grad $\Phi = 0$, div **curl** $\underline{A} = 0$ and **curl curl** $\underline{A} =$ **grad** div $\underline{A} - \nabla^2 \underline{A}$

1.26 If $\underline{G} = 2xye^z\hat{i} + e^z x^2 \hat{j} + (x^2 y \, e^z + z^2)\hat{k}$ show that $\underline{\nabla} \wedge \underline{G} = 0$ and find a function $\Phi(x, y, z)$ such that $\underline{G} = \underline{\nabla}\Phi$.

1.27 If $\underline{F} = 2x \; \hat{i} + x^2 \; y \; \hat{j} + z^2 \; \hat{k}$ and $\underline{G} = x \; \hat{i} + y \; \hat{j}$ verify the identities:
(i) $\underline{\nabla}.(\underline{F} \wedge \underline{G}) = \underline{G}.(\underline{\nabla} \wedge \underline{F}) - \underline{F}.(\underline{\nabla} \wedge \underline{G})$
(ii) $\underline{\nabla}(\underline{F}.\underline{G}) = (\underline{F}.\underline{\nabla})\underline{G} + (\underline{G}.\underline{\nabla})\underline{F} + \underline{F} \wedge (\underline{\nabla} \wedge \underline{G}) + \underline{G} \wedge (\underline{\nabla} \wedge \underline{F})$
Using indices verify the identities in (i) and (ii) in general.

1.8 Vector Integral Calculus

- **Line Integrals**

A line integral is obtained as a limiting case of discrete summation. To take a specific example, consider the problem of moving an object up an inclined plane from point P_1 to P_2 through a distance ℓ (see Figure 1.22a).

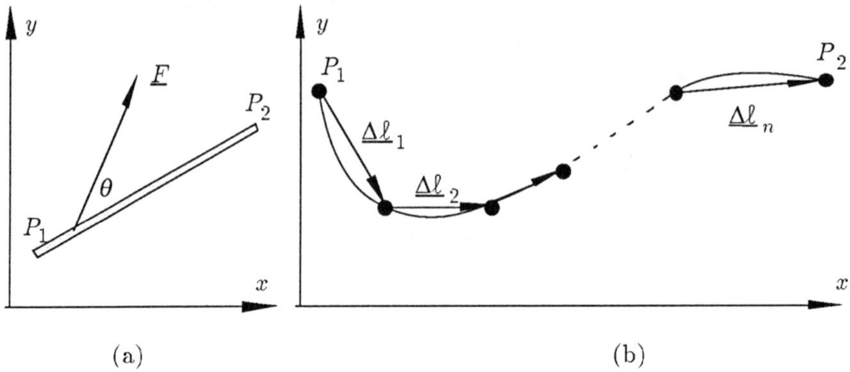

Figure 1.22

If a constant force $\underline{F}$ is applied to the object then the work done is given by $|\underline{F}|\ell\cos\theta$ or in vector form $\underline{F}.\underline{\ell}$, where $\underline{\ell}$ denotes the vector $\overrightarrow{P_1P_2}$. Now if the force $\underline{F}$ varies in magnitude and direction (in which case the angle θ varies) this expression will be incorrect. Also if P_1P_2 is not a straight line some modification of approach is required.

What we can do in all cases (see Figure 1.22b), is to divide the path P_1P_2 into small segments which can be approximated by straight lines and over which $\underline{F}$ can be assumed constant, say at a value $\underline{F}_i$ for the ith segment. Then if $\underline{\Delta\ell}_i$ denotes the ith segment, the work done in moving the object along this segment is approximately $\underline{F}_i.\underline{\Delta\ell}_i$ and for the whole path of N segments the total work done is approximately $\sum_{i=1}^{N}\underline{F}_i.\underline{\Delta\ell}_i$. The exact work done can be obtained by taking smaller and smaller segments and letting N increase. The limit of this sum as $N\to\infty$ and all $\underline{\Delta\ell}_i \to 0$ is called the **line integral** of $\underline{F}$ along the path C from P_1 to P_2 and is written $\int_C \underline{F}.\underline{d\ell}$. Thus a line integral, like the more familiar definite integral, is defined as the limit of a sum.

We should note that we use the symbol $\oint_C$ to denote a line integral over a **closed** path; that is, one where the initial point P_1 and the end point P_2 coincide.

The essential ingredients then for a line integral of the form $\int_C \underline{F}.\underline{d\ell}$ are a vector field $\underline{F}$ and a path C traversed in a definite sense. Despite the term 'line integral' the path C is not necessarily a straight line.

The result of a line integral is a scalar number. If $\underline{F}$ denotes a physical quantity then the units of $\int_C \underline{F}.\underline{d\ell}$ are the units of $\underline{F}$ multiplied by the units of distance, for example if $\underline{F}$ is a force (in newtons) and $\underline{d\ell}$ is a length in metres then the units of $\int_C \underline{F}.\underline{d\ell}$ are Nm or Joules J, the units of **work.**

The notation $\int_C \underline{F}.\underline{d\ell}$ is a common one for a line integral and is clearly neat and concise. Some authors use $\underline{dr}$ instead of $\underline{d\ell}$ for the 'line element'. In cartesian coordinates this quantity has the form

$$\underline{d\ell}\ (\text{or } \underline{dr}) = dx\hat{\imath} + dy\hat{\jmath} + dz\hat{k}$$

and it can be thought of as a vector between points with coordinates (x, y, z) and $x + dx, y + dy, z + dz)$.

- **Surface Integrals**

There is often a need to sum (or, in the limiting case, integrate) a physical quantity across a surface and such a calculation arises in a number of areas of Engineering and Science.

Consider, for example, conduction of heat. Heat will be conducted through a medium if a temperature gradient exists. In a simple one-dimensional case, such as heat flowing in a thin rod with temperature $T(x)$ whose surfaces are insulated and which is lying along the x-direction, the basic physical law is that the rate Q at which heat flows across unit area is proportional to the temperature gradient $\frac{dT}{dx}$. Mathematically we can express this law by the simple scalar equation

$$Q = -\kappa \frac{dT}{dx} \qquad \textbf{(1.58)}$$

where the constant of proportionality κ is called the thermal conductivity of the medium.

In the more general three-dimensional case we can represent the heat flow by a vector field $\underline{Q}$ whose direction at any point gives the direction of the heat flow and whose magnitude (as in one-dimensional case) gives the rate at which heat crosses a surface of unit area normal to the flow. The three-dimensional equivalent of (1.58) is, assuming that κ is a constant and independent of direction,

$$\underline{Q} = -\kappa \underline{\nabla} T \qquad \textbf{(1.59)}$$

where $\underline{\nabla} T$ is the gradient of the scalar field T.

Now if $\underline{Q}$ is constant in direction and magnitude then, by its definition, the rate at which heat crosses the flat surface of area A normal to the direction of flow is $f = |\underline{Q}|A$, see Figure 1.23a. If the area A is inclined at an angle θ (Figure 1.23b) then the rate of heat flow is $f = |\underline{Q}| \cos\theta\ A$, i.e. we take the component of $\underline{Q}$ normal to A.

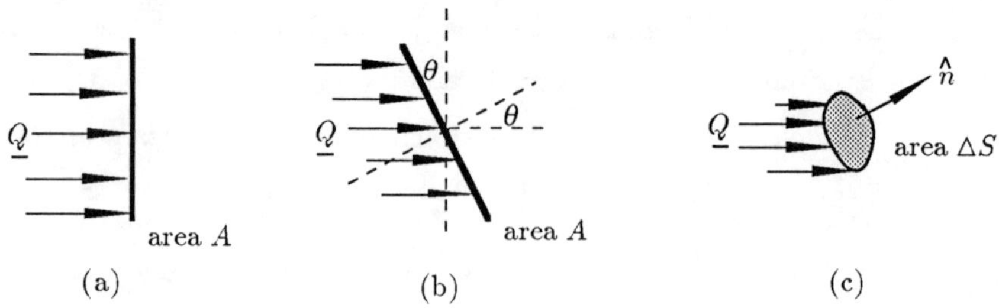

Figure 1.23

If $\underline{Q}$ is varying across A or if A is not flat these simple results will not be valid. However, if we take a small enough surface element of area say ΔS (a more common notation that ΔA) then the surface element can be assumed approximately flat and any variation of $\underline{Q}$ across ΔS will be small. Hence $\Delta F = |\underline{Q}| \cos\theta \ \Delta S$ will give us the approximate rate of heat flow across ΔS.

It is convenient to re-write this equation in vector form. To do this we associate with the area ΔS a direction specified by a unit vector $\hat{n}$ taken perpendicular or **normal** to ΔS (Figure 1.23c) and define a **vector area**

$$\underline{\Delta S} = \Delta S \ \hat{n}$$

(See also Section 1.3)

The expression $|\underline{Q}| \cos\theta \ \Delta S$ then becomes $\underline{Q}. \ \hat{n}\Delta S$ or $\underline{Q}.\underline{\Delta S}$. (Note that there are two possible unit normals $\hat{n}_1$ and $\hat{n}_2$ to a given surface element ΔS. However $\hat{n}_1 = -\hat{n}_2$ so the only possible ambiguity is one of sign rather than magnitude).

For a general surface S (Figure 1.24a) the total heat flow crossing S is approximated by adding the flows across the N elements ΔS_i $(i = 1, 2, \ldots, N)$ which make up S. For the ith such element ΔS_i:

$$\text{rate of heat flow } \Delta f_i \approx \underline{Q}_i.\hat{n}_i \Delta S_i \qquad \text{or} \qquad \underline{Q}_i.\underline{\Delta S}_i$$

so that the total rate of heat flow across S is

$$f \approx \sum_{i=1}^{N} \underline{Q}.\hat{n}_i \Delta S_i \qquad \text{or} \qquad \sum_{i=1}^{N} \underline{Q}_i.\underline{\Delta S}_i$$

If we now increase N and correspondingly let the elements ΔS_i become smaller and smaller the sum will tend to a limit and the rate of heat flow will be given by $f = \lim_{N\to\infty} \sum_{i=1}^{N} \underline{Q}_i.\hat{n} \ \Delta_i S$ which is written as $\iint_S \underline{Q}.\hat{n} \ dS$. The integral is called the **surface integral** of $\underline{Q}$ over S. The alternative notation $\iint_S \underline{Q}.\underline{dS}$ is also used. If the surface integral S is **closed** then, by convention, we use for $\hat{n}$ the unit normal which is directed **outwards** from the volume enclosed by S (Figure 1.24b) and the surface integral is written as $\oiint_S \underline{Q}.\hat{n} \ dS$.

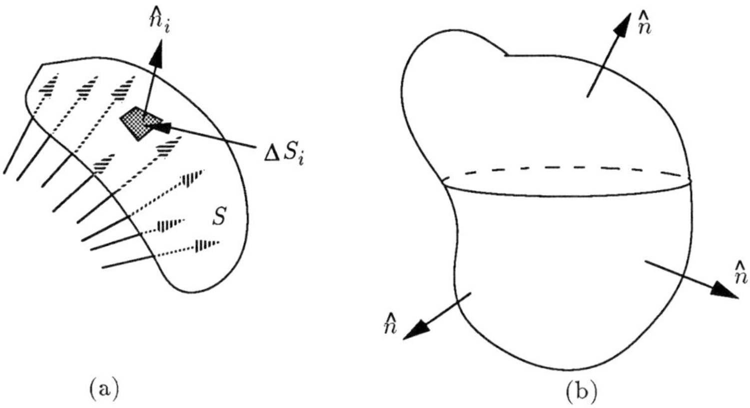

Figure 1.24

In general for any vector field $\underline{F}$ we can denote the surface integral of $\underline{F}$ across a surface S by either of the representations

$$\iint_S \underline{F}.\hat{n}\, dS \qquad \text{or} \qquad \iint_S \underline{F}.\underline{dS}$$

where the integrals are limiting cases of summations over S as already discussed.

- **Volume Integrals**

A volume integral is one taken over a closed region or volume in space. Also volume integrals, like their line and surface counterparts, are limiting cases of summations.

To begin with a physical example, we learn in elementary science that the mass of a body occupying a volume V is obtained by multiplying V by the mass density ρ_v of the body. However, this is actually only valid if ρ_v is constant throughout the body. If ρ_v is variable then to find the total mass we first divide the volume V into N small elements, $\Delta V_1,\ \Delta V_2,\ \ldots,\ \Delta V_N$ (where V is the sum of all these small volumes i.e. $V = \sum_{i=1}^{N} \Delta V_i$). We can then say that if (x_i, y_i, z_i) are the coordinates of a point within the volume element ΔV_i the mass of the element ΔV_i is $\Delta M_i \approx \rho_v(x_i, y_i, z_i)\Delta V_i$ and the total mass of the whole body is given approximately by the summation

$$M \approx \sum_{i=1}^{N} \rho_v(x_i, y_i, z_i)\Delta V_i \qquad \mathbf{(1.60)}$$

To improve the approximation we take smaller and smaller elements and correspondingly let $N \to \infty$. Assuming that ρ_v is a well-behaved function of the coordinates x, y and z it can be shown that the sum in (1.60) tends to a finite limit which is called **volume integral** of ρ_v over the volume V. The mass M is then given exactly by

$$M = \iiint_V \rho_v dV \qquad \mathbf{(1.61)}$$

In a similar vein we can define the volume integral of any scalar field $f(x, y, z)$:

$$\iiint_V f(x, y, z)\ dV = \lim_{N\to\infty} \sum_{i=1}^{N} f(x_i, y_i, z_i)\Delta V_i$$

EXERCISES

1.28 If $\underline{F} = x \sin y\ \hat{\imath} + \cos y\ \hat{\jmath}$ is a force field, calculate the work done in moving a particle along the following paths
(i) a straight line path from (0,0) to (0,1) and then another straight line path to (1,1)
(ii) a straight line path from (0,0) to (1,0) and then to (1,1)
(iii) from (0,0) to (1,1) along the parabola $y = x^2$.

1.29 Distinguish between the area of a surface and a surface integral.

1.30 (a) A vector field $\underline{F}$ is given by $\underline{F} = \dfrac{y^2}{2}\hat{\imath} + x\ \hat{\jmath}$. Find the surface integral of $\underline{\nabla} \wedge \underline{F}$ over the hemisphere $z = +\sqrt{4 - x^2 - y^2}$.
(b) Evaluate the line integral $\oint_C \underline{F}.\underline{d\ell}$ over the circle $x^2 + y^2 = 4,\ z = 0$.
(You should obtain the same answer as for the surface integral).

1.31 Given the vector field $\underline{A} = (2x^2 - 3z)\hat{\imath} - 2xy\ \hat{\jmath} - 4x\ \hat{k}$ evaluate the volume integral $\iiint_V \underline{\nabla}.\underline{A}\ dV$ where V is the region bounded by the planes $x = 0,\ y = 0,\ z = 0$ and $2x + 2y + z = 4$.

• Integral Theorems of Vector Analysis

At the heart of the theory of vector fields are two powerful theorems. These are the **divergence theorem** and **Stokes' theorem** and they are a major unifying influence in the subject in that they provide connections between the various types of integral – line, surface and volume – that we have introduced above.

The divergence theorem and Stokes' theorem are generalisations of more elementary theorems in the integral calculus. We now list the various types of integral that we are likely to encounter, together with their associated theorems, and note the common threads which connect them together.

- **Definite Integral**

The basic result here is familiar enough

$$\int_a^b \frac{df}{dx} = f(b) - f(a) \tag{1.62}$$

(this is the fundamental theorem of calculus)

- **Line Integrals**

If Φ is a scalar field then a fundamental result for line integrals is

$$\int_{P_1}^{P_2} \frac{d\Phi}{ds}\, ds = \Phi(P_2) - \Phi(P_1) \tag{1.63}$$

This result can readily be seen as a generalization of (1.62) (to the case in which the path of integration is not a straight line). Indeed we can extend (1.63) to the case of a **vector line integral**, defined by

$$\int_{P_1}^{P_2} \frac{d\underline{F}}{ds}\, ds = \hat{\imath} \int_{P_1}^{P_2} \frac{dF_x}{ds}\, ds + \hat{\jmath} \int_{P_1}^{P_2} \frac{dF_y}{ds}\, ds + \hat{k} \int_{P_1}^{P_2} \frac{dF_z}{ds}\, ds$$

Using (1.63) for each component

$$\int_{P_1}^{P_2} \frac{d\underline{F}}{ds}\, ds = \hat{\imath}\{F_x(P_2) - F_x(P_1)\} + \hat{\jmath}\{F_y(P_2) - F_y(P_1)\} + \hat{k}\{F_z(P_2) - F_z(P_1)\}$$

that is

$$\int_{P_1}^{P_2} \frac{d\underline{F}}{ds}\, ds = \underline{F}(P_2) - \underline{F}(P_1) \tag{1.64}$$

If we imagine the path of integration C from P_1 to P_2 to be a part of a longer path (Figure 1.25) then the points P_1 and P_2 separate the set of points on this longer path into two regions – points on C and points not on C. We can say that P_1 and P_2 constitute the **boundary** of C. With respect to the longer path the boundary points of C allows a natural definition of the terms **inside** and **outside**. We can also define, again quite naturally, an **intrinsic** positive direction at the boundary points to be that direction which points from **inside** to **outside** the path C. The positive direction for the arc-length parameter is that shown in Figure 1.25 and is clearly opposite to the intrinsic positive direction P_1

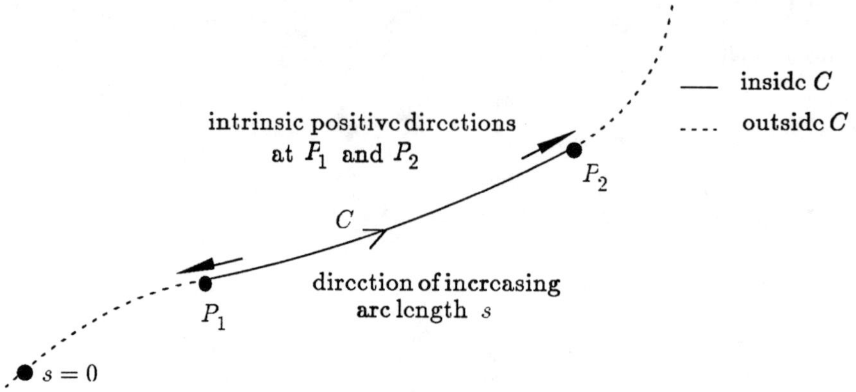

Figure 1.25

The result (1.64) then tells us, that the integral over a region, of a differential operator $\frac{d}{ds}$ acting on $\underline{F}$ is equal to the sum $(-\underline{F}(P_1) + \underline{F}(P_2))$ evaluated at the boundary of the region. (The negative sign is included in this 'sum' since at P_1 the positive intrinsic direction is in the negative direction for the measurement of arc-length).

- **Stokes' Theorem**

This theorem can be thought of as similar in form to (1.64) but of dimension one higher:

$$\iint_S \underline{\nabla} \wedge \underline{F}.\hat{n}\ dS = \oint_C \underline{F}.\underline{d\ell} \tag{1.65}$$

Here the 'region' of integration is a surface S and the 'boundary' is the closed curve C. The surface S may be regarded as part of a larger surface (Figure 1.26a).

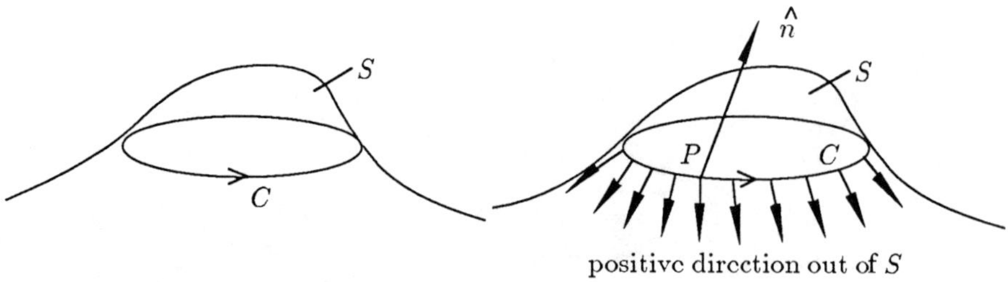

(a) (b)

Figure 1.26

There is a naturally defined 'inside' and 'outside' and a naturally defined positive direction (on S) pointing outwards (Figure 1.26b). The positive outward direction (perpendicular to C) together with the tangent direction to C naturally defines a positive normal to S according to the right hand rule. (See Figure 1.26b at the point P). The similarity of (1.65) to (1.64) is clear:- the operator $\frac{d}{ds}$ is now the curl $\underline{\nabla}\wedge$. The integral of $\underline{\nabla}\wedge\underline{F}$ over a region S is equal to the 'sum' evaluated on the boundary (not surprisingly this 'sum' is now a line integral as the boundary is now comprised of the infinite number of points C).

Of course, the direction chosen for positive $\hat{n}$ is intimately connected to the 'positive' direction on C. If C is traversed in the opposite direction then $\hat{n}$ also must be reversed.

In terms of indices, Stokes' theorem takes the form:

$$\iint_S \eta_{ijk}F_{k,j}n_i dS = \oint_C F_i dx_i$$

- **The Divergence Theorem**

The divergence theorem too, can be thought of as similar in form to (1.64) but of dimension two higher:

$$\iiint_V \underline{\nabla}.\underline{F}dV = \oiint_S \underline{F}.\hat{n}dS \qquad \textbf{(1.66)}$$

Here the 'region' of integration is a volume V and the boundary is the closed surface S. The region V may be regarded as part of a larger volume allowing a natural definition for the 'positive' normal to S (pointing from 'inside' to the 'outside' as in Figure 1.27)

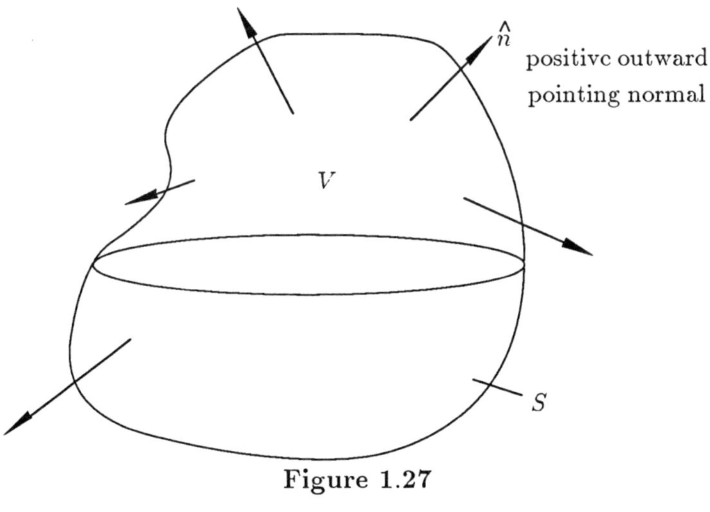

Figure 1.27

In index notation the divergence theorem may be written:

$$\iiint_V F_{i,i}dV = \oiint_S F_i n_i dS$$

Other vector integral identities may be obtained from Stokes' theorem or the divergence theorem. For example Green's first identity:

$$\text{(i)} \quad \oiint_S u\underline{\nabla}v.\underline{dS} = \iiint_V (u\nabla^2 v + \underline{\nabla}u.\underline{\nabla}v)dV$$

may be derived using the divergence theorem with integrand $u\underline{\nabla}v$ whilst Green's second identity

$$\text{(ii)} \quad \oiint_S (u\underline{\nabla}v - v\underline{\nabla}u).\underline{dS} = \iiint_V (u\nabla^2 v - v\nabla^2 u)dV$$

may be achieved from the first by interchanging u with v in (i) and then subtracting the two identities obtained.

More generally if $F_{ij...p}$ is an indexed object which is a function of position then a more general form of the divergence theorem is:

$$\oiint_S F_{ij...p}n_i dS = \iiint_V F_{ij...p,i}dV \qquad \textbf{(1.67)}$$

and a more general form of Stokes' theorem is

$$\oint_C F_{ij...p}dx_i = \iint_S \eta_{ikm}F_{mj...p,k}n_i dS \qquad \textbf{(1.68)}$$

EXERCISES

1.32 If $\underline{r}$ denotes a position vector from the origin to a point with coordinates (x, y, z) show that

(i) $\oint_C \underline{r}.\underline{d\ell} = 0$ for any closed surface C

(ii) $\oiint_S \underline{r}.\hat{n}dS = 3$ (volume enclosed by S)

(iii) If S is the surface bounded by the paraboloid $z = 4 - (x^2 + y^2)$ and the xy plane show that $\oiint_S \underline{r}.\hat{n}dS = 24\pi$

1.33 (a) Show that the volume enclosed by any closed surface S is $V = \dfrac{1}{6}\oiint_S \underline{\nabla}r^2.\hat{n}dS$

(b) If there exists a scalar field $h(x, y, z)$ such that $\nabla^2\Phi = h\Phi$ and $\nabla^2\vartheta = h\vartheta$ in a region R bounded by a closed surface S, prove that

$$\oiint_S (\Phi\underline{\nabla}\vartheta - \vartheta\underline{\nabla}\Phi).\hat{n}\ dS = 0$$

(c) If Φ is harmonic in a region R show that

$$\oiint_S \Phi\underline{\nabla}\Phi.\hat{n}dS = \iiint_R |\underline{\nabla}\Phi|^2 dV$$

(d) Show that if $\underline{A} = \underline{\nabla}\Phi$ and Φ is harmonic then

$$\iiint_R |\underline{A}|^2 dV = \oiint_S \Phi\underline{A}.\hat{n}\ dS$$

1.34 Verify the divergence theorem for the vector field $\underline{F} = xy\hat{i} - yz\hat{j} + 3\hat{k}$ and the hemisphere $x^2 + y^2 + z^2 = 1,\ z \geq 0$.

1.35 Verify Stokes' Theorem for the vector field $\underline{F} = x\hat{i} + (2z - x)\hat{j} + y^2\hat{k}$ using two surfaces each bounded by the circle $x^2 + y^2 = 4,\ z = 1$

Chapter 2

CARTESIAN TENSORS

2.1 Introduction

In this text we shall deal, almost exclusively, with Cartesian coordinates. They are certainly the most 'familiar' coordinates and, mathematically, the easiest to deal with. This is not to imply that they are always the most convenient to use. In physical situations which display certain symmetries then it may be appropriate to suspend the use of Cartesian coordinates and to use a coordinate system more attuned to the natural symmetry inherent in the problem. However, there is often a hefty price to be paid when using 'curvilinear' coordinates – particularly when dealing with vectors. The base vectors $\hat{\imath}$, $\hat{\jmath}$, $\hat{k}$ in Cartesians are constant in both magnitude and direction. This is not the case for curvilinear coordinates; at least one of the base vectors $\hat{e}_1$, $\hat{e}_2$, $\hat{e}_3$ has a direction which is *position dependent.* This considerably complicates the treatment of vector differentiation.

So, unless special circumstances arise, we will use Cartesian coordinates. However, we will not *demand* that all observers (experimenters) use exactly the same system of Cartesian coordinates. That would be too restrictive. To accomodate the freedom as to the choice of *which* Cartesian coordinates one may use, we must develop a formalism so as to be able to compare objects (they may be scalars or vectors or some other object) in different coordinate systems.

Consider, for example, two Cartesian coordinate systems x_i and x'_i (referred to as *unprimed* and *primed* systems) with origins at O and O' respectively. (See Figure 2.1). A free vector $\underline{v}$ is 'observed' from both coordinate systems.

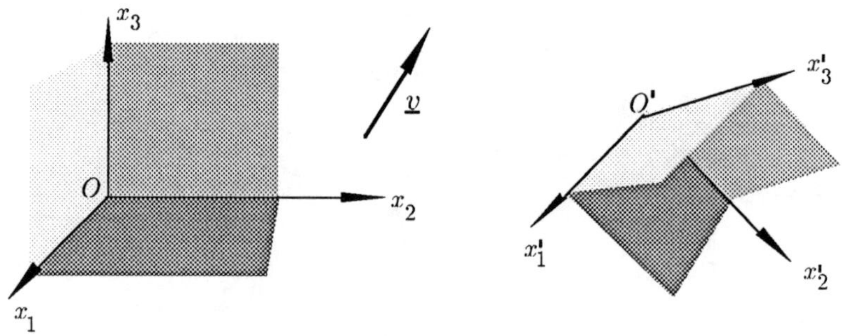

Figure 2.1

With respect to x_i and x'_i the vector will appear to be 'different'. Generally, it will certainly have different components in the two coordinate systems. Each observer would transfer the vector (keeping it pointing in the same direction) so that it points out of *their* origin. The components of the vector with respect to each observer can then be determined.

An observer at O labels this vector $\underline{v}_O$ whilst an observer at O' labels this vector $\underline{v}_{O'}$. How can the two observers agree that $\underline{v}_O$ and $\underline{v}_{O'}$ are in fact the same vector?

We note first that if the two coordinate systems are the same apart from a translation (see Figure 2.2) then $\underline{v}_O$ and $\underline{v}_{O'}$ would have *exactly* the same components in both coordinate systems and both observers would agree that *their* vectors are identical.

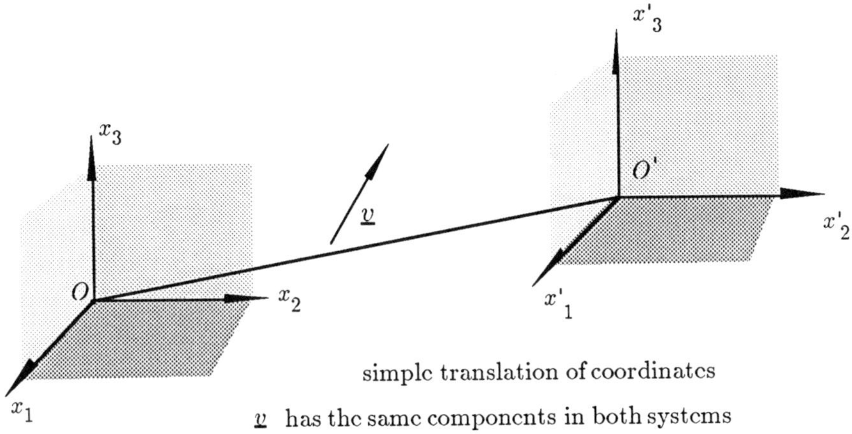

simple translation of coordinates

$\underline{v}$ has the same components in both systems

Figure 2.2

Thus, as far as the observation (or measurement) of a free vector is concerned we can ignore a simple translation of coordinates.

In the more general case (a combined translation and rotation) the observer at O can rotate his coordinate system so as to be aligned with the coordinate system at O' (this is only possible if both coordinate systems are right-handed or if both are left-handed). As the rotation at O takes place the components of $\underline{v}_O$ will change. If, when the alignment is complete, the components of $\underline{v}_O$ are the same as the components of $\underline{v}_{O'}$ then both observers can agree that they are viewing the same vector.

This problem of comparing objects in different coordinate systems must be understood before we proceed as it is fundamental to the study of any physical science. When we consider the comparison mathematically, we shall be led to re-define what we mean by a vector (in terms of how its components transform from one coordinate system to another) and also to generalise the vector to produce new objects called tensors.

Since simple translations of coordinate systems are of no consequence in the description of free vectors we shall only consider the relation between the components of a vector in Cartesian coordinate systems which differ by a rotation alone. Here, and throughout this text, we shall only use right-handed coordinate systems.

2.2 Rotation of Cartesian Coordinates

Consider two coordinate systems x_i and x'_i, both with the same origin (see Figure 2.3).

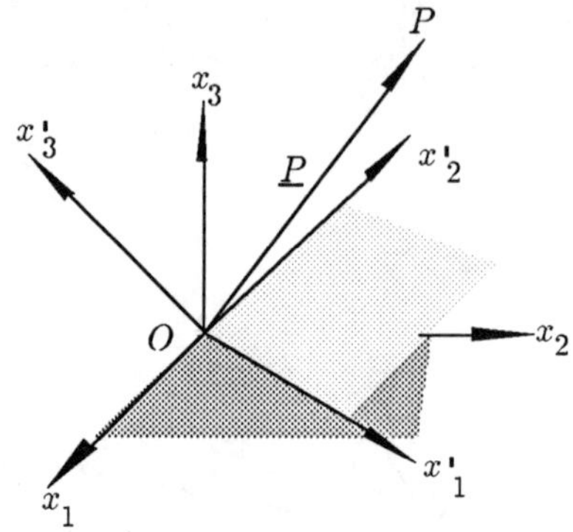

Figure 2.3

A point P in space may be specified by listing its coordinates in either system. Let the coordinates of $\underline{P}$ be P_i in the unprimed system and P'_i in the primed system. Then we have, for any vector, (see (1.20))

$$\underline{P} = \begin{cases} (\underline{P}.\hat{\imath})\hat{\imath} + (\underline{P}.\hat{\jmath})\hat{\jmath} + (\underline{P}.\hat{k})\hat{k} & \text{unprimed system} \\ (\underline{P}.\hat{\imath}')\hat{\imath}' + (\underline{P}.\hat{\jmath}')\hat{\jmath}' + (\underline{P}.\hat{k}')\hat{k}' & \text{primed system} \end{cases} \qquad \textbf{(2.1)}$$

In particular we can *choose* $\underline{P} \equiv \hat{\imath}$ and use the second of (2.1) to find the relation:

$$\hat{\imath} = (\hat{\imath}.\hat{\imath}')\hat{\imath}' + (\hat{\imath}.\hat{\jmath}')\hat{\jmath}' + (\hat{\imath}.\hat{k}')\hat{k}' \qquad \textbf{(2.2a)}$$

Similarly, by choosing $\underline{P} \equiv \hat{\jmath}$ and $\underline{P} \equiv \hat{k}$ in turn, we find:

$$\hat{\jmath} = (\hat{\jmath}.\hat{\imath}')\hat{\imath}' + (\hat{\jmath}.\hat{\jmath}')\hat{\jmath}' + (\hat{\jmath}.\hat{k}')\hat{k}' \qquad \textbf{(2.2b)}$$

and

$$\hat{k} = (\hat{k}.\hat{\imath}')\hat{\imath}' + (\hat{k}.\hat{\jmath}')\hat{\jmath}' + (\hat{k}.\hat{k}')\hat{k}' \qquad \textbf{(2.2c)}$$

Introducing the short-hand notation

$$\begin{array}{lll} \alpha_{11} = \hat{\imath}.\hat{\imath}' & \alpha_{21} = \hat{\imath}.\hat{\jmath}' & \alpha_{31} = \hat{\imath}.\hat{k}' \\ \alpha_{12} = \hat{\jmath}.\hat{\imath}' & \alpha_{22} = \hat{\jmath}.\hat{\jmath}' & \alpha_{32} = \hat{\jmath}.\hat{k}' \\ \alpha_{13} = \hat{k}.\hat{\imath}' & \alpha_{23} = \hat{k}.\hat{\jmath}' & \alpha_{33} = \hat{k}.\hat{k}' \end{array} \qquad \textbf{(2.3)}$$

and, interpreting the scalar products (2.3) in the usual way we see that

$$\alpha_{ij} = \text{cosine of the angle between the positive } x'_i \text{ and } x_j \text{ axes} \qquad \textbf{(2.4)}$$

We refer to these quantities as **rotation cosines.**

Now, in (2.1), using an obvious notation for the components of $\underline{P}$ in primed and unprimed systems, we shall write $\underline{P}.\hat{\imath} = P_1$, $\underline{P}.\hat{\jmath} = P_2$, $\underline{P}.\hat{k} = P_3$, $\underline{P}.\hat{\imath}' = P_1'$ etc and so

$$P_1\hat{\imath} + P_2\hat{\jmath} + P_3\hat{k} = P_1'\hat{\imath}' + P_2'\hat{\jmath}' + P_3'\hat{k}' \qquad \textbf{(2.5)}$$

Using (2.2) and (2.3) in (2.5) we have

$$P_1(\alpha_{11}\hat{\imath}' + \alpha_{21}\hat{\jmath}' + \alpha_{31}\hat{k}') + P_2(\alpha_{12}\hat{\imath}' + \alpha_{22}\hat{\jmath}' + \alpha_{32}\hat{k}') \\ + P_3(\alpha_{13}\hat{\imath}' + \alpha_{23}\hat{\jmath}' + \alpha_{33}\hat{k}') = P_1'\hat{\imath}' + P_2'\hat{\jmath}' + P_3'\hat{k}'$$

Equating coefficients implies

$$P_1' = \alpha_{11}P_1 + \alpha_{12}P_2 + \alpha_{13}P_3 = \alpha_{1j}P_j$$
$$P_2' = \alpha_{21}P_1 + \alpha_{22}P_2 + \alpha_{23}P_3 = \alpha_{2j}P_j$$
$$P_3' = \alpha_{31}P_1 + \alpha_{32}P_2 + \alpha_{33}P_3 = \alpha_{3j}P_j$$

or, more succinctly,

$$P_i' = \alpha_{ij}P_j \qquad \textbf{(2.6)}$$

Alternatively, by choosing $\underline{P} \equiv \hat{\imath}'$, $\hat{\jmath}'$ and $\hat{k}'$ in turn in (2.1) would provide expressions (similar to those in (2.2)) for $\hat{\imath}'$, $\hat{\jmath}'$ and $\hat{k}'$ in terms of $\hat{\imath}$, $\hat{\jmath}$ and $\hat{k}$. Substituting these expressions into (2.5) implies

$$P_i = \alpha_{ji}P_j' \qquad \textbf{(2.7)}$$

Equations (2.6) (or equivalently (2.7)) are the equations which relate the components of a vector with respect to two Cartesian coordinate systems, one being rotated with respect the other.

Example 2.1

Determine the coordinates of a vector $\underline{P}$ referred to a primed system of coordinates obtained by rotating the x_1, x_2 axes about the x_3 axis through an angle θ.

Solution

The situation is described in Figure 2.4.
From the definition of α_{ij} in (2.4) we find

$$\alpha_{11} = \cos\theta \qquad \alpha_{21} = \cos(90+\theta) = -\sin\theta \qquad \alpha_{31} = 0$$
$$\alpha_{12} = \cos(90-\theta) = \sin\theta \qquad \alpha_{22} = \cos\theta \qquad \alpha_{32} = 0$$
$$\alpha_{13} = 0 \qquad \alpha_{23} = 0 \qquad \alpha_{33} = 1$$

implying, from (2.6)

$$P_1' = \cos\theta\ P_1 + \sin\theta P_2$$
$$P_2' = -\sin\theta P_1 + \cos\theta P_2$$

$$P_3' = P_3$$

These relations are the familiar ones connecting two representations of a *single* vector when referred to two coordinate systems differing by a planar rotation.

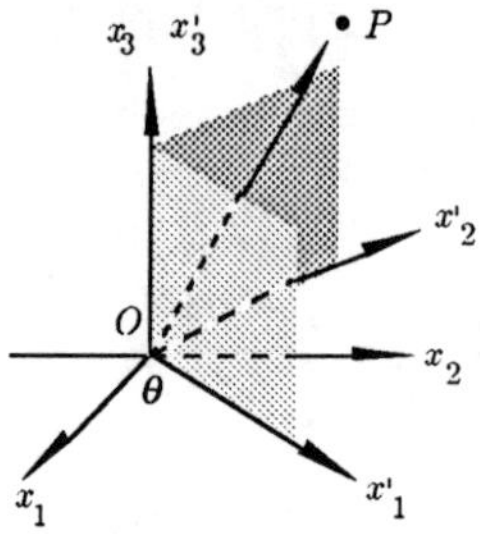

Figure 2.4

The nine rotation cosines (the α_{ij}'s) are not all independent. They satisfy a number of identities, resulting from the vector identities enjoyed by the unit vectors:

$$\hat{\imath}.\hat{\imath} = 1, \quad \hat{\imath}.\hat{\jmath} = 0, \; ..... \; \hat{\imath}'.\hat{\imath}' = 1, \; .... \; \hat{\imath}'.\hat{k}' = 0, \; ... \text{ etc}$$

By taking the scalar product of both sides of (2.2a) with the vector $\hat{\imath}$ we find

$$1 = (\hat{\imath}.\hat{\imath}')^2 + (\hat{\imath}.\hat{\jmath}')^2 + (\hat{\imath}.\hat{k}')^2$$

or

$$1 = \alpha_{11}^2 + \alpha_{21}^2 + \alpha_{31}^2 = \alpha_{k1}\alpha_{k1}$$

Repeating this process with (2.2b) we obtain

$$0 = (\hat{\jmath}.\hat{\imath}')(\hat{\imath}'.\hat{\imath}) + (\hat{\jmath}.\hat{\jmath}')(\hat{\jmath}'.\hat{\imath}) + (\hat{\jmath}.\hat{k}')(\hat{k}'.\hat{\imath})$$

or

$$0 = \alpha_{12}\alpha_{11} + \alpha_{22}\alpha_{21} + \alpha_{32}\alpha_{31} = \alpha_{k2}\alpha_{k1} \qquad \text{etc}$$

Proceeding in this way (taking the scalar product of (2.2a,b,c) with $\hat{\jmath}$ and $\hat{k}$ in turn) we find the full set of relations can be conveniently written in the form

$$\alpha_{ji}\alpha_{jk} = \delta_{ik} \tag{2.8a}$$

Also, proceeding to utilise the identities $\hat{\imath}'.\hat{\imath}' = 1, \quad \hat{\imath}'.\hat{\jmath}' = 0$ etc we find

$$\alpha_{ij}\alpha_{kj} = \delta_{ik} \tag{2.8b}$$

In (2.8a) the summation is on the first index of the rotation cosines whilst in (2.8b) the summation is on the second index.

• The Gradient Vector

Thus far we have been dealing with *free* vectors. Under a combined translation and rotation of coordinate systems only the rotation affects the components of such a vector; the precise relationship between the two sets of components being detailed in (2.6) (or (2.7)). However, the position vector of a point is *not* a free vector. For this kind of vector translations *are* important. To see this consider Figure 2.5.

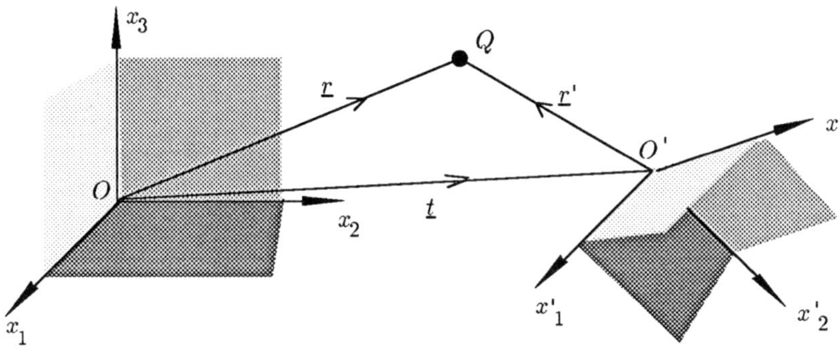

Figure 2.5

If $\underline{r}$ and $\underline{r}'$ are the position vectors of Q with respect to primed and unprimed systems then, using the parallelogram rule:

$$\underline{r}' = \underline{r} - \underline{t} \qquad \text{or} \qquad \underline{r} = \underline{r}' + \underline{t}$$

where $\underline{t}$ is the (constant) vector $\overrightarrow{OO'}$. Let

$$\underline{r}' = x'_1\hat{\imath}' + x'_2\hat{\jmath}' + x'_3\hat{k}' \qquad \text{and} \qquad \underline{r} = x_1\hat{\imath} + x_2\hat{\jmath} + x_3\hat{k}$$

then

$$x'_1 = x_1\hat{\imath}.\hat{\imath}' + x_2\hat{\jmath}.\hat{\imath}' + x_3\hat{k}.\hat{\imath}' - \underline{t}.\hat{\imath}' \qquad \text{etc}$$

These relations between the position vector components can be written more concisely using indices:

$$x'_i = \alpha_{ij}x_j - t'_i \tag{2.9a}$$

or alternatively

$$x_i = \alpha_{ji}x'_j + t_i \tag{2.9b}$$

So, unless $\underline{t} = 0$, the position vector x_i does not transform according to (2.6) (or (2.7)) on changing Cartesian coordinate systems. However, derivatives with respect to the position vector *do* transform according to (2.6) (since they eliminate the *constant* vector $\underline{t}$).

If Φ is a scalar function of position then the gradient of Φ calculated in the unprimed system is the vector $\dfrac{\partial \Phi}{\partial x_i}(\equiv \underline{\nabla}\Phi)$ whilst the gradient of Φ in the primed system would be $\dfrac{\partial \Phi}{\partial x'_i}$. Now from (2.9b)

$$\frac{\partial x_i}{\partial x'_k} = \alpha_{ji}\frac{\partial x'_j}{\partial x'_k} = \alpha_{ji}\delta_{jk} = \alpha_{ki} \tag{2.10}$$

(since $\underline{t}$ is a constant vector), and so

$$\frac{\partial \Phi}{\partial x'_i} = \frac{\partial \Phi}{\partial x_k}\frac{\partial x_k}{\partial x'_i} = \frac{\partial \Phi}{\partial x_k}\alpha_{ik} \tag{2.11a}$$

and hence the vector $\dfrac{\partial \Phi}{\partial x_i}$ *does* transform according to the rule for any *free* vector (2.6), irrespective of whether a translation has or has not occurred.

As in Section 1.6 we shall normally write a derivative using the 'comma' notation and so (2.11a) may be rewritten as

$$(\Phi_{,i})' = \alpha_{ik}\Phi_{,k} \tag{2.11b}$$

In an exactly similar way one may show that

$$\Phi_{,i} = \alpha_{ki}(\Phi_{,k})' \tag{2.12}$$

(This last result could be obtained by multiplying both sides of (2.11b) by α_{ij} and then using the identity (2.8a)).

EXERCISES

2.1 Verify the identities in (2.8) for the rotation cosines (listed in matrix form):

$$\alpha_{ij} = \begin{bmatrix} \frac{1}{3} & \frac{1}{3}+\frac{1}{\sqrt{3}} & \frac{1}{3}-\frac{1}{\sqrt{3}} \\ \frac{1}{3}-\frac{1}{\sqrt{3}} & \frac{1}{3} & \frac{1}{3}+\frac{1}{\sqrt{3}} \\ \frac{1}{3}+\frac{1}{\sqrt{3}} & \frac{1}{3}-\frac{1}{\sqrt{3}} & \frac{1}{3} \end{bmatrix}$$

Show also that the point with coordinates (0,1,1) in the unprimed system has coordinates $(\frac{2}{3}, \frac{2}{3}, \frac{2}{3} - \frac{1}{\sqrt{3}})$ with respect to the primed system.

2.2 (a) Let P be a fixed point with coordinates x_i in an unprimed system. On rotating the coordinate system through an angle θ about a direction n_i then (see Section 2.5)

$$x'_i = x_i \cos\theta + (1-\cos\theta)x_k n_k n_i + \sin\theta\, \eta_{ijk} x_j n_k$$

Show that $\alpha_{ik} = \delta_{ik}\cos\theta + (1-\cos\theta)n_i n_k + \sin\theta\, \eta_{ikm} n_m$

(b) Show that if a rotation of 180^0 about a line through the origin and the point with coordinates (p_1, p_2, p_3) is performed then $\alpha_{ik} = -\delta_{ik} + \dfrac{2p_i p_k}{p_m p_m}$

2.3 Cartesian Tensors

If A_i and B_i are two free vectors then we can imagine forming the 'product' A_iB_j (we remember that A_iB_i would be the scalar product; $\eta_{ijk}A_jB_k$ would be the vector product). This is a new product which has not arisen in ordinary vector algebra. It has nine 'components'; A_1B_1, A_1B_2, ..., A_3B_3. How does this object transform under a change of coordinate system? In the primed system the corresponding object (defined in the same way) would be $A'_iB'_j$.

Now since A_i, B_i are vectors:

$$A'_i = \alpha_{ij}A_j \qquad \text{and} \qquad B'_i = \alpha_{ij}B_j$$

and so

$$A'_iB'_j = (\alpha_{ik}A_k)(\alpha_{jm}B_m) = \alpha_{ik}\alpha_{jm}A_kB_m \tag{2.13}$$

The quantity A_iB_j transforms in a similar way to a free vector in the sense that the transformation rule (2.13) contains a separate set of rotation cosines α_{ij} for each free index. We are now in a position to define what is meant by a tensor:

- An object which transforms according to the rule $\Phi' = \Phi$ under a change of coordinate systems is called a **zeroth order tensor** or scalar.

- Any one-indexed object that transforms according to (2.6) is called a **first order tensor** or vector.

- More generally, an object with n free indices, denoted by $A_{ijk...p}$ in the unprimed system and by $A'_{ijk...p}$ in the primed system and which transforms according to the rule:

$$A'_{ijk...p} = \alpha_{il}\alpha_{jm}\alpha_{kn}.....\alpha_{pq}A_{lmn...q} \tag{2.14}$$

is called an **nth order tensor**.

Example 2.2
If Φ is a scalar show that $\Phi_{,ij}$ is a second order tensor.

Solution

From (2.11a) the gradient of Φ transforms according to the rule:

$$\frac{\partial \Phi}{\partial x_i'} = \alpha_{ik}\frac{\partial \Phi}{\partial x_k}$$

hence

$$\frac{\partial^2 \Phi}{\partial x_j' \partial x_i'} = \frac{\partial}{\partial x_j'}(\alpha_{ik}\frac{\partial \Phi}{\partial x_k}) = \frac{\partial}{\partial x_m}(\alpha_{ik}\frac{\partial \Phi}{\partial x_k})\frac{\partial x_m}{\partial x_j'}$$

since α_{ik} are constants and again using (2.10)

$$\frac{\partial^2 \Phi}{\partial x_j' \partial x_i'} = \alpha_{ik}\frac{\partial^2 \Phi}{\partial x_m \partial x_k}\alpha_{jm}$$

and so we can write

$$(\Phi_{,ij})' = \alpha_{ik}\alpha_{jm}\Phi_{,km}$$

Thus $\Phi_{,ij}$ transforms according to (2.14) and so is a second order tensor.

Example 2.3
Show that δ_{ij} is a second order tensor.

Solution

In a primed or unprimed system the definition of the Kronecker delta is the same

$$\delta_{ij}' \equiv \delta_{ij} \equiv \begin{cases} 1 & \text{if} \quad i = j \\ 0 & \text{if} \quad i \neq j \end{cases}$$

Let the components of δ_{ij} in the primed system be B_{ij}. Then if δ_{ij} is a tensor

$$B_{ij} = \alpha_{ik}\alpha_{jm}\delta_{km} = \alpha_{ik}\alpha_{jk}$$

using the substitution property of the Kronecker delta. Hence using (2.8b),

$$B_{ij} = \delta_{ij} = \begin{cases} 1 & \text{if} \quad i = j \\ 0 & \text{if} \quad i \neq j \end{cases}$$

implying $B_{ij} = \delta_{ij}'$ showing that the Kronecker delta components in primed and unprimed systems are related through

$$\delta_{ij}' = \alpha_{ik}\alpha_{jm}\delta_{km}$$

and so δ_{ij} *is* a second order tensor.
One can show that the permutation symbol η_{ijk} is a third order tensor. (This requires a sophisticated knowledge of the properties of determinants and so will be omitted from a text of this level).

2.4 Properties of Tensors

In this section we discuss some of the major properties of tensors; how they can be combined, how they can be differentiated, what is the meaning of a tensor equation and so on. Behind each property or construction is the fundamental law (2.14) defining how tensor components in different coordinate systems are related.

- **Null Tensor**

An nth order tensor $A_{ij\ldots p}$ is said to be a null (or zero) tensor if each component is zero. That is if

$$A_{ij\ldots p} = 0$$

From the tensor transformation law (2.14), if a tensor is null in one coordinate system then it is null in *all* coordinate systems.

- **Equality of Tensors**

Two nth order tensors $A_{ij\ldots p}$ and $B_{ij\ldots p}$ are said to be equal if and only if their components with respect to the same coordinate system satisfy

$$A_{ij\ldots p} = B_{ij\ldots p}$$

Two tensors which can be shown to be equal in one coordinate system are equal in all coordinate systems.

- **Addition of Tensors**

Addition is only defined between tensors of the same order. The sum of two nth order tensors $A_{ij\ldots p}$ and $B_{ij\ldots p}$ is defined as $C_{ij\ldots p}$ through

$$C_{ij\ldots p} = A_{ij\ldots p} + B_{ij\ldots p} \tag{2.15}$$

and is also an nth order tensor. The tensor character of $C_{ij\ldots p}$ is easily established by defining $C'_{ij\ldots p} = A'_{ij\ldots p} + B'_{ij\ldots p}$ in a primed coordinate system. Since $A_{ij\ldots p}$ and $B_{ij\ldots p}$ are both tensors satisfying (2.14) the result easily follows.

- **Multiplication of Tensors**

The multiplication of an nth order tensor $A_{ij\ldots k}$ with an mth order tensor $B_{pq\ldots r}$ is an $(m+n)$th order tensor $C_{ij\ldots kpq..r}$ defined by:

$$C_{ij\ldots kpq\ldots r} = (A_{ij\ldots k})(B_{pq\ldots r}) \tag{2.16}$$

As with addition, the tensor character of $C_{ij\ldots kpq\ldots r}$ is verified by defining $C'_{ij\ldots kpq\ldots r}$ as the product $(A'_{ij\ldots k})(B'_{pq\ldots r})$ and proceeding as above.

- **Contraction of Tensors**

If $A_{ij..p..q..r}$ is a kth order tensor then the result of contracting any two indices (see Section 1.6), say index p with index q, produces a tensor $A_{ij..p..p..r}$ of order $(k-2)$.

Proof

Since $A_{ij..p..q..r}$ is a tensor of order k then

$$A'_{ij..p..q..r} = \alpha_{ia}\alpha_{jm}\ldots\alpha_{pn}..\alpha_{qs}..\alpha_{rt}A_{am..n..s..t}$$

in which there are k α-terms.

Contracting p with q and using the identity $\alpha_{pn}\alpha_{ps} = \delta_{ns}$ we obtain

$$A'_{ij..p..p..r} = \alpha_{ia}\alpha_{jm}...\delta_{ns}.....\alpha_{rt}A_{am..n..s..t}$$

$$= \alpha_{ia}\alpha_{jm}........\alpha_{rt}A_{am..n..n..t}$$

verifying that $A_{am..p..p..t}$ transforms as a tensor of order $(k-2)$. (The contraction process has removed two of the α-terms on the right hand side).

For example, since $\Phi_{,ij}$ is a second order tensor then $\Phi_{,ii}$ (the Laplacian of Φ) is a scalar (zeroth order tensor). Also $A_{i,j}$ is a second order tensor and so $A_{i,i}$ (the divergence of A_i) is a scalar.

As we have noted earlier in Section 1.6 contraction of the indices p and q in a tensor can be effected by multiplying the tensor by δ_{pq}.

- **Tensor Equations**

An nth order tensor equation is simply a rearrangement of an nth order null tensor. For example if $(A_{ij} - B_{ikkj} + C_{ij})$ is a second order null tensor then

$$A_{ij} - B_{ikkj} + C_{ij} = 0$$

We say that

$$A_{ij} = B_{ikkj} - C_{ij}$$

or

$$B_{ikkj} = A_{ij} + C_{ij}$$

are second order tensor equations.

If the validity of a tensor equation can be determined in one (convenient) coordinate system then its validity is guaranteed in *all* coordinate systems. Tensor equations are *independent* of the coordinate system. If there is a single overriding reason for using and working with tensors, it is this.

A law of physics (or of elasticity in the case of this text) must be capable of being expressed in a form which is not dependent on any coordinate system, if it is to be valid. Consider, for example, Newton's second law (before Newton). Two budding physicists each perform an experiment; both apply a force to a mass m and then measure the corresponding acceleration. Physicist 1 happens to choose a coordinate system (x, y, z) such that the force points along his 'y-axis'. He then formulates his second law of motion in terms of the three equations:

$$0 = a_x$$

$$F_y = ma_y$$

$$0 = a_z$$

Physicist 2 chooses a more general coordinate system (x', y', z') and formulates his law in the form

$$F_{x'} = ma_{x'}$$

$$F_{y'} = ma_{y'}$$

$$F_{z'} = ma_{z'}$$

Both of these 'laws' are clearly coordinate dependent. Both formulations are valid only for the particular coordinate systems each experimenter has chosen. If these Physicists are to communicate their results to the outside world and if they are to make a claim that they are truly universal laws of physics, they must either

- re-express their results in coordinate independent form (if this is possible)

or

- Explain how the quantities within their equations transform under a change of coordinate system.

In this case the first approach is easier. Either physicist could rephrase their results by stating that 'the applied force is equal to the mass times the acceleration' (which is clearly coordinate independent) or more concisely, in mathematical terms, reformulate their results in terms of vectors:

$$\underline{F} = m\underline{a}$$

In elasticity theory we shall find that objects (tensors) more general than vectors occur quite naturally. If we are to formulate the laws of elasticity in coordinate independent form we shall need to develop the relations that arise in terms of *tensor equations*, since these are transparently independent of the particular coordinate system chosen.

- **Symmetric and Skew-Symmetric Tensors**

An nth order tensor $A_{ij..p..q..r}$ which satisfies

$$A_{ij..p..q..r} = A_{ij..q..p..r} \tag{2.17}$$

is said to be symmetric in the indices p and q whilst if

$$A_{ij..p..q..r} = -A_{ij..q..p..r} \tag{2.18}$$

then the tensor is said to be skew-symmetric in the indices p and q.

Evidently if one can establish the symmetry properties of a tensor in one coordinate system they will be valid in all coordinate systems since (2.17) and (2.18) are tensor equations.

As we have already noted in Section 1.6 the tensor δ_{ij} is symmetric in i and j (or simply symmetric) whereas the permutation symbol η_{ijk} is skew-symmetric in *all* pairs of indices.

Example 2.4

Show that if $\eta_{ijk}\sigma_{jk} = 0$ then σ_{jk} is symmetric

Solution

The simplest way to exhibit the symmetry of σ_{jk} is to list the three equations

$$\eta_{1jk}\sigma_{jk} = 0 \qquad \eta_{2jk}\sigma_{jk} = 0 \qquad \eta_{3jk}\sigma_{jk} = 0$$

that is (noting the values of the permutation symbol listed in (1.40a,b))

$$\eta_{123}(\sigma_{23} - \sigma_{32}) = 0 \qquad \eta_{213}(\sigma_{13} - \sigma_{31}) = 0 \qquad \eta_{312}(\sigma_{12} - \sigma_{21}) = 0$$

from which of course we deduce

$$\sigma_{23} = \sigma_{32} \qquad \sigma_{13} = \sigma_{31} \qquad \sigma_{12} = \sigma_{21}$$

hence σ_{ij} is symmetric.
(More formally: multiply both sides of $\eta_{ijk}\sigma_{jk} = 0$ by η_{iam} and use the $\eta - \delta$ identity, first introduced in Section 1.6, to give

$$\eta_{iam}\eta_{ijk}\sigma_{jk} = 0 \qquad \text{or} \qquad (\delta_{aj}\delta_{mk} - \delta_{ak}\delta_{jm})\sigma_{jk} = 0$$

which implies $\sigma_{am} - \sigma_{ma} = 0$ showing that σ_{am} is symmetric).

- **The Derivative of a tensor**

If $A_{ij...p}$ is an nth order tensor field (i.e. the components are functions of position) then the operation of taking a partial derivative with respect to one of the coordinates produces a new tensor of order $(n+1)$ with components defined by $\dfrac{\partial}{\partial x_s}(A_{ij...p})$ or simply $A_{ij...p,s}$.

Proof

Let $B_{ij...ps} \equiv \dfrac{\partial}{\partial x_s}(A_{ij...p})$ and assume the same definition in all coordinate systems. Then

$$\begin{aligned} B'_{ij...ps} &\equiv \frac{\partial}{\partial x'_s}(A'_{ij...p}) \\ &= \frac{\partial}{\partial x'_s}(\alpha_{iq}\alpha_{jr}.....\alpha_{pt}A_{qr...t}) \qquad \text{since } A_{qr...t} \text{ is a tensor} \\ &= \frac{\partial}{\partial x_m}(\alpha_{iq}\alpha_{jr}.....\alpha_{pt}A_{qr...t})\frac{\partial x_m}{\partial x'_s} \\ &= \alpha_{iq}\alpha_{jr}.....\alpha_{pt}(\frac{\partial}{\partial x_m}(A_{qr...t}))\alpha_{sm} \qquad \text{from (2.10)} \end{aligned}$$

Hence

$$A'_{ij...p,s} = \alpha_{iq}\alpha_{jr}......\alpha_{pt}\alpha_{sm}A_{qr...t,m} \tag{2.19}$$

verifying that $A_{qr...t,m}$ transforms as an $(n+1)$th order tensor.

The partial derivative, denoted by a comma, can be interpreted as the derivative of a tensor with respect to a vector x_i (a first order tensor).

EXERCISES

2.3 Show that the curl of a vector is a first order tensor, whilst the divergence is a zeroth order tensor.

2.4 The instantaneous velocity v_i of a rigid body is $v_i = t_i + \eta_{ijk}\omega_j x_k$ in which t_i and ω_i are constant vectors. Show that the curl of this vector is $2\omega_i$ whilst the divergence vanishes.

2.5 Show that *any* second order tensor may be written as the sum of a symmetric and a skew-symmetric tensor.

2.6 If T_{ij} is a second order tensor and n_i is a vector then show that $T_{ij}n_in_j$ is a scalar.

2.7 Determine the components of a symmetric second order tensor S_{ij} when referred to a primed system of coordinates x_i' obtained from x_i by a planar rotation through an angle θ about the x_3-axis (see (4.16) of Section 4.4).

2.5 The Rotation Tensor

The reader may not be fully convinced that there is any need for tensors. To add further to the argument as to the usefulness of tensors we shall show that the commonly occurring geometrical operation of performing a *finite* rotation is most simply described, mathematically, in terms of a second order tensor.

A single finite rotation from P to P' (see Figure 2.6) is completely specified by (i) the direction and location in space of the axis of rotation and (ii) the angle θ turned through. We use the word 'finite' when θ is not small (usually $\theta > 5^0$).

A natural question to ask is whether a finite rotation can be represented by a vector. It is tempting to think that such a representation is possible – after all rotation *seems* vectorial in nature and it is uniquely characterised by three pieces of information.

Whether or not a finite rotation can be represented by a vector is easily checked. We know that vector addition is commutative i.e. if $\underline{a}$ and $\underline{b}$ are vectors then $\underline{a} + \underline{b} = \underline{b} + \underline{a}$.

We investigate, therefore, whether a similar result holds for rotations, i.e. if we perform two successive rotations R_1 and R_2 in that order do we get the same result if we apply R_2 first and then R_1?

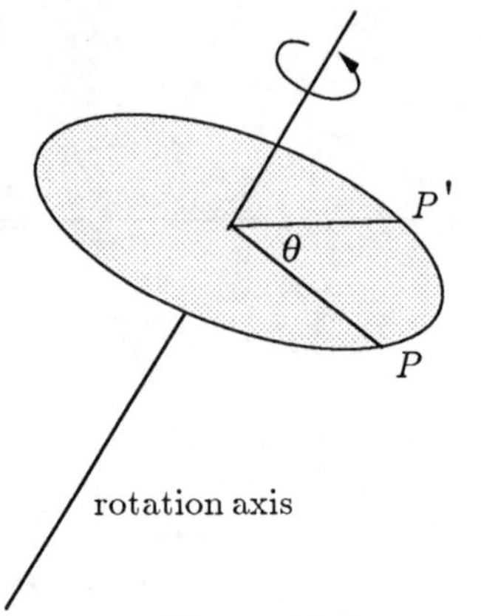

Figure 2.6

To answer this question we set up a Cartesian coordinate system and consider an initial point P with coordinates $(0,0,1)$ as in Figure 2.7a

We apply two simple right-handed rotations.

- R_1 : about the $x-$axis through angle 90^0
- R_2 : about the $y-$axis through angle 90^0

(The term 'right-handed' here simply means that the sense of the rotation is the same as that of a right handed screw advancing in the positive direction along a coordinate axis).

Applying R_1 first takes P into the point $P'(0,-1,0)$ on the y-axis as shown in Figure 2.7b. Application of R_2 will then leave P' unchanged.

Conversely, applying R_2 first takes P into $P''(1,0,0)$ on the x-axis as shown in Figure 2.7c. Application of R_1 will then leave P'' unchanged.

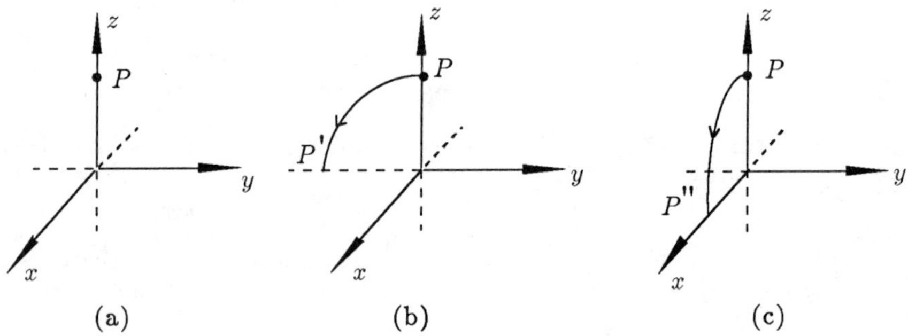

Figure 2.7

We can see clearly that the order in which the two rotations, about different axes, are performed is significant i.e. R_1 followed by R_2 has a differenent resultant from R_2 followed by R_1. In other words the commutative law **fails** for finite rotations. That is, contrary perhaps to our intuition a finite rotation **cannot** be fully characterised by a single vector. We shall see however, that it can be characterised by a second order tensor. We shall derive this basic result using classical vector methods and then simply convert to tensor notation.

Consider a rigid body undergoing a finite rotation through an angle θ about an axis in the direction of the unit vector $\hat{n}$. If $\underline{r}$ denotes the position vector of a particle originally at P (choosing an arbitrary point O on the rotation axis as origin) we shall determine an expression for the position vector $\underline{r}'$ of this particle after rotation. See Figure 2.8a

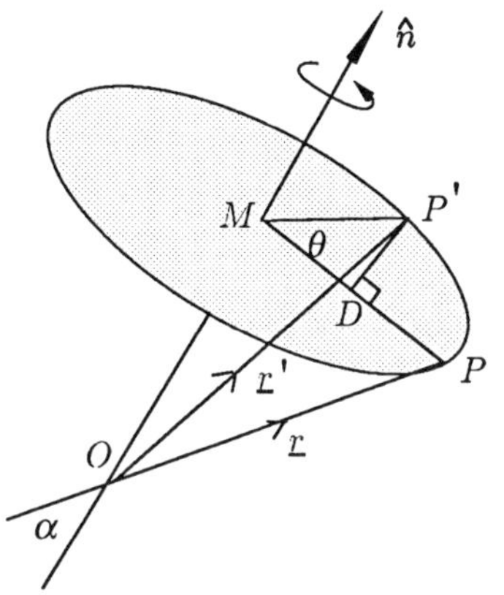

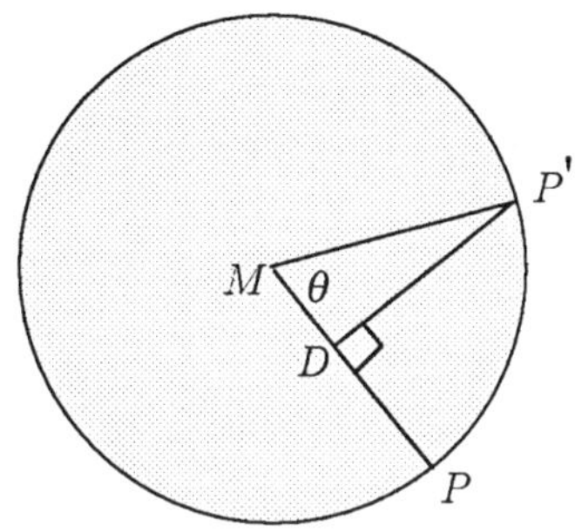

Figure 2.8

Clearly

$$\underline{r}' = \underline{r} + \overrightarrow{PD} + \overrightarrow{DP'} \tag{2.20}$$

where D is the point on MP such that $P'D$ and MP are perpendicular. (see Figure 2.8b).
Also

$$\overrightarrow{OM} = |\overrightarrow{OM}|\hat{n} = (\underline{r}.\hat{n})\hat{n}$$

and so

$$\overrightarrow{MP} = \underline{r} - \overrightarrow{OM} = \underline{r} - (\underline{r}.\hat{n})\hat{n}$$

Now, since P and P' lie on a circle with centre on the rotation axis at M

$$|\overrightarrow{P'M}| = |\overrightarrow{PM}|$$

Using this we find

$$\overrightarrow{PD} = |\overrightarrow{PD}|\frac{\overrightarrow{PM}}{|\overrightarrow{PM}|} = |\overrightarrow{PM}|(1-\cos\theta)\frac{\overrightarrow{PM}}{|\overrightarrow{PM}|} = \overrightarrow{PM}(1-\cos\theta)$$

$$= ((\underline{r}.\hat{n})\hat{n} - \underline{r})(1-\cos\theta) \qquad \text{(Note the sign change)}$$

Also $|\overrightarrow{DP'}| = |\overrightarrow{PM}|\sin\theta$ and $\overrightarrow{DP'}$ is in the direction of $\hat{n}\wedge\underline{r}$ and so

$$\overrightarrow{DP'} = |\overrightarrow{PM}|\sin\theta\frac{\hat{n}\wedge\underline{r}}{|\hat{n}\wedge\underline{r}|} = |\overrightarrow{PM}|\sin\theta\frac{\hat{n}\wedge\underline{r}}{|\underline{r}|\sin\alpha}$$

But from triangle OPM, 0 $\quad |\overrightarrow{PM}| = |\underline{r}|\sin\alpha$ and therefore

$$\overrightarrow{DP'} = (\hat{n}\wedge\underline{r})\sin\theta$$

Finally substituting for $\overrightarrow{PD}$ and $\overrightarrow{DP'}$ in (2.20)

$$\underline{r}' = \underline{r} + ((\underline{r}.\hat{n})\hat{n} - \underline{r})(1-\cos\theta) + (\hat{n}\wedge\underline{r})\sin\theta$$

or

$$\underline{r}' = \underline{r}\cos\theta + (\underline{r}.\hat{n})\hat{n}(1-\cos\theta) + (\hat{n}\wedge\underline{r})\sin\theta \qquad \textbf{(2.21)}$$

Rewriting this in index notation with the ith component of $\underline{r}$, $\underline{r}'$ and $\hat{n}$ denoted by x_i, x'_i and n_i respectively we find

$$x'_i = R_{ik}x_k \qquad \textbf{(2.22)}$$

in which

$$R_{ik} = \cos\theta\delta_{ik} + (1-\cos\theta)n_i n_k + \eta_{ijk}n_j\sin\theta \qquad \textbf{(2.23)}$$

Clearly, by its construction, R_{ik} is a second order tensor (it is a sum of terms each of which is a second order tensor). It is neither symmetric nor skew-symmetric.

According to (2.22) (x'_1, x'_2, x'_3) are the coordinates of the end point of a vector after it has been rotated through an angle θ about an axis in the direction of n_i. For example, if we rotate the vector with end point coordinates (p_1, p_2, p_3) through an angle θ about the x_3-axis then its final coordinates are $p'_i = R_{ik}p_k$ in which $n_i = (0,0,1) = \delta_{i3}$ and so

$$R_{ik} = \cos\theta\delta_{ik} + (1-\cos\theta)\delta_{i3}\delta_{k3} + \eta_{i3k}\sin\theta$$

Hence, from (2.22)

$$p'_1 = \cos\theta p_1 - \sin\theta p_2$$

$$p'_2 = \sin\theta p_1 + \cos\theta p_2$$

$$p'_3 = p_3$$

These relations should be compared with those determined in Example 2.1. The sets of equations are similar but not identical. In fact rotating a coordinate system (as in Example 2.1) through an angle θ is equivalent to rotating all the vectors of interest through an angle $-\theta$. The above transformation equations will be identical to those of Example 2.1 if θ is replaced by $-\theta$.

Although there is a very close correspondence between R_{ik} and α_{ik} (in fact α_{ik} is identical to R_{ik} except that θ is replaced by $-\theta$) there is one important difference. The indices on R_{ik} (as with any tensor indices) refer to a *single* coordinate system whereas the indices on α_{ik} refer to two distinct coordinate systems. Whilst R_{ik} is a second order tensor; α_{ik} cannot even be considered as a tensor – it is simply a collection of nine numbers defined according to (2.3) and which satisfy the relations (2.8).

- **Infinitessimal Rotations**

In elasticity theory we shall encounter so-called infinitessimal rotations, in which the angle θ (in radians) is very small; typically $\theta < 5^0$ will suffice (corresponding to a radian measure of approximately 0.1). In this case we have

$$\cos\theta \approx 1 - \frac{\theta^2}{2} \qquad \text{and} \qquad \sin\theta \approx \theta - \frac{\theta^3}{6}$$

and so, using these approximations we can write (2.22) as

$$x'_i \approx x_i + \eta_{ijk} x_k n_j \theta \qquad \text{ignoring non–linear terms in } \theta$$

so that the change in position (the displacement) as a result of an infinitessimal rotation is

$$x'_i - x_i \approx \eta_{ijk} x_k \omega_j$$

where $\omega_j = n_j\theta$ is a vector called the angular rotation vector. It specifies both the angle of rotation θ and the axis n_i about which the rotation takes place.

If a translation along vector t_i of the body as a whole is effected in addition to this infinitessimal rotation then the displacement is given by:

$$t_i + \eta_{ijk} x_k \omega_j \qquad (2.24)$$

EXERCISES

2.8 Determine the final coordinates of a point P initially at $(1,0,0)$ after successive rotations of 45^0 about the x_1-axis followed by 90^0 about the x_2-axis. What would be the position if the order of rotations was reversed?

2.9 Explain carefully the differences between finite rotations and infinitessimal rotations. When can finite rotations be treated as if they were infinitessimal rotations?

2.10 Determine the primed coordinates x'_i of a fixed point P which has coordinates x_i in the unprimed system if x'_i and x_i are related via a rotation through an angle θ about a direction n_i (see Exercise 2.2(a))

2.6 Isotropic Tensors

We have already noted that the Kronecker delta δ_{ij} and the permutation symbol η_{ijk} are tensors. However they have the unusual property that each of the components of these tensors have the same values in all coordinate systems. That is, they transform into themselves:

$$\delta'_{ij} = \delta_{ij} \qquad \text{and} \qquad \eta'_{ijk} = \eta_{ijk}$$

Tensors which have unchanging component values in all coordinate systems are called *isotropic*

We can attempt to determine, methodically, all the isotropic tensors of order zero, of order one and so on. Obviously, according to our definition, any scalar is a zeroth order isotropic tensor. It is also easy to show that there are no non-null first order isotropic tensors. For if we assume that v_i is an isotropic tensor then, according to (2.6)

$$v'_i \equiv v_i = \alpha_{ij} v_j \tag{2.25}$$

This equation, (2.25) has to be true in *all* coordinate systems. Choose α_{ij} to correspond to a rotation of 180^0 about the x_3-axis. Then from the results deduced in Example 2.1

$$\begin{array}{lll} \alpha_{11} = -1 & \alpha_{21} = 0 & \alpha_{31} = 0 \\ \alpha_{12} = 0 & \alpha_{22} = -1 & \alpha_{32} = 0 \\ \alpha_{13} = 0 & \alpha_{23} = 0 & \alpha_{33} = 1 \end{array} \tag{2.26}$$

and so (2.25) implies

$$v_1 = -v_1 \qquad v_2 = -v_2 \qquad v_3 = v_3$$

from which we conclude that $v_1 = v_2 = 0$. By choosing a rotation of 180^0 about the x_1-axis we can deduce that v_3=0. So that the only first order isotropic tensor is null which proves our claim.

A similar technique is used to show that an isotropic tensor of second order is proportional to δ_{ij}. To show this let v_{ij} be a second order isotropic tensor; then

$$v'_{ij} = v_{ij} = \alpha_{ik}\alpha_{jm} v_{km} \tag{2.27}$$

If we choose the rotation defined in (2.26) we find

$$\begin{array}{lll} v_{11} = v_{11} & v_{12} = v_{12} & v_{13} = -v_{13} \\ v_{21} = v_{21} & v_{22} = v_{22} & v_{23} = -v_{23} \\ v_{31} = -v_{31} & v_{32} = -v_{32} & v_{33} = v_{33} \end{array}$$

from which we deduce $v_{13} = v_{31} = v_{23} = v_{32} = 0$. By similar transformations about the x_1-axis we can show that $v_{12} = v_{21} = 0$.

If we now consider a planar transformation through an angle θ about the x_3-axis we have (again refer to Example 2.1)

$$\begin{array}{lll} \alpha_{11} = \cos\theta & \alpha_{21} = -\sin\theta & \alpha_{31} = 0 \\ \alpha_{12} = \sin\theta & \alpha_{22} = \cos\theta & \alpha_{32} = 0 \\ \alpha_{13} = 0 & \alpha_{23} = 0 & \alpha_{33} = 1 \end{array}$$

then (2.27) implies, for example

$$\begin{aligned} v_{11} &= \alpha_{1k}\alpha_{1m}v_{km} \\ &= \alpha_{11}(\alpha_{11}v_{11} + \alpha_{12}v_{12}) + \alpha_{12}(\alpha_{11}v_{21} + \alpha_{12}v_{22}) \\ &= \cos^2\theta\ v_{11} + \sin^2\theta\ v_{22} \end{aligned}$$

leading to $v_{11} = v_{22}$. Similar transformations about the x_2-axis shows that $v_{11} = v_{33}$. Hence we can write the second order isotropic tensor in the form

$$v_{ij} = K\delta_{ij}$$

where K is a constant, which proves our claim.

Further, it may be shown that any third order isotropic tensor must be proportional to η_{ijk}.

As we shall see in Chapter 5, fourth order isotropic tensors are of fundamental importance in elasticity theory and so of particular interest to this text. Using a rather tedius procedure similar to that described above (which we shall omit) it can be shown that the most general fourth order isotropic tensor has the form:

$$v_{ijkm} = \kappa\delta_{ij}\delta_{km} + \beta\delta_{ik}\delta_{mj} + \gamma\delta_{im}\delta_{jk} \tag{2.28}$$

in which κ, β and γ are constants.

2.7 Second Order Symmetric Tensors

Symmetric second order tensors arise quite naturally in the static and dynamic analysis of massive bodies. Consider a body V of density ρ (possibly variable) rotating with angular velocity $\underline{\omega}$ about a fixed point O. We choose this fixed point to be the origin of coordinates. We focus attention on an elementary region δV of mass $\delta m = \rho\delta V$ located at position vector $\underline{x}$. The velocity of this mass is (as usual) denoted by $\underline{\dot{x}}$ and the angular momentum $\delta\underline{h}$ is, by definition $\delta\underline{h} = \delta m\underline{x} \wedge \underline{\dot{x}}$. Now a simple calculation shows that:

$$\frac{d}{dt}(\delta m\underline{x} \wedge \underline{\dot{x}}) = \delta m\underline{x} \wedge \underline{\ddot{x}} = \underline{x} \wedge \delta\underline{F}$$

in which we have used Newton's second law that the applied force ($\delta\underline{F}$) equals the mass times the acceleration. This result states that the rate of change of angular momentum equals the applied couple. The total angular momentum $\underline{h}$ is $\iiint_V \rho\underline{x} \wedge \underline{\dot{x}}dV$. This may be expressed directly in terms of the rotation vector $\underline{\omega}$. Since $\underline{x} \wedge \underline{\dot{x}} = \underline{x} \wedge (\underline{\omega} \wedge \underline{x})$ then:

$$\underline{h} = \iiint_V \rho\underline{x} \wedge (\underline{\omega} \wedge \underline{x})dV$$

transforming to indices

$$h_i = \iiint_V \rho\eta_{ijk}x_j(\underline{\omega} \wedge \underline{x})_k dV = \iiint_V \rho\eta_{ijk}\eta_{kmp}x_jx_p\omega_m dV$$

But, using the $\eta - \delta$ identity described in Section 1.6, this can be rewritten:

$$h_i = \iiint_V \rho(\delta_{im}\delta_{jp} - \delta_{ip}\delta_{jm})x_j x_p \omega_m dV = \mathrm{I}_{im}\omega_m$$

in which

$$\mathrm{I}_{im} = \iiint_V \rho(\delta_{im}x_p x_p - x_i x_m)dV$$

The object I_{im} is, by construction, a second order symmetric tensor. It is called the inertia tensor of the body. The terms I_{11}, I_{22}, I_{33} are called the moments of inertia about the x_1-, x_2- and x_3-axes respectively. The terms I_{12}, I_{13}, I_{23} are called the products of inertia. The relation 'angular momentum equals inertia times the rotation vector' is akin to the corresponding relation for linear motion 'momentum equals mass times the velocity vector', though the multiplication is tensorial in the case of rotational motion as compared to scalar multiplication in the case of linear motion.

If the body is rigid, the inertia tensor components are characteristics of the body of alone. They depend only on the orientation of the coordinate system and the mass distribution. As we shall show below, one can always choose coordinates (which imply, in the case of a moving body, a *moving* set of axes) for which the products of inertia vanish. This considerably simplifies the analysis of rotating bodies. (For example the components of the angular momentum vector would take the simple form $(\mathrm{I}_{11}\omega_1,\ \mathrm{I}_{22}\omega_2,\ \mathrm{I}_{33}\omega_3)$). In static situations we normally orientate axes (almost without thought) so that products of inertia vanish.

As well as being important in the static and dynamic analysis of rigid bodies, second order symmetric tensors play a central role in elasticity theory also and so some of the properties of such tensors in general will now be considered. The most important of these properties is that there exists a coordinate system x'_j, referred to which the components of the symmetric tensor S_{ij} satisfy

$$S'_{ij} = 0 \qquad i \neq j$$

We say S_{ij} has been diagonalised. (Much of the terminology in this section is taken from matrix theory since the theory of second order tensors and of matrices is, in many respects, identical).

We first need some preliminary results concerning the eigenvalues and eigenvectors of the tensor S_{ij}. This rather theoretical exercise turns out to be extremely useful, not just in elasticity theory, but in many other branches of engineering and science.

The scalar λ is said to be an eigenvalue and n_i the corresponding eigenvector of the tensor S_{ij} if they satisfy the system of linear homogeneous equations:

$$S_{ij}n_j = \lambda n_i \tag{2.29}$$

In this system S_{ij} is assumed given and λ and n_i are to be determined. From the standard theory of linear equations a non-trivial solution for n_i only exists if

$$det(S_{ij} - \lambda\delta_{ij}) = 0 \tag{2.30}$$

that is

$$\begin{vmatrix} S_{11}-\lambda & S_{12} & S_{13} \\ S_{21} & S_{22}-\lambda & S_{23} \\ S_{31} & S_{32} & S_{33}-\lambda \end{vmatrix} = 0$$

When this is expanded we obtain a cubic in λ. This may be solved to give three solutions which we shall denote by $\lambda^{(1)}$, $\lambda^{(2)}$ and $\lambda^{(3)}$. We also assume, for the present, that the three roots are distinct:

$$\lambda^{(1)} \neq \lambda^{(2)} \qquad \lambda^{(1)} \neq \lambda^{(3)} \qquad \lambda^{(2)} \neq \lambda^{(3)}$$

For each of $\lambda^{(1)}$, $\lambda^{(2)}$ and $\lambda^{(3)}$ the corresponding eigenvectors $n_i^{(1)}$, $n_i^{(2)}$ and $n_i^{(3)}$ respectively, may then be determined from (2.29).

Example 2.5
Determine the eigenvalues and eigenvectors of the tensor S_{ij} given in matrix form:

$$\begin{bmatrix} 2 & 0 & 0 \\ 0 & 2 & 2 \\ 0 & 2 & -1 \end{bmatrix}$$

Verify that the eigenvectors are mutually orthogonal

Solution

The eigenvalues of S_{ij} are the solutions of the equation (from (2.30))

$$\begin{vmatrix} 2-\lambda & 0 & 0 \\ 0 & 2-\lambda & 2 \\ 0 & 2 & -1-\lambda \end{vmatrix} = 0$$

Expanding we find

$$(2-\lambda)(\lambda-3)(\lambda+2) = 0$$

Hence the three eigenvalues are $\lambda^{(1)} = -2$, $\lambda^{(2)} = 2$ and $\lambda^{(3)} = 3$. For each eigenvalue we determine the corresponding eigenvector according to (2.29).

- **case 1**

$\lambda^{(1)} = -2$. Then $n_i^{(1)}$ satisfies $S_{ij}n_i^{(1)} = \lambda^{(1)}n_i^{(1)}$. That is

$$2n_1^{(1)} = -2n_1^{(1)}$$

$$2n_2^{(1)} + 2n_3^{(1)} = -2n_2^{(1)}$$

$$2n_2^{(1)} - n_3^{(1)} = -2n_3^{(1)}$$

The solution of these equations is

$$n_1^{(1)} = 0 \qquad n_3^{(1)} = -2n_2^{(1)} \qquad \text{or} \;\; n_i^{(1)} = (0, a, -2a)$$

where $a \neq 0$ is an arbitrary constant.

- **case 2**

$\lambda^{(2)} = 2$. Then $n_i^{(2)}$ satisfies $S_{ij}n_i^{(2)} = \lambda^{(2)}n_i^{(2)}$. That is

$$2n_1^{(2)} = 2n_1^{(2)}$$
$$2n_2^{(2)} + 2n_3^{(2)} = 2n_2^{(2)}$$
$$2n_2^{(2)} - n_3^{(2)} = 2n_3^{(2)}$$

The solution of these equations is

$$n_i^{(2)} = (b, 0, 0)$$

where $b \neq 0$ is an arbitrary constant.

- **case 3**

$\lambda^{(3)} = 3$. Then $n_i^{(3)}$ satisfies $S_{ij}n_i^{(3)} = \lambda^{(3)}n_i^{(3)}$. Solving for $n_i^{(3)}$ as in cases 1 and 2 we find

$$n_i^{(3)} = (0, 2c, c)$$

where $c \neq 0$ is an arbitrary constant.
An easy calculation verifies that these three eigenvectors are mutually orthogonal:

$$n_i^{(1)}n_i^{(2)} = 0 \qquad n_i^{(1)}n_i^{(3)} = 0 \qquad n_i^{(2)}n_i^{(3)} = 0$$

This is not coincidental. For symmetric second order tensors the eigenvectors corresponding to *distinct* eigenvectors are always mutually orthogonal. We shall take this opportunity to make further enquiries into this interesting result.
Since we have three orthogonal directions we can orient a coordinate system x'_m so that x'_1 points along $\hat{n}_i^{(1)}$; x'_2 points along $\hat{n}_i^{(2)}$ and x'_3 points along $\hat{n}_i^{(3)}$. Taking unit vectors along the eigenvector directions obtained in this example we have

$$\hat{\imath}' = (0, \frac{1}{\sqrt{5}}, -\frac{2}{\sqrt{5}}) \qquad \hat{\jmath}' = (1, 0, 0) \qquad \hat{k}' = (0, \frac{2}{\sqrt{5}}, \frac{1}{\sqrt{5}})$$

and so from (2.3)

$$\alpha_{11} = 0 \qquad \alpha_{21} = 1 \qquad \alpha_{31} = 0$$
$$\alpha_{12} = \frac{1}{\sqrt{5}} \qquad \alpha_{22} = 0 \qquad \alpha_{32} = \frac{2}{\sqrt{5}}$$
$$\alpha_{13} = -\frac{2}{\sqrt{5}} \qquad \alpha_{23} = 0 \qquad \alpha_{33} = \frac{1}{\sqrt{5}}$$

Since S_{ij} is a tensor we can determine its components when referred to this new system of coordinates:

$$S'_{ij} = \alpha_{ik}\alpha_{jm}S_{km}$$

As symmetry is preserved under coordinate transformations only the diagonal and upper diagonal elements of S'_{ij} need be determined. We find, after a short calculation (which the reader should verify) for each component:

$$S'_{ij} = \begin{bmatrix} -2 & 0 & 0 \\ 0 & 2 & 0 \\ 0 & 0 & 3 \end{bmatrix}$$

This is quite remarkable. Referred to a system of coordinates with axes aligned along the eigenvector directions the tensor S_{ij} is in diagonal form ($S'_{ij} = 0, \quad i \neq j$). Also the diagonal elements are the eigenvalues of the tensor!

The results obtained in the previous example will now be confirmed to apply generally using a more formal approach. In particular we shall show that for a second order symmetric tensor S_{ab}

- The eigenvalues of S_{ab} are real.
- Eigenvectors corresponding to distinct eigenvectors are orthogonal.
- In the case that S_{ab} has three distinct eigenvalues, there exists three mutually orthogonal eigenvectors. If 'primed' coordinates are chosen so that the x'_k-axis is aligned along the direction of $n_a^{(k)}$ then, when referred to 'primed' axes, S_{ab} is diagonal (that is $S'_{ab} = 0, \quad a \neq b$)

To prove the first claim let λ be eigenvalue and n_i corresponding eigenvector of S_{ij}; then

$$S_{ij}n_j = \lambda n_i \tag{2.31}$$

Taking complex conjugates (denoted by a star *) of both sides of (2.31)

$$S_{ij}n_j^* = \lambda^* n_i^* \qquad \text{since } S_{ij} \text{ is real} \tag{2.32}$$

Now multiply (2.31) by n_i^* and (2.32) by n_i and subtract to give

$$S_{ij}n_jn_i^* - S_{ij}n_j^*n_i = (\lambda - \lambda^*)n_in_i^* \tag{2.33}$$

But, for the second term on the left hand side of (2.33)

$$S_{ij}n_j^*n_i = S_{ji}n_i^*n_j = S_{ij}n_i^*n_j$$

where the symmetry of S_{ij} has been used. From (2.33) we then deduce

$$(\lambda - \lambda^*)n_in_i^* = 0$$

As each term of the scalar $n_in_i^*$ involves the product of a number with its complex conjugate (and as such can never be negative) it follows that $\lambda = \lambda^*$ which proves our first claim that eigenvalues of a second order symmetric tensor are real. This is important since, as we shall see later in discussions on the stress and the strain tensors (both symmetric), the eigenvalues of these tensors have direct physical significance.

The proof of the second claim runs along similar lines. If $\lambda^{(1)}$ and $\lambda^{(2)}$ are distinct eigenvalues and $n_i^{(1)}$, $n_i^{(2)}$ the corresponding eigenvectors then

$$S_{ij}n_j^{(1)} = \lambda^{(1)}n_i^{(1)} \qquad \text{and} \qquad S_{ij}n_j^{(2)} = \lambda^{(2)}n_i^{(2)}$$

Multiplying the first of these by $n_i^{(2)}$ and the second by $n_i^{(1)}$ and subtracting we obtain

$$S_{ij}n_j^{(1)}n_i^{(2)} - S_{ij}n_j^{(2)}n_i^{(1)} = (\lambda^{(1)} - \lambda^{(2)})n_i^{(1)}n_i^{(2)}$$

But again the left hand side is zero due to the symmetry of S_{ij} and so

$$(\lambda^{(1)} - \lambda^{(2)})n_i^{(1)}n_i^{(2)} = 0$$

However, $\lambda^{(1)} \neq \lambda^{(2)}$ as the eigenvalues are distinct and so the only possibility is that

$$n_i^{(1)}n_i^{(2)} = 0$$

That is, the eigenvectors corresponding to distinct eigenvalues are orthogonal.

To prove our final claim we rotate axes so that x'_1 points along $n_i^{(1)}$, x'_2 points along $n_i^{(2)}$ and x'_3 points along $n_i^{(3)}$. (We assume the eigenvectors are *normalised* $n_i^{(\alpha)}n_i^{(\alpha)} = 1$ for $\alpha = 1, 2$ or 3).

With respect to the *new* system of coordinates $n_i^{(1)}$, $n_i^{(2)}$ and $n_i^{(3)}$ will have components

$$(n_i^{(1)})' = (1,0,0) \qquad (n_i^{(2)})' = (0,1,0) \qquad (n_i^{(3)})' = (0,0,1)$$

so that we can write

$$(n_i^{(\alpha)})' = \delta_{i\alpha} \tag{2.34}$$

(we emphasize that α is not a tensor index – it is merely a label).

Now consider the scalar $S_{ij}n_i^{(\alpha)}n_j^{(\beta)}$ in which α and β are one of 1,2 or 3. Since it is a scalar it is unchanging in the primed system and so:

$$S'_{ij}(n_i^{(\alpha)})'(n_j^{(\beta)})' = S_{ij}n_i^{(\alpha)}n_j^{(\beta)} = \lambda^{(\alpha)}n_j^{(\alpha)}n_j^{(\beta)}$$

However, from (2.34)

$$S'_{ij}\delta_{i\alpha}\delta_{j\beta} = \lambda^{(\alpha)}n_j^{(\alpha)}n_j^{(\beta)}$$

That is

$$S'_{\alpha\beta} = \lambda^{(\alpha)}n_j^{(\alpha)}n_j^{(\beta)}$$

Thus if α is the same as β

$$S'_{\alpha\alpha} = \lambda^{(\alpha)} \qquad \text{since} \quad n_j^{(\alpha)}n_j^{(\alpha)} = 1 \quad (\alpha \text{ not summed})$$

showing that the diagonal components of S_{ij} with respect to the primed system are the eigenvalues. On the other hand if α is distinct from β then

$$S'_{\alpha\beta} = \lambda^{(\alpha)} n_j^{(\alpha)} n_j^{(\beta)} = 0$$

due to the orthogonality of 'distinct' eigenvectors. This proves our third claim made concerning the properties of second order symmetric tensors.

When a system of coordinates x'_i is chosen in this way (that is, aligned along the eigenvectors) it is called a *Principal axis system* for S_{ij}. The eigenvalues are called the *Principal values* of S_{ij}. When the eigenvalues are distinct the Principal axis system is unique.

If an eigenvalue is repeated once (that is the cubic equation (2.30) has a double root) then there are just two distinct eigenvalues. For the eigenvalue which is not repeated, $\lambda^{(1)}$ say, a corresponding eigenvector $n_i^{(1)}$ can be found in the usual way. For the repeated eigenvalue an infinity of eigenvectors can be found, each lying in a plane orthogonal to $n_i^{(1)}$. In this case a Principal axis system exists (with $n_i^{(1)}$ and *any* two perpendicular directions in the plane orthogonal to $n_i^{(1)}$) though, obviously, it is not unique.

If an eigenvalue is repeated twice (that is the cubic equation (2.30) has a triple root) then there is just one distinct eigenvalue. In this case *any* three mutually orthogonal set of directions constitutes a Principal axis system.

EXERCISES

2.11 Show that the angle θ between two vectors $\underline{u}$ and $\underline{v}$ is given by the expression

$$\cos\theta = \frac{u_i v_i}{\sqrt{u_j u_j}\sqrt{v_k v_k}}$$

2.12 An object is said to be **invariant** if it transforms as a scalar on change of coordinate system. Show that u_i is a vector if, for arbitrary vectors v_i the object $u_i v_i$ is an invariant

2.13 If T_{ijk} is skew-symmetric with respect to the first and third indices and is symmetric with respect to the first and last pair of indices then $T_{ijk} = 0$.

2.14 The constitutive equations of linear elasticity are $\sigma_{ij} = 2\mu\varepsilon_{ij} + \lambda\delta_{ij}\varepsilon_{kk}$ (see Chapter 5) where μ, λ are scalars and σ_{ij} and ε_{ij} are **symmetric** second-order tensors. Show that

(i) $\sigma_{kk} = (2\mu + 3\lambda)\varepsilon_{kk}$

(ii) $\varepsilon_{kk} = \dfrac{1}{2\mu}\sigma_{ij} - \dfrac{\lambda}{2\mu(2\mu+3\lambda)}\delta_{ij}\sigma_{kk}$

(iii) $\sigma_{ij}\varepsilon_{ij} = 2\mu\varepsilon_{ij}\varepsilon_{ij} + \lambda\varepsilon_{kk}^2$

2.15 Determine the eigenvalues and eigenvectors of the symmetric tensor S_{ij}:

$$S_{11} = 3 \quad S_{12} = -6 \quad S_{13} = -4 \quad S_{22} = 4 \quad S_{23} = 2 \quad S_{33} = -1$$

2.16 Show that the tensor S_{ij}:

$$S_{11} = 4 \quad S_{12} = -2 \quad S_{13} = 4 \quad S_{22} = 1 \quad S_{23} = -2 \quad S_{33} = 4$$

has only two distinct eigenvalues. Find three mutually orthogonal eigenvectors and hence 'diagonalise' this tensor

Chapter 3

THE ANALYSIS OF STRESS

3.1 Introduction

The analysis of stress is fundamental to the study of the mechanical behaviour of all materials whether they behave plastically, viscoelastically, or, as in the considerations of this text, in a linearly elastic manner.

The occurrence of stress is commonplace. For example, if we apply equal and opposite forces to either end of a straight bar of uniform (though arbitrary) cross-section then it is a matter of common experience that after a very short time during which it suffers a slight elongation the bar will be in equilibrium (see Figure 3.1). If the bar is made of 'stiff' material like metal or wood then the elongation is not likely to be detectable to the eye. Sensitive equipment is required to measure the extension. More 'obvious' results are obtained with a highly elastic material such as rubber. An elastic band extended between each hand provides an excellent example.

Figure 3.1

The result of applying these forces is to *stress* the bar (or band). If we touch the exterior curved part of the bar we feel no force. Even if we cut the bar in the direction of F no change in the configuration of the bar will be observed. However, if we cut the bar transversely at any point then it will immediately begin to tear apart. The forces have been transmitted through the bar to the point of the cut.

We are also familiar with the fact that if this same force F is applied to a bar of larger area of cross-section (take a number of elastic bands between your hands) then its effect in deforming the bar is reduced in proportion. Indeed the deformation produced (elongation and thinning) is directly related to the 'normal' stress σ which is defined, for this simple system, as the magnitude of the applied force divided by the cross-sectional area. That is:

$$\sigma = \frac{F}{A} \tag{3.1}$$

Now consider a non-uniform bar shown in Figure 3.2. It is again a matter of common experience that if the force F is large enough and if the bar ruptures then it is very likely to rupture at a point near P. It is at this point that the normal stress is largest since the same force F is transmitted across each cross-section but the area through which it is transmitted is much reduced in the vicinity of P.

We now extend these ideas to more generally shaped bodies.

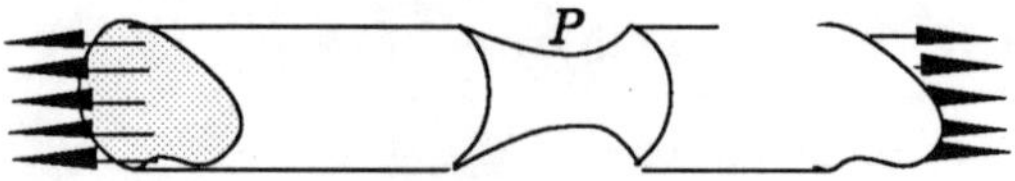

Figure 3.2

3.2 The Stress Tensor

As we have noted, the analysis of stress applies to all continua. Generally it depends only on geometry and on Newtonian dynamics. However, in our case, since only static situations are considered, we shall only require a use of Newton's equations of equilibrium.

Consider a continuous body acted on by forces as shown in Figure 3.3. Within the body we examine a sub-region V which is separated into two parts by a dividing plane S of area A.

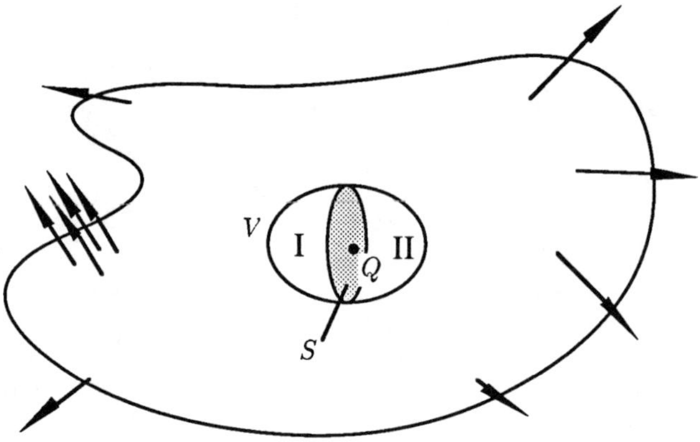

Figure 3.3

The forces exerted by region I on region II are transmitted across S and, by Newton's third law, are equal and opposite to the forces exerted by region II on region I.

Let Q be a point of S. The forces acting in region II may be resolved (see Section 1.4) and are equivalent to a single force F_i and a single couple H_i acting at Q. Now shrink V (and hence S) always ensuring that Q lies on S.

It is a basic assumption of stress analysis that, as the volume V decreases and the resolved force F_i and the resolved couple H_i also decrease in magnitude, that the ratio of force to area tends to a definite limit and the ratio of couple to area tends to zero. Mathematically, we write:

$$\lim_{A \to 0} \frac{F_i}{A} = T_i \qquad \text{and} \qquad \lim_{A \to 0} \frac{H_i}{A} = 0 \tag{3.2}$$

(In the absence of so-called body couples the second limit is essentially a consequence of the first since, roughly:

$$\text{couple} = \text{force} \times \text{dimension}$$

$$\text{and if} \quad \frac{\text{force}}{\text{area}} \to \text{definite limit} \qquad \text{from the first of (3.2)}$$

$$\text{then} \quad \frac{\text{couple}}{\text{area}} = \frac{\text{force} \times \text{dimension}}{\text{area}} \to 0$$

since the dimension (on A) tends to zero.

Body couples are not commonplace though they may be produced by the effects of a magnetic field on a magnetic material. The theory developed in this text will therefore not be applicable to such materials).

The vector T_i is called the stress vector at Q and its components have the dimensions of (force$\times$area^{-1}). The stress vector will depend not only on the coordinates of Q (that is the position within the body) but also upon the orientation of the plane S. If n_i is the unit normal to S then we emphasise this dependence by writing $T_i^{(n)}$ instead of T_i. (The superscript (n) is simply a label. It is *not* a tensor index).

The fundamental result of stress analysis is that there exists, at each point Q of the material, a second order tensor σ_{ij}, independent of n_i, such that

$$T_i^{(n)} = \sigma_{ij} n_j \tag{3.3}$$

The proof of this relation, known as the Cauchy relation, hinges on the use of Newton's second law that the forces on an element of the body which is in equilibrium, should sum to zero.

- **Proof**

Consider any plane S containing an arbitrary point Q of the medium. The three mutually perpendicular planes $S^{(1)}$, $S^{(2)}$ and $S^{(3)}$ form, with S, a tetrahedron as shown in Figure 3.4. Let n_i, $n_i^{(1)}$, $n_i^{(2)}$ and $n_i^{(3)}$ respectively be the unit normals to these planes. (I apologise in advance for the somewhat cumbersome notation).

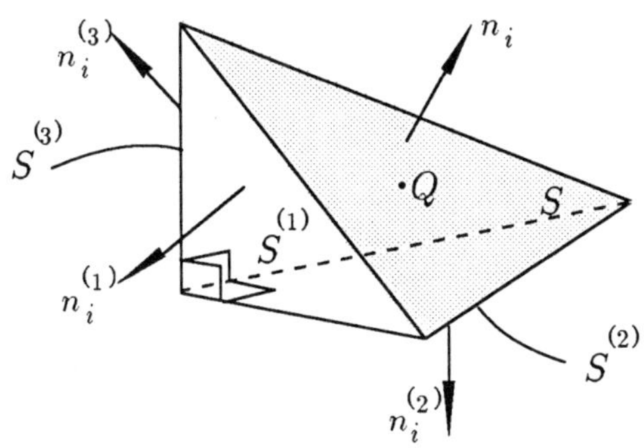

Figure 3.4

If A, $A^{(1)}$, $A^{(2)}$ and $A^{(3)}$ are the areas of the triangles so formed then (using the vector property of areas of a closed polygonal region; see (1.27) of Section 1.3)

$$An_i + A^{(1)}n_i^{(1)} + A^{(2)}n_i^{(2)} + A^{(3)}n_i^{(3)} = 0 \tag{3.4}$$

Hence, taking scalar products of (3.4) with $n_i^{(1)}$, $n_i^{(2)}$ and $n_i^{(3)}$ in turn we obtain

$$A^{(1)} = -An_in_i^{(1)} \qquad A^{(2)} = -An_in_i^{(2)} \qquad A^{(3)} = -An_in_i^{(3)} \tag{3.5}$$

At each point on the faces of the tetrahedron a stress vector is defined. Its value may change from point to point. To obtain the total force acting on each face we consider an infinitessimal region on each plane of area dS. The total force on this region is obtained by multiplying the area by the stress vector at the point. The total force on the face is then the sum over the face (or integral, if the stress vector varies from point to point) of all such contributions. This will account for the forces over each face of the tetrahedron.

We must also take account of body forces in the application of Newton's equations of equilibrium. The typical body force is that due to gravity which affects every material particle of the tetrahedron not just those on the bounding faces. If B_i denotes the total body force (units are force×mass^{-1}) acting on a volume dV in the interior of the tetrahedron then Newton's second law takes the form

$$\begin{aligned} 0 = \iiint_V \rho B_i dV + \iint_S T_i^{(n)} dS + \iint_{S^{(1)}} T_i^{(n^{(1)})} dS \\ + \iint_{S^{(2)}} T_i^{(n^{(2)})} dS + \iint_{S^{(3)}} T_i^{(n^{(3)})} dS \end{aligned} \tag{3.6}$$

where $T_i^{(n)}$, $T_i^{(n^{(1)})}$,... are the stress vectors on S, $S^{(1)}$, ... respectively and ρ is the density of the material.

The equilibrium of the tetrahedron under the action of the applied forces, as expressed in (3.6) is valid irrespective of size. We now shrink the tetrahedron to infinitessimal size so that volume and the surface faces are so small that the integrals may be approximated by discrete sums.

Let l be the longest dimension on the tetrahedron. Using Taylor's theorem we may express the stress vector (each component is a function of the coordinates) on each face in terms of its value at the point Q. This assumes that the stress vector components are continuous. Then

$$\begin{aligned} T_i^{(n)} &= (T_i^{(n)})_Q + O(l) \\ T_i^{(n^{(k)})} &= (T_i^{(n^{(k)})})_Q + O(l) \quad k = 1, 2, 3 \end{aligned} \tag{3.7}$$

The $O(l)$ symbol is shorthand notation for any collection of terms each of which has *at least* a single factor l and as such will vanish as $l \to 0$.

The first term on the right hand sides of the expansions in (3.7) is the value of the stress vector at a point Q and is therefore a constant.

Now, using (3.7)

$$\begin{aligned} \iint_S T_i^{(n)} dS &= (T_i^{(n)})_Q \iint_S dS + \iint_S O(l) dS \\ &= (T_i^{(n)})_Q A + O(l^3) \end{aligned} \tag{3.8}$$

With similar results for the other integrals. Also

$$\iiint_V \rho B_i dV = O(l^3) \tag{3.9}$$

Substituting these relations, (3.8), (3.9) into (3.6) and dividing by A we obtain

$$(T_i^{(n)})_Q + (T_i^{(n^{(1)})})_Q \frac{A^{(1)}}{A} + (T_i^{(n^{(2)})})_Q \frac{A^{(2)}}{A} + (T_i^{(n^{(3)})})_Q \frac{A^{(3)}}{A} + O(l) = 0$$

since $A = O(l^2)$. (The body force terms being proportional to the volume of the tetrahedron are of order l smaller than the surface force terms on the tetrahedron which, of course, are proportional to area).

Since Q is an arbitrary point we can drop the suffix and utilising (3.5) we obtain, in the limit as $l \to 0$:

$$T_i^{(n)} = T_i^{(n^{(1)})} n_k n_k^{(1)} + T_i^{(n^{(2)})} n_k n_k^{(2)} + T_i^{(n^{(3)})} n_k n_k^{(3)} \equiv \sigma_{ik} n_k \tag{3.10}$$

in which the two-indexed object σ_{ik} is defined by this relation; that is

$$\sigma_{ik} \equiv T_i^{(n^{(1)})} n_k^{(1)} + T_i^{(n^{(2)})} n_k^{(2)} + T_i^{(n^{(3)})} n_k^{(3)} \tag{3.11}$$

By the manner in which σ_{ik} is defined, being the sum of terms each of which is a second order tensor, it is evident that σ_{ik} is a second order tensor. It is called the **stress tensor**.

To show that σ_{ik} is independent of n_i we simply rotate the plane S (thus changing n_i) but keep $S^{(1)}$, $S^{(2)}$ and $S^{(3)}$ fixed. The tensor σ_{ik} is unchanged and $T_i^{(n)}$ will still be defined as in (3.10). Thus σ_{ik} is independent of n_i. This completes the proof.

The tensorial nature of 'stress' has been deduced by utilising Newton's second law applied to an infinitessimal tetrahedron and using continuity arguments. If we apply Newton's second law applied to an arbitrarily shaped region of a body which has been stressed as a result of the application of body forces to the interior and of surface forces acting on the boundary we find that the 'stress' represented by the tensor σ_{ik} is further constrained. If the portion of the body has volume V and is bounded by a surface S then for the forces to be in equilibrium we must have

$$\iiint_V \rho B_i dV + \iint_S T_i^{(n)} dS = 0 \tag{3.12}$$

where n_i (on the superscript of T) is a unit normal to S. Using (3.10) in (3.12) implies

$$\iiint_V \rho B_i dV + \iint_S \sigma_{ij} n_j dS = 0 \tag{3.13}$$

Using the divergence theorem (see Section (1.8)) on the second integral of (3.13) implies

$$\iiint_V \rho B_i dV + \iiint_V \sigma_{ij,j}\, dV = 0 \tag{3.14}$$

that is

$$\iiint_V (\rho B_i + \sigma_{ij},_j)dV = 0 \tag{3.15}$$

But since V is arbitrary we can conclude that

$$\sigma_{ij},_j + \rho B_i = 0 \qquad \text{throughout } V \tag{3.16}$$

(If this were not zero, but positive say at a particular point then by continuity it would be positive at all points in a region V'. Choosing V to be V' would then contradict (3.15)).

In a similar way we demand that the moments of the forces about any point should vanish. (When the forces are in equilibrium, as they are in this case due to (3.16), we can choose *any* convenient point for the calculation of the moment since the total moment is independent of positition; see Section 1.4). We choose the origin of coordinates for convenience.

For a volume element dV, with mass ρdV, the moment of the body force B_i about the origin is $\eta_{ijk}x_jB_k\rho dV$ (corresponding to $\underline{r} \wedge \underline{B}\rho dV$). Similarly for a surface element dS the moment of the surface force, on this element, about the origin is $\eta_{ijk}x_jT_k^{(n)}dS$. Thus over the complete region we demand:

$$\iiint_V \eta_{ijk}x_jB_k\rho dV + \iint_S \eta_{ijk}x_jT_k^{(n)}dS = 0 \tag{3.17}$$

(Again, we are assuming that there are no body couples, as might arise in magnetic materials).

Using (3.10) in (3.17) implies

$$\iiint_V \eta_{ijk}x_jB_k\rho dV + \iint_S \eta_{ijk}x_j\sigma_{km}n_m dS = 0 \tag{3.18}$$

Also if we use (3.16) and apply the divergence theorem to the surface integral in (3.18) we obtain:

$$-\iiint_V \eta_{ijk}x_j(\sigma_{km},_m)dV + \iiint_V (\eta_{ijk}x_j\sigma_{km}),_m dV = 0 \tag{3.19}$$

that is

$$\iiint_V \eta_{ijk}x_j,_m \sigma_{km}dV = 0$$

But $x_j,_m = \delta_{jm}$ and so, finally

$$\iiint_V \eta_{imk}\sigma_{km}dV = 0 \tag{3.20}$$

or $\eta_{imk}\sigma_{km} = 0$ at each point of V. As already noted in Example 2.4 this simply implies that

$$\sigma_{km} = \sigma_{mk} \tag{3.21}$$

that is, in the absence of body couples, the stress tensor is *symmetric*.

The stress vector $T_i^{(n)}$ at each point, Q say, within a body is fully defined by (3.10). It is associated with an infinitessimal plane area with direction n_i. We should note that the direction of $T_i^{(n)}$ is not necessarily in the direction of n_i (see

Figure 3.5). However, $T_i^{(n)}$ has a component along n_i called the **normal stress** at Q and a component tangential to the plane area called the **shear stress** at Q.

Clearly the component of $T_i^{(n)}$ in the direction of n_i is $T_i^{(n)} n_i$ and so, using vector addition:

$$T_i^{(n)} = (T_k^{(n)} n_k) n_i + (T_i^{(n)} - (T_k^{(n)} n_k) n_i) \tag{3.22}$$

the first term on the right hand side of (3.22) being parallel to n_i and the second being perpendicular to n_i. (The scalar product of the second term with n_i is zero).

The magnitude of the normal stress is $T_k^{(n)} n_k$ and that of the shear stress is (by Pythagoras) $\{T_i^{(n)} T_i^{(n)} - (T_i^{(n)} n_i)^2\}^{1/2}$

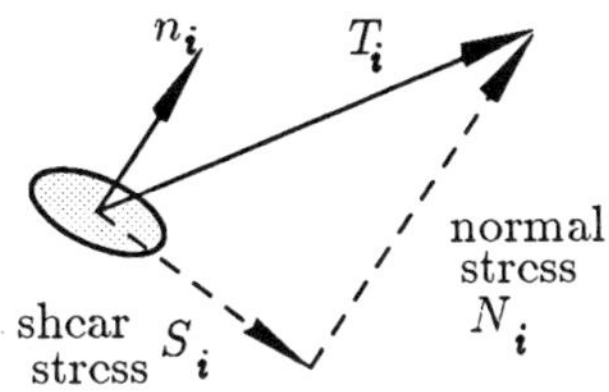

Figure 3.5

Example 3.1
A uniform bar of cross-sectional area A has its axis aligned along the x_1-axis and is subjected to equal and opposite forces F as shown in Figure 3.6. If

$$\sigma_{11} = \frac{F}{A} \qquad \text{other} \quad \sigma_{ij} = 0$$

determine the stress vector at any point on a plane at an angle α to the axis of the bar. Determine the magnitudes of the normal and shear stress on this plane and, finally, verify that the total force on this plane is F.

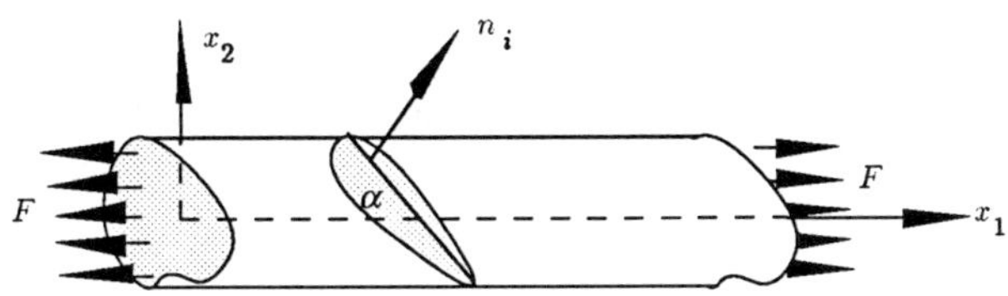

Figure 3.6

Solution

It is clear from the diagram that $n_i = (\sin\alpha,\ \cos\alpha,\ 0)$ and so, from (3.10)

$$T_i^{(n)} = \sigma_{i1}n_1 + \sigma_{i2}n_2 + \sigma_{i3}n_3$$
$$= \sigma_{i1}\sin\alpha + \sigma_{i2}\cos\alpha$$

giving $T_i^{(n)} = (\frac{F}{A}\sin\alpha,\ 0,\ 0)$. The normal component of $T_i^{(n)}$ on this plane is

$$T_i^{(n)}n_i = \frac{F}{A}\sin^2\alpha$$

which is the magnitude of the normal stress. The magnitude of the shear stress is

$$\{T_i^{(n)}T_i^{(n)} - (T_i^{(n)}n_i)^2\}^{1/2} = \{\frac{F^2}{A^2}\sin^2\alpha - \frac{F^2}{A^2}\sin^4\alpha\}^{1/2}$$
$$= \frac{F}{A}\sin\alpha\cos\alpha$$

The total force on the inclined plane is $(\frac{F}{A}\sin\alpha)A'$ in the x_1-direction, where A' is the area of the inclined element. If A'_k is the vector area of this element then $A'_k = A'n_k$ (see Section 1.3). But the projection of A'_k onto the x_3x_2 plane is A'_1 and this is equal to A. Therefore

$$A = A'_1 = A'\sin\alpha$$

Therefore the total force on the inclined element is $(\frac{F}{A}\sin\alpha)\frac{A}{\sin\alpha} = F$ as expected.

Example 3.2

The torsion problem of a circular cylinder is concerned with the deformation produced when the two ends of the cylinder are twisted by equal and opposite couples. A circular cylinder of radius a has its axis aligned along the x_3-axis. Verify that if at each point $(x_1,\ x_2,\ x_3)$ the stress tensor is given by

$$\sigma_{31} = \sigma_{13} = -kx_2 \qquad \sigma_{23} = \sigma_{32} = kx_1 \qquad \text{other } \sigma_{ij} = 0$$

where k is a constant then this fully specifies a situation in which the cylinder is subjected to torsion by equal and opposite couples applied to either end.

Solution

We shall simply verify that the forces which result from the given stress tensor are compatible with the torsion problem. We firstly show that the lateral (i.e. curved) surface of the cylinder is stress free (that is, no forces are applied to this part of the boundary). Secondly we shall show that the forces applied to the ends of the cylinder constitute a torsional couple.
From Figure 3.7a we can see that a normal to the lateral surface has the form $n_i = (\alpha,\ \beta,\ 0)$ and clearly has a zero x_3-component.

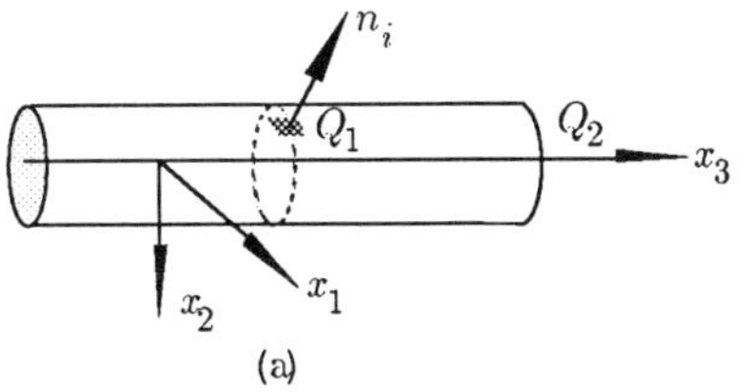

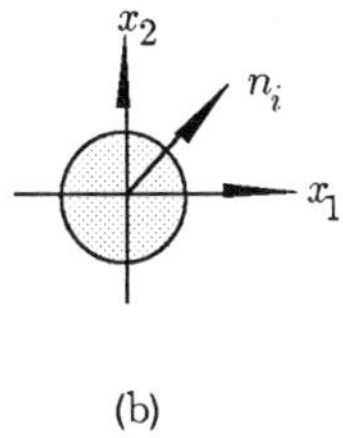

Figure 3.7

Hence, if we therefore consider a point Q_1 on the lateral surface then the stress vector at that point has components

$$T_1^{(n)} = \sigma_{1j}n_j = 0 \quad T_2^{(n)} = \sigma_{2j}n_j = 0 \quad T_3^{(n)} = \sigma_{3j}n_j = -k\alpha x_2 + k\beta x_1$$

But on a circle in the x_1x_2 plane (centred on the x_3-axis) the unit normal at any point on the boundary (with equation $x_1^2 + x_2^2 = a^2$) has the form $n_i = (\frac{x_1}{a}, \frac{x_2}{a}, 0)$ since the normal is always in the same direction as the position vector of the point under consideration (see Figure 3.7b). Thus

$$T_3^{(n)} = -kx_2(\frac{x_1}{a}) + kx_1(\frac{x_2}{a}) = 0$$

so that the lateral surface is indeed stress free since $T_i^{(n)}$ is zero there.
If we now choose a point Q_2 on the right hand end of the cylinder then the normal at Q_2 is $m_i = (0,0,1)$ and so, at this point, the stress vector has the form:

$$T_i^{(m)} = \sigma_{ij}m_j = (-kx_2, kx_1, 0)$$

But this vector is always tangential to the circle $x_1^2 + x_2^2 = r^2$ since $T_i^{(m)}n_i = 0$ where n_i is the normal to the circle (obtained above). It is in the direction shown in Figure 3.8a since the cross product of the normal with the stress vector: $\eta_{ijk}n_jT_k^{(m)} = (0,0,k(x_1^2 + x_2^2))$ which is a vector in the positive x_3-direction. Of course had $T_i^{(m)}$ been in the opposite direction to that shown then this vector product would have been in the negative x_3-direction.
It is clear that the resultant of all these forces (due to the shearing stress at each point) is a pure couple in the x_3-direction. (By symmetry the resultant force is zero). We can readily determine the magnitude of the resultant couple as follows.
An elementary area at Q_2 (see Figure 3.8b) has an area $\approx r\delta r\delta\theta$ and so the force acting over this area has magnitude

$$|T_i^{(m)}|r\delta r\delta\theta = k\sqrt{x_1^2 + x_2^2}\, r\delta r\delta\theta = kr^2\delta r\delta\theta$$

This force has a couple about the origin of magnitude $kr^3\delta r\delta\theta$ and adding all couples due to forces at a distance r from the axis we obtain a couple of magnitude $kr^3\delta r(2\pi)$ due to the 'ring' of forces. Now summing all couples as r changes from 0 to a we find the total couple acting over the end of the cylinder to have components

$$(0,0, \int_0^a kr^3(2\pi)dr) = (0,0, \frac{\pi ka^4}{2})$$

The calculation at the other end of the cylinder is identical except that we must use $m_j = (0,0,-1)$. For this end the resultant couple is of equal magnitude but in the opposite direction, that is $(0,0,-\frac{\pi k a^4}{2})$ which is required for the overall equilibrium of the body.

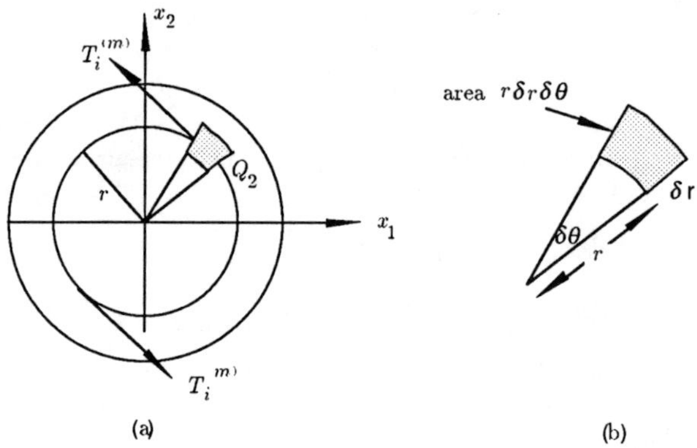

Figure 3.8

EXERCISES

3.1 A square plate is acted on by a shearing stress τ on each of its edges as shown in Figure 3.9. Determine the tensile and shearing stresses on planar elements at P and Q (inclined at 45^0 and 135^0 respectively to the x_1-axis). Assume that the stress tensor components are constant throughout the plate.

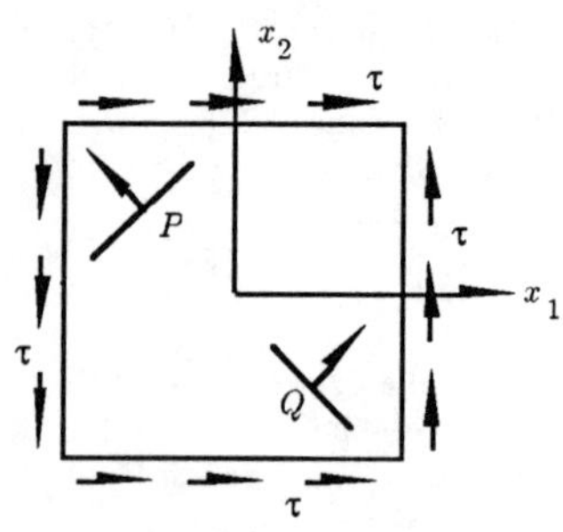

Figure 3.9

3.2 The state of stress at a point in a body is

$$\sigma_{11} = 200, \quad \sigma_{12} = 400, \quad \sigma_{13} = 300, \quad \sigma_{22} = 0 \quad \sigma_{23} = 0, \quad \sigma_{33} = -100$$

Determine the stress vector acting on a plane with normal (1,2,2) passing through the point. Find also the magnitudes of the normal and shear stresses on the plane.

3.4 Show that if $T_i^{(n)}$ is the stress vector acting on a plane P, then the stress vector acting across any plane that contains $T_i^{(n)}$ lies in the plane P.

3.3 Principal Axes

In Section 2.7 the properties of second order symmetric tensors were discussed. In this section we shall apply some of the results obtained there to the particular case of the stress tensor.

Consider a point Q within (or on the surface) of a body throughout which there is defined a stress tensor σ_{ij}. For a specific plane area through Q with normal n_i the stress vector $T_i^{(n)}$ is related to the stress tensor through:

$$T_i^{(n)} = \sigma_{ij} n_j \tag{3.23}$$

As already noted the normal stress points in the normal direction with magnitude $T_i^{(n)} n_i$. That is, if N_i denotes the normal stress vector then

$$N_i = (T_k^{(n)} n_k) n_i \tag{3.24}$$

whilst if S_i denotes the shear stress vector then, by the parallelogram law of vector addition, (see Figure 3.5)

$$S_i = T_i^{(n)} - (T_k^{(n)} n_k) n_i \tag{3.25}$$

As the orientation of the plane, through Q, changes then so the stress vector $T_i^{(n)}$ changes according to (3.23), inducing corresponding changes on N_i and S_i according to (3.24) and (3.25).

For a given stress tensor at Q we might hope to find a particular plane through Q (that is, a *specific* n_i) across which the shear stress is zero. For such a plane, the stress vector is a pure normal stress. On such a plane $S_i = 0$, that is:

$$T_i^{(n)} - (T_k^{(n)} n_k) n_i = 0$$

or, in terms of the stress tensor,

$$\sigma_{ij} n_j - (\sigma_{jk} n_k n_j) n_i = 0 \tag{3.26}$$

If we denote the magnitude of the normal stress $\sigma_{jk} n_k n_j$ by σ then

$$\sigma_{ij} n_j - \sigma n_i = 0 \tag{3.27}$$

But this is a standard eigenvalue problem first introduced in Section 2.7. For a non-trivial solution for the three unknowns (n_1, n_2, n_3) we require

$$det(\sigma_{ij} - \sigma\delta_{ij}) = 0 \qquad \textbf{(3.28)}$$

As we have seen, this is a cubic in σ with, in general, three distinct roots. These three values of σ, denoted by $\sigma^{(1)}$, $\sigma^{(2)}$ and $\sigma^{(3)}$ are the eigenvalues of the stress tensor and the corresponding values of n_i, denoted by $n_i^{(1)}$, $n_i^{(2)}$ and $n_i^{(3)}$ (found from (3.27) when σ is known) are the eigenvectors. Since σ_{ij} is symmetric these three roots are real and furthermore the eigenvectors corresponding to distinct eigenvalues are mutually orthogonal.

In the case of distinct eigenvalues the coordinate system may be rotated so that the axes coincide with the three mutually orthogonal eigenvectors. The rotated coordinate system is called the Principal axis system. When referred to this system we know (see Section 2.7) that the stress tensor σ'_{ij} only has non-zero diagonal components with values:

$$\sigma'_{11} = \sigma^{(1)} \qquad \sigma'_{22} = \sigma^{(2)} \qquad \sigma'_{33} = \sigma^{(3)}$$

That is, in a principal axis system the diagonal terms of the stress tensor are equal to its eigenvalues which are the the magnitudes of the normal stresses on the three planes with normals $n_i^{(1)}$, $n_i^{(2)}$ and $n_i^{(3)}$. The normal stresses in these special directions are called the Principal stresses of σ_{ij}.

Example 3.3
Suppose that the stress state at Q is given in matrix form by

$$\sigma_{ij} = \begin{bmatrix} 3 & -6 & -4 \\ -6 & 4 & 2 \\ -4 & 2 & -1 \end{bmatrix}$$

Determine the principal stresses and principal directions for this system and verify that with respect to the principal axis system the stress tensor is diagonal.

Solution

The principal stresses (the eigenvalues of σ_{ij}) are obtained by solving (3.28). For this stress tensor we have

$$\begin{vmatrix} 3-\sigma & -6 & -4 \\ -6 & 4-\sigma & 2 \\ -4 & 2 & -1-\sigma \end{vmatrix} = 0$$

Expanding the determinant implies

$$(\sigma + 1)(\sigma + 4)(\sigma - 11) = 0$$

so that $\sigma^{(1)} = -1$, $\sigma^{(2)} = -4$ and $\sigma^{(3)} = 11$ are the eigenvalues and hence the values of the principal stresses. (The negative values indicate compressive stresses, the positive value is a tensile stress).

The principal directions are obtained by solving (3.27) for each eigenvalue.

- **case 1** $\sigma = \sigma^{(1)} = -1$. Then writing out the three equations in (3.27) we find

$$\begin{aligned} 4n_1 - 6n_2 - 4n_3 &= 0 \\ -6n_1 + 5n_2 + 2n_3 &= 0 \\ -4n_1 + 2n_2 \qquad &= 0 \end{aligned}$$

Solving this system implies $n_2 = 2n_1$, $n_3 = -2n_1$. However, since n_k is a unit vector we obtain the principal direction corresponding to principal stress $\sigma^{(1)}$ as

$$n_i^{(1)} = \frac{1}{3}(1, 2, -2)$$

- **case 2** $\sigma = \sigma^{(2)} = -4$. Proceeding as in case 1 we find

$$n_i^{(2)} = \frac{1}{3}(2, 1, 2)$$

- **case 3** $\sigma = \sigma^{(3)} = 11$. Here we find the corresponding principal direction as

$$n_i^{(3)} = \frac{1}{3}(-2, 2, 1)$$

A simple calculation verifies that the three principal directions are mutually orthogonal.
If we now choose a new coordinate system (x_1', x_2', x_3') whose positive directions are aligned along $n_i^{(1)}, n_i^{(2)}$ and $n_i^{(3)}$ then, with respect to this primed system,

$$(n_i^{(1)})' = (1, 0, 0) \qquad (n_i^{(2)})' = (0, 1, 0) \qquad (n_i^{(3)})' = (0, 0, 1)$$

Now for any two vectors $n_i^{(\alpha)}$, $n_i^{(\beta)}$ the object $\sigma_{ij} n_i^{(\alpha)} n_j^{(\beta)}$ is a scalar and so

$$\sigma_{ij}'(n_i^{(\alpha)})'(n_j^{(\beta)})' = \sigma_{ij} n_i^{(\alpha)} n_j^{(\beta)}$$

By assigning values to α and β we can isolate particular components of σ_{ij}'. For example, taking $\alpha = 1$ and $\beta = 1$ we obtain

$$\begin{aligned} \sigma_{11}' &= \sigma_{ij} n_i^{(1)} n_j^{(1)} \\ &= \frac{1}{9}[3(1) + 2(-6)(2) + 2(-4)(-2) + 4(4) + 2(2)(-4) - (4)] \\ &= -1 \qquad (= \sigma^{(1)} \quad \text{as expected}) \end{aligned}$$

Similarly, choosing $\alpha = 2$ and $\beta = 2$ we find

$$\sigma'_{22} = \sigma_{ij} n_i^{(2)} n_j^{(2)}$$
$$= \frac{1}{9}[3(4) + 2(-6)(2) + 2(-4)(4) + 4(1) + 2(2)(2) - (4)]$$
$$= -4 \qquad (= \sigma^{(2)})$$

Choosing $\alpha = 3$ and $\beta = 3$ shows that $\sigma'_{33} = 11$ $(= \sigma^{(3)})$. Other combinations of α and β confirm the 'off-diagonal' components of σ'_{ij} are zero

---

It is interesting to note what happens at a stress free point on the boundary of a body. If Q is such a point then $T_i^{(m)} = 0$ there (in which m_i is a unit normal to the boundary). Thus plane elements tangential to the boundary are free of shearing stresses:

$$S_i = T_i^{(m)} - (T_k^{(m)} m_k) m_i = 0$$

Hence the direction m_i *must* be one of the principal directions $n_i^{(1)}$, $n_i^{(2)}$ or $n_i^{(3)}$. Further, one of the principal stresses $\sigma^{(1)}$, $\sigma^{(2)}$ or $\sigma^{(3)}$ *must* be zero since its magnitude is

$$\sigma = \sigma_{ij} m_j m_i = T_i^{(m)} m_i = 0$$

Since principal directions are mutually perpendicular the other two principal directions *must* be tangential to the boundary surface. As we shall see in our discussion, in Section 3.7 these results are particularly useful in the application of the photoelastic technique to the measurement of stress.

3.4 Maximum Normal and Shear Stresses

Although the determination of principal axes and principal stresses is an interesting mathematical problem its fundamental importance in stress analysis is due to the following result:

At a point Q within a body the algebraically largest and the algebraically least normal stress at Q are principal stresses. Depending on the sign of the principal stresses these may be tensile or compressive.

Proof

The magnitude of the normal stress on an area element with unit normal n_i is $T_i^{(n)} n_i$ (or equivalently $\sigma_{ij} n_i n_j$). Our aim is to find that *particular* unit normal n_i so that the function G:

$$G = \sigma_{ij} n_i n_j \tag{3.29}$$

attains its maximum or minimum value. This is a relatively standard max/min problem requiring the optimisation of the function $G = \sigma_{ij} n_i n_j$ subject to the constraint $n_i n_i = 1$.

Using the Lagrange multiplier technique we know that the turning points of G are exactly the same as the turning points of the function F:

$$F = \sigma_{ij}n_in_j - \lambda(n_in_i - 1) \quad \textbf{(3.30)}$$

where λ is the Lagrange multiplier. Optimising F with respect to each component of n_i and λ we find

$$\frac{\partial F}{\partial n_k} = \sigma_{ij}(\delta_{ik}n_j + n_i\delta_{jk}) - \lambda(2\delta_{ik}n_i) = 0$$

that is,

$$\sigma_{kj}n_j - \lambda n_k = 0 \quad \textbf{(3.31)}$$

and

$$\frac{\partial F}{\partial \lambda} = -(n_in_i - 1) = 0 \quad \textbf{(3.32)}$$

But equations (3.31) (together with (3.32)) are precisely the equations seen earlier (see (3.27)). Thus the solution for n_i in (3.31) are the principal directions and, taking the scalar product of (3.31) with n_k,

$$\lambda = \sigma_{kj}n_jn_k \qquad (\equiv \sigma \text{ previously})$$

which of course, we recognise as the principal stresses. The algebraically largest and least values of $G = \sigma_{ij}n_in_j$ must therefore be principal stresses and are achieved when n_i is a principal direction.

(An alternative method of proof hinges on the fact that $\sigma_{ij}n_in_j$ is a scalar and as such is independent of the choice of coordinate system at Q. If a coordinate system x'_i is chosen to coincide with the principal axes at Q then, with respect to the primed system $\sigma'_{ij} = 0 \;\; i \neq j$ and so

$$\sigma_{ij}n_in_j = \sigma'_{ij}n'_in'_j = \sigma^{(1)}(n'_1)^2 + \sigma^{(2)}(n'_2)^2 + \sigma^{(3)}(n'_3)^2$$

Thus we need to optimise

$$H = \sigma^{(1)}(n'_1)^2 + \sigma^{(2)}(n'_2)^2 + \sigma^{(3)}(n'_3)^2$$

subject to the constraint that

$$(n'_1)^2 + (n'_2)^2 + (n'_3)^2 = 1$$

Proceed as before to show that the max/min values of H are, depending on the values of n'_i, just $\sigma^{(1)}$, $\sigma^{(2)}$ or $\sigma^{(3)}$. This is exactly the same solution as that obtained above).

These extreme values of the normal stress at a point are obviously important in the study of the failure of materials.

In a similar way one may determine the maximum shear stress magnitude at a point Q by optimising (see (3.25))

$$S_iS_i = \{T_i^{(n)}T_i^{(n)} - (T_i^{(n)}n_i)^2\} \qquad \text{subject to } n_in_i = 1$$

If we choose a coordinate system aligned along the principal directions at Q this is equivalent to optimising

$$(\sigma^{(1)})^2n_1^2 + (\sigma^{(2)})^2n_2^2 + (\sigma^{(3)})n_3^2 - \{\sigma^{(1)}n_1^2 + \sigma^{(2)}n_2^2 + \sigma^{(3)}n_3^2\}^2$$

subject to $n_in_i = 1$. The result of this calculation is recorded in Table 3.1

We observe that the maximum shearing stress magnitude at a point Q is equal to half the difference between the greatest and least normal stresses at Q. The directions n_i, of course, are with reference to the principal axis system at Q.

n_1	n_2	n_3	max/min shear stress $(S_iS_i)^{\frac{1}{2}}$
0	0	± 1	0
0	± 1	0	0
± 1	0	0	0
0	$\pm\sqrt{2}/2$	$\pm\sqrt{2}/2$	$\lvert\sigma^{(2)} - \sigma^{(3)}\rvert/2$
$\pm\sqrt{2}/2$	0	$\pm\sqrt{2}/2$	$\lvert\sigma^{(3)} - \sigma^{(1)}\rvert/2$
$\pm\sqrt{2}/2$	$\pm\sqrt{2}/2$	0	$\lvert\sigma^{(1)} - \sigma^{(2)}\rvert/2$

Table 3.1

3.5 Plane Stress

Thus far we have, quite properly, considered the analysis of stress from a three dimensional point of view but, for a significant class of physical situations, the dimensionality of the problem may be, for all practical applications, taken as two. This applies particularily to a thin plate loaded within its own plane (see Figure 3.10) in such a way that those stress elements with a component in a direction normal to the plate may be taken as 'small' in comparison to the other stress elements.

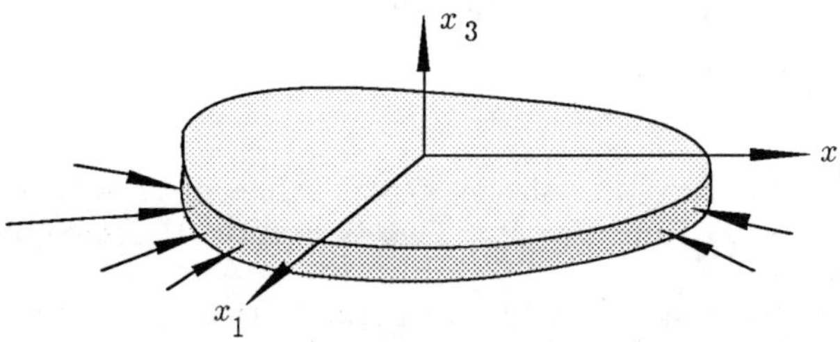

Figure 3.10

More precisely, if the plate lies in the x_1x_2 plane then under many loading conditions it is a reasonable assumption to take σ_{13}, σ_{23} and σ_{33} as small in comparison to σ_{11}, σ_{12} and σ_{22}. (A simple physical argument supports this view. The top and bottom faces of the plate are stress free and so, on these faces $T_i^{(n)} = 0$ in which $n_i = (0, 0, \pm 1)$. From the relation $T_i^{(n)} = \sigma_{ij}n_j$ it follows that σ_{13}, σ_{23} and σ_{33} are *identically* zero at these points. If the plate is thin it is a reasonable assumption to take these stress components as being small *throughout* the plate).

Some situations arise in which σ_{13}, σ_{23} and σ_{33} are *identically* zero throughout the body. In this case the state of stress in the body is referred to as **plane stress**, and what we outline below applies particularily to this case.

Of course generally, at each point, there will be three principal directions defined by (3.27). In particular (3.28) may be expanded as:

$$\begin{vmatrix} \sigma_{11}-\sigma & \sigma_{12} & \sigma_{13} \\ \sigma_{21} & \sigma_{22}-\sigma & \sigma_{23} \\ \sigma_{31} & \sigma_{32} & \sigma_{33}-\sigma \end{vmatrix} = 0$$

which becomes

$$(\sigma_{11}-\sigma)[(\sigma_{22}-\sigma)(\sigma_{33}-\sigma)-\sigma_{32}\sigma_{23}] - \sigma_{12}[\sigma_{21}(\sigma_{33}-\sigma)-\sigma_{31}\sigma_{23}] + \sigma_{13}[\sigma_{21}\sigma_{32}-\sigma_{31}(\sigma_{22}-\sigma)] = 0$$

Now ignoring quadratic terms in σ_{13}, σ_{23} and σ_{33} as being very small in comparison to the other terms this equation may be simplified to:

$$(\sigma_{33}-\sigma)[(\sigma_{11}-\sigma)(\sigma_{22}-\sigma)-\sigma_{12}^2] = 0$$

which is easily solved to give

$$\sigma^{(1)},\ \sigma^{(2)} = \frac{1}{2}[(\sigma_{11}+\sigma_{22}) \pm \sqrt{(\sigma_{11}-\sigma_{22})^2+4\sigma_{12}^2}] \qquad \textbf{(3.33)}$$

and

$$\sigma^{(3)} = \sigma_{33}$$

The eigenvector $\sigma^{(3)}$ gives rise to an eigenvector $n_i^{(3)}$ satisfying (from (3.27))

$$\begin{aligned} \sigma_{11}n_1^{(3)} + \sigma_{12}n_2^{(3)} + \sigma_{13}n_3^{(3)} &= \sigma_{33}n_1^{(3)} \\ \sigma_{21}n_1^{(3)} + \sigma_{22}n_2^{(3)} + \sigma_{23}n_3^{(3)} &= \sigma_{33}n_2^{(3)} \\ \sigma_{31}n_1^{(3)} + \sigma_{32}n_2^{(3)} \qquad &= 0 \end{aligned}$$

A 'good' approximation to this system of equations, when σ_{13}, σ_{23} are small is $n_k^{(3)} = (0,0,1)$, so that one of the principal directions is normal to the plane of the plate.

In a practical problem σ_{33} is of little interest since it is small in comparison to the *maximum* principal stress and indeed the physical situation may be regarded, for all practical purposes, as a two dimensional stress problem with a two dimensional stress tensor:

$$\sigma_{ij} = \begin{bmatrix} \sigma_{11} & \sigma_{12} \\ \sigma_{21} & \sigma_{22} \end{bmatrix} \qquad \textbf{(3.34)}$$

and principal stresses given by (3.33). Of course, if Q is a stress free point on the boundary edge of the plate, then using the arguments of the previous section, one of the principal stress magnitudes $\sigma^{(1)}$ or $\sigma^{(2)}$ will be zero there and the non-zero principal stress vector will be tangential to the boundary. Note also that in these two dimensional problems the maximum shear stress is

$$\frac{1}{2}|\sigma^{(2)} - \sigma^{(1)}| = \sqrt{\frac{1}{4}(\sigma_{11} - \sigma_{22})^2 + \sigma_{12}^2} \qquad (3.35)$$

(We must choose $n_3 = 0$ in Table 3.1 since now only plane elements orthogonal to the plane of the plate are allowed – physically we are treating the plate as having no thickness).

EXERCISES

3.3 (a) At a point, the stress tensor is given by

$$\sigma_{11} = 0, \quad \sigma_{12} = 5, \quad \sigma_{13} = 5, \quad \sigma_{22} = 0 \quad \sigma_{23} = -5, \quad \sigma_{33} = -10$$

Find the principal stresses and the principal directions. Determine the magnitude of the maximum shearing stress.

3.4 Prove that, if with respect to an unprimed Cartesian coordinate system

$$\sigma_{13} = 0, \quad \sigma_{23} = 0 \text{ and } \sigma_{33} = 0$$

then under a coordinate transformation to a primed system:

$$\sigma'_{11} + \sigma'_{22} = \sigma_{11} + \sigma_{22}$$

(See Exercise 2.7). That is, in plane stress situations, the sum of the 'longitudinal' stresses is an invariant.

3.5 At a point in a plate in plane stress the non-zero stress components are

$$\sigma_{11} = -24000 \qquad \sigma_{12} = -8000 \qquad \sigma_{22} = -12000$$

Determine the three principal stresses, the principal shear stresses and their corresponding principal directions.

3.6 Photoelastic Measurement of Principal Stresses

In 1816 Brewster communicated his discovery of the photoelastic effect in the London Transactions of the Royal Society. He observed that certain transparent materials, when in a state of plane stress, are birefringent. That is, a beam of light incident on a specimen of such material is refracted within the material into *two* rays called an ordinary and an extraordinary ray. See Figure 3.11

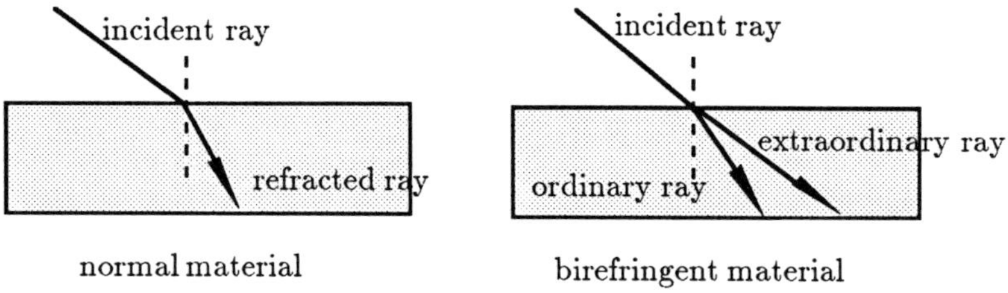

Figure 3.11

Even when the incident ray is normal to the specimen the two rays still exist. They travel at different speeds (depending upon the stresses within the material) and give rise to an interference pattern. These patterns may be analysed and throw light (excuse the pun) on the levels of the internal stresses.

Before proceeding in more detail we shall briefly discuss the nature of polarised light which plays a fundamental role in the phenomenon.

A point source of monochromatic light consists of waves (transverse waves to be precise) in randomly oriented planes. If the rays from a point source are passed through a polarising filter which has, for example its optical axis in the vertical plane then only those rays vibrating in the vertical plane will be allowed to pass through. The emergent light is called *plane polarised light.* A polariscope is an arrangement of two identical polarising filters. The first filter is called the polariser and the second is called the analyser. In the arrangement shown in Figure 3.12 the optical axis of the analyser is rotated through an angle θ from the optical axis of the polariser.

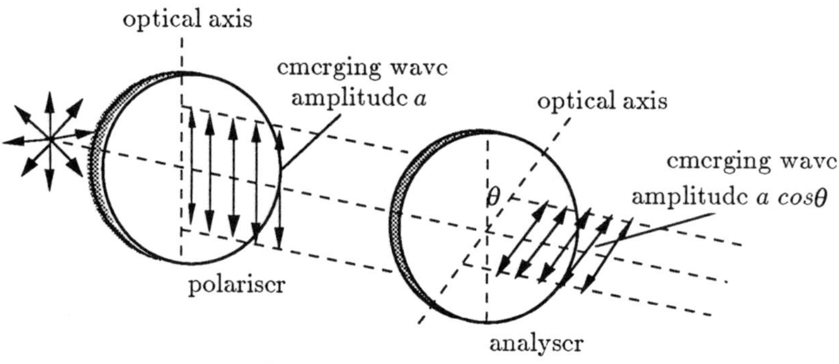

POLARISCOPE

Figure 3.12

It is clear that if $\theta = \pi/2$ (so that the polariser and analyser are *crossed*) then no light will emerge from the analyser.

Polarised light which is incident normally on a transparent plate of birefringent material in a state of plane stress splits into two rays. The motion of these rays are governed by the so-called photoelastic laws first formulated by James Clerk Maxwell (Sci Papers, Vol 1 p 30):

- The two rays vibrate transversely along the two principal stress directions of the material (see Figure 3.13).

• The phase difference between the two defracted rays of light is proportional to the difference in the magnitude of the principal stresses and also proportional to the thickness of the plate.

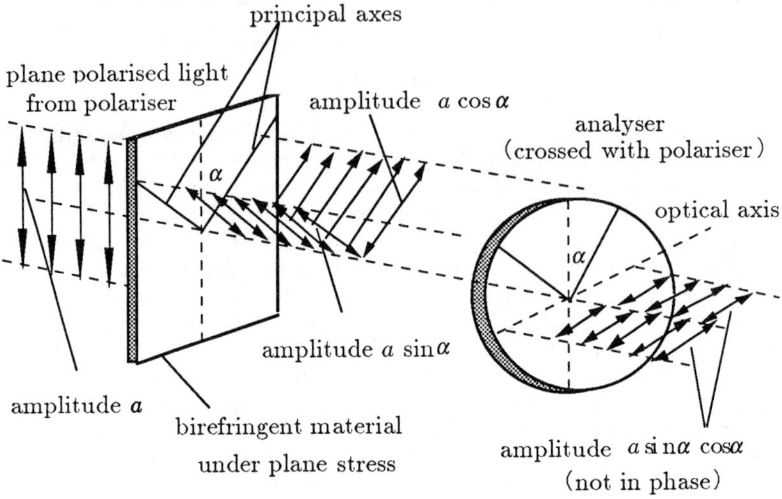

Figure 3.13

When a piece of photoelastc material under plane stress is placed in a polariscope between crossed polariser and analyser filters then the two refracted waves emerging from the material are brought together again as they pass through the analyser. Although the amplitude of these waves are equal their phase angles are not and so produce interference fringes. These fringes are called *isochromatics.* (We emphasise that Figure 3.13 refers to a specific point in the birefringent material. At other points, the direction of the principal axes and the values of the principal stresses may change leading to differing levels of interference)

Other dark regions, called *isoclinics*, appear in the field of view. These are due to a coincidental alignment between the optical axis of the polariser and the principal axes in the birefringent material ($\alpha = \pi/2$ in Figure 3.13). When this occurs only light in the direction of the optical axis will be transmitted. This plane polarised light will be unable to pass through the analyser since the polariser is crossed. At this point a dark region will be seen in the field of view. The locus of all such dark points constitute the isoclinics and, typically, take the form of dark bands often masking much of the field of view. These bands, though providing valuable information on the direction of the principal stresses, may be filtered out without affecting the isochromatic fringes through the use of a pair of crossed quarter-wave plates placed in the polariscope on either side of the photoelastic specimen. For fuller details the reader is referred to various textbooks on the subject most notable amongst these being by Coker and Filon ('Photo-elasticity', Cambridge University Press, New York 1931).

The polarised wave which leaves the polariser may be represented by the simple harmonic form $a \cos pt$, where a is the amplitude and p is related to the frequency of the wave that is used. The two waves emerging from the birefringent material have amplitudes $a\cos\alpha$ and $a\sin\alpha$ (see Figure 3.13). However, due to the effect of the principal stresses, one of these waves is now retarded with respect to the other so that each may be represented by

$$a\cos\alpha\cos pt \quad \text{and} \quad a\sin\alpha\cos(pt-\delta)$$

where the phase difference δ may be expressed in the form (due to the second photoelastic law)

$$\delta = Ch(\sigma^{(2)} - \sigma^{(1)}) \tag{3.36}$$

in which h is the thickness of the material, $\sigma^{(1)}$, $\sigma^{(2)}$ are the principal stresses and C is a constant depending upon the temperature, the photoelastic material and the wavelength of the light being used.

The amplitude of each of these waves emerging from the analyser is simply the respective component of the incident waves in the direction of the optical axis. The $\sigma^{(1)}$-wave is modified by a factor $\cos\alpha$ and the $\sigma^{(2)}$-wave by a factor of $\sin\alpha$. That is, on emergence we have two waves of equal amplitude differing only by a phase shift δ. The resultant wave is therefore (taking account of direction)

$$\begin{aligned} a\cos\alpha\sin\alpha\cos pt &- a\sin\alpha\cos\alpha\cos(pt-\delta) \\ &= a\cos\alpha\sin\alpha[\cos pt - \cos pt\cos\delta - \sin pt\sin\delta] \\ &= a\cos\alpha\sin\alpha[\cos pt(2\sin^2\frac{\delta}{2}) - 2\sin pt\sin\frac{\delta}{2}\cos\frac{\delta}{2}] \\ &= a\sin 2\alpha[\sin\frac{\delta}{2}][\sin(pt-\frac{\delta}{2})] \end{aligned}$$

The amplitude of the emergent wave is $a\sin 2\alpha\sin\frac{\delta}{2}$. When this is zero, at a pont, the corresponding point in the field of view will be dark. The locus of these points give rise to the isoclinics (being produced when $\alpha = 0$ or when $\alpha = \pi/2$; implying one of the principal axes at the point is aligned along the optical axis of the polariser) and to the isochromatic fringes, arising when

$$\sin\frac{\delta}{2} = 0$$

that is,

$$\delta = 2\pi n \qquad n = 0, 1, 2, ...$$

or, from (3.36)

$$\sigma^{(2)} - \sigma^{(1)} = \frac{kn}{h} \qquad (k = \frac{2\pi}{C}) \tag{3.37}$$

k is called the material fringe value and n is the fringe order. k is a constant of the birefringent material and of the wavelength of light being used. It may be determined by using (3.37) in a problem in which the principal stresses are *known.*

By knowing k, h and n the difference between the principal stresses (twice the maximum shear stress) at a point may be inferred. In particular at a stress free point on the boundary one of the principal stresses is zero (see previous section) and so, depending as the point is in compression or in tension

$$\sigma^{(2)} = \frac{kn}{h} \qquad \text{or} \qquad -\sigma^{(1)} = \frac{kn}{h} \qquad (\text{assuming } \sigma^{(2)} > 0)$$

That is, knowing the fringe order at a point (by counting fringes from the zero-order fringe) the principal stresses at the boundary points in the specimen may be determined directly. This is extremely useful since in many instances the maximum principal stresses occur on the boundary of the stressed object. (See, in particular, the analysis of Torsion in a Chapter 7).

To apply this analysis to real objects made from structural materials like steel and concrete (which of course are not transparent) in a state of plane stress, a scaled model of the object is constructed in birefringent material and a similarly scaled-down loading is applied. The stress levels measured in the model can be related to the actual stress levels observed in the structural object through the relation:

$$(\text{stress})_o = (\text{stress})_m \left(\frac{L_o}{L_m}\right)\left(\frac{l_m}{l_o}\right)\left(\frac{h_m}{h_o}\right)$$

in which the subscripts o and m refer to the structural object and model respectively and

$L =$ loading applied

$l =$ typical lateral dimension

$h =$ thickness

As a specific example we shall consider the analysis of a ring of thickness 0.005m of bifringent material (Araldite CT-200) subjected to diametral compression as shown in Figure 3.14 (I am most grateful to K Sharples, founder of Sharples Stress Engineers Ltd, Bamber Bridge, Preston England for supplying this example)

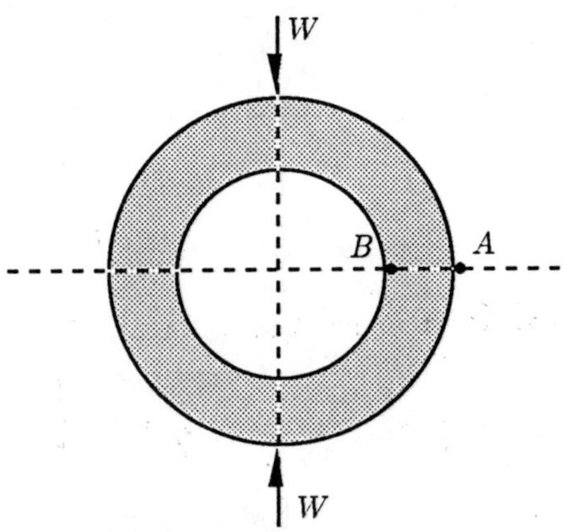

Figure 3.14

We shall use the photoelastic technique to measure the principal stresses at points on the stress-free boundaries.

When placed in a circular polariscope (this is a polariscope which has been modified by the addition of quarter-wave plates so as to exclude the isoclinics)

illuminated with sodium light the interference pattern shown in the right half of Figure 3.15 is obtained. We can now begin counting fringes from the zeroth order fringe. The zeroth fringe may sometimes difficult to isolate. However, if the model is viewed in white light these zero fringes are black in colour since, at these points, the principal stresses are equal and so the waves emerging from the analyser interfere destructively *irrespective* of the wavelength.

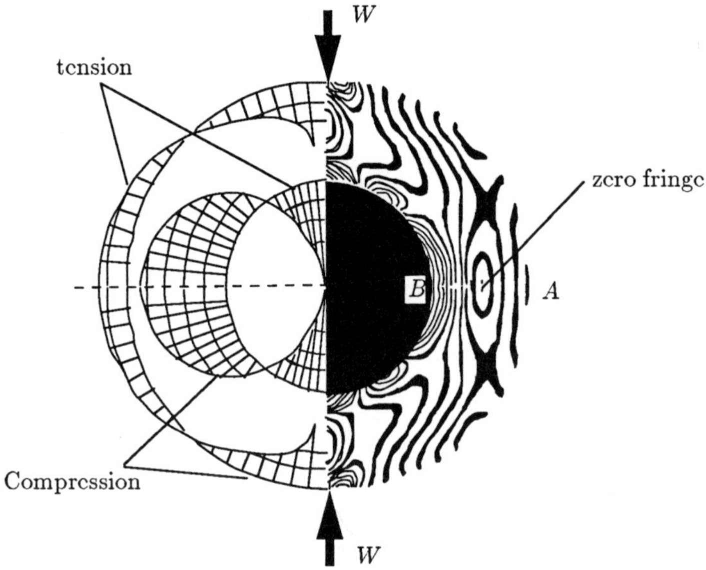

Figure 3.15

Having found the zero fringe we can then count fringes (using magnifying glasses if necessary). For this example, we conclude that at point A the order is $n = 4$ whilst at point B the order is $n = 12$.

For this material it is known that $k = 10\text{KN m}^{-1}$. Hence at A, where it is clear that the stress is tensile

$$\sigma^{(2)} = \frac{10(4)}{0.005} = 8000 \text{ KN m}^{-2}$$

Also at point B where it is physically obvious that the stress is compressive

$$\sigma^{(1)} = -\frac{10(12)}{0.005} = -24000 \text{ KN m}^{-2}$$

By counting fringe orders at other points on the outer and inner boundaries complete stress plots can be obtained as shown in the left half of Figure 3.15. These compare very favourably with the exact solution to this problem.

In this brief introduction to photoelasticity we have covered only the elementary applications. More specific texts and publications should be consulted if further information on the practical applications of the technique is required.

EXERCISES

3.6 Let S_{ij} be a symmetric tensor. If axes are rotated by an angle θ in the positive sense about the x_3-axis show that there are two values of θ for which S'_{ij} is zero where

$$\tan 2\theta = \frac{2S_{12}}{S_{11} - S_{22}}$$

3.7 The stress tensor at the origin in a certain fluid is given by

$$\sigma_{ij} = \begin{bmatrix} 1 & 2 & 0 \\ 2 & 1 & 0 \\ 0 & 0 & 2 \end{bmatrix}$$

Find the normal stress on the plane $2x_1 + 3x_2 + 4x_3 = 0$. Find also the equation of the plane through the origin on which the normal stress is a maximum

3.8 Show, for a two dimensional elastic system that the stress components in primed and unprimed coordinate systems, related one to the other by a positive rotation through angle θ about the x_3-axis, are related by the complex equations

$$\sigma'_{11} + \sigma'_{22} = \sigma_{11} + \sigma_{22}$$

$$\sigma'_{11} - \sigma'_{22} + 2i\sigma'_{12} = e^{-2i\theta}(\sigma_{11} - \sigma_{22} + 2i\sigma_{12})$$

$$\sigma'_{13} + i\sigma'_{23} = e^{-i\theta}(\sigma_{13} + i\sigma_{23})$$

3.9 The stress components at a point are

$$\sigma_{11} = 0 \quad \sigma_{22} = 0 \quad \sigma_{33} = 3 \quad \sigma_{12} = -2 \quad \sigma_{23} = 3 \quad \sigma_{13} = 1$$

Show that the three principal stresses and their corresponding directions are:

$$\sigma_A = 0.49 \qquad n_i = (0.049, -0.54, -0.84)$$

$$\sigma_A = 1.48 \qquad n_i = (0.83, -0.45, -0.33)$$

$$\sigma_A = -3.34 \qquad n_i = (0.55, 0.72, -0.43)$$

Chapter 4

THE ANALYSIS OF STRAIN

4.1 The Strain Tensor

The concept of strain is applicable to any material. The material may flow like water or be as rigid as iron. In this text we shall be concerned with *structural materials* which are able to sustain or support large loads. The material may take many geometrical forms: it might, for example, be in the form of a reinforced concrete girder used in the construction industry. It may be formed with great precision as in an engine block which is able to sustain extremely large pressure loads resulting from the controlled explosion of petrol with air. More mundanely it may simply be a large mass of concrete used as a stable 'bed' for a sensitive optical experiment.

Consider a body of such structural material. This *continuum* is comprised of a large number of particles. A change in the relative position between particles is called a **deformation**. When deformations occur the body is said to be *strained*. The strain may result from the application of compressive or tensile forces applied to its surface or from the action of 'body' forces (such as gravity) acting throughout the body. As well as the action of forces, straining can occur simply by heating the body. In this introductory text, we shall ignore temperature effects.

In many treatments, especially in classical mechanics, bodies are treated as rigid in which all particles are assumed to remain a fixed distance apart. In elasticity theory it is the very fact that bodies *are* deformable that is of interest. Most structural materials *appear* not to deform unless the loads applied to them are extremely large. In fact under 'normal' loads all bodies (fixed in position so as not to be movable as a rigid body) deform though the deformations are extremely small. In small-displacement elasticity theory we exploit this natural property of structural materials to our advantage.

Of course, we are all familiar with substances like soft putty or rubber which easily deform under the action of normal loads. These, and substances like them, cannot be analysed by the theory outlined in this text. Clearly, they are not the materials one would use in any significant way in building structures which are able to support large loads without sustaining unduly large deformations.

Bodies deform when the relative position between particles change. However, forces may be applied to a body *without* producing deformation. The intention may simply be to move the body from one point to another. Such a movement may be translation or a rotation or a combination of both. Under such movements we say the body undergoes a **rigid-body motion**.

To obtain a quantative measure of strain we introduce a mathematical description of deformation.

Consider a body as shown in Figure 4.1. Each particle, at A say, in the body has a specific coordinate position x_i (or (x_1, x_2, x_3)) with respect to the Cartesian system with an origin at O.

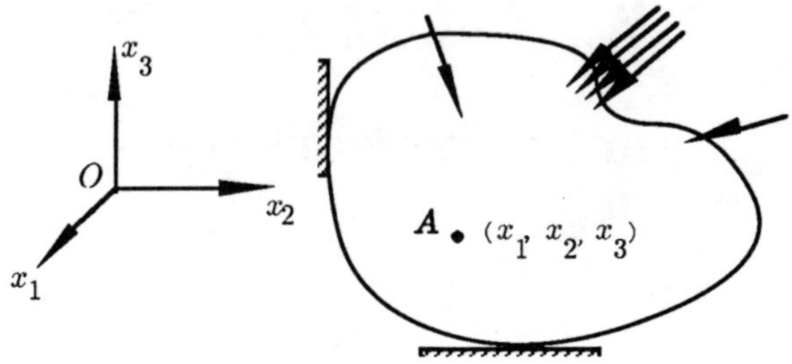

Figure 4.1

Under the action of applied forces the body is deformed resulting in a movement or *displacement* of the particle at A to a new point A^*. If the coordinates of A^* are x_i^* then by simple vector addition the displacement of A, denoted by u_i, is given by

$$u_i = x_i^* - x_i \tag{4.1}$$

As we have already noted, for most structural materials the magnitude of this displacement $|u_i u_i|^{1/2}$ will be extremely small for particles under normal loads. To obtain a measure of the deformation or strain induced in the body by the applied loads we concentrate on the change in length of an elementary line of particles AB. This line is deformed to the line A^*B^* after the loads are applied. We shall develop a measure of strain which is appropriate to a very small region within the body and so shall take B, and hence B^*, as being very close to A and A^* respectively. Anticipating this we desiginate the coordinates of B as $x_i + dx_i$ and those of B^* as $x_i^* + dx_i^*$ in which the 'd' notation is used to denote a small quantity. For clarity, the deformation has been highly exaggerated in Figure 4.2.

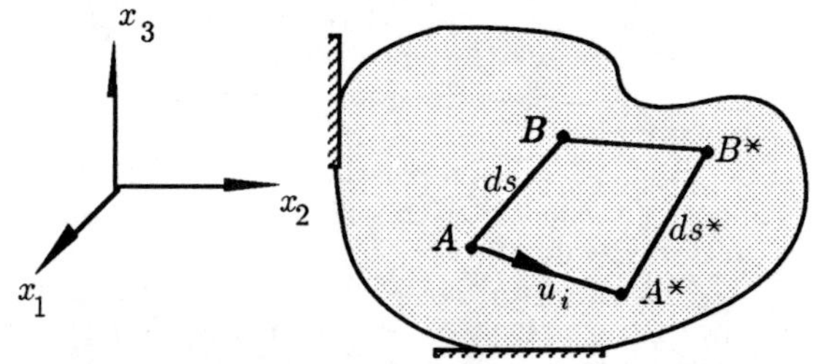

Figure 4.2

The length of the element before and after deformation is ds and ds^* respectively and, using the scalar product to determine the magnitude, we obtain:

$$ds^2 = (x_i + dx_i - x_i)(x_i + dx_i - x_i) = dx_i dx_i$$

and

$$(ds^*)^2 = dx_i^* dx_i^*$$

An obvious measure of strain would be to record the change in length $(ds^* - ds)$ of such a line of particles in each small region of the body. Mathematically this measure, involving, as it does, square roots is difficult to deal with. Instead the measure that has been universally adopted is $(ds^*)^2 - ds^2$:

$$(ds^*)^2 - ds^2 = dx_i^* dx_i^* - dx_i dx_i \tag{4.2}$$

For a given deformation we shall assume that the final coordinates x_i^* of a particle are smooth functions of the original coordinates x_i so that the displacement vector u_i of the particle may also be considered to be a smooth function of the original coordinates.

Since $x_i^* = u_i + x_i$ then

$$dx_i^* = \frac{\partial x_i^*}{\partial x_j} dx_j = (\frac{\partial u_i}{\partial x_j} + \delta_{ij}) dx_j$$
$$= (u_{i,j} + \delta_{ij}) dx_j$$

giving

$$\begin{aligned}(ds^*)^2 - ds^2 &= (u_{i,j} + \delta_{ij})(u_{i,k} + \delta_{ik}) dx_j dx_k - dx_i dx_i \\ &= (u_{i,k} u_{i,j} + u_{j,k} + u_{k,j} + \delta_{jk}) dx_j dx_k - dx_i dx_i \\ &= (u_{i,k} u_{i,j} + u_{j,k} + u_{k,j}) dx_j dx_k\end{aligned} \tag{4.3}$$

As well as assuming that displacements are small we shall also assume that displacement derivatives $u_{i,k}$ $i = 1,2,3;$ $k = 1,2,3$ are also so small so that products of these terms may be neglected in comparison with linear terms. The overriding reason for doing so is mathematical. The theory that results from this assumption is a *linear* theory and the governing equations of the theory (systems of simultaneous and partial differential equations) are also linear allowing analytical solutions to be developed. The predictions (on deformations and stresses) that are based on this linear theory compare well with the deformations and stresses actually measured in structural materials; thereby justifying the assumption of linearity.

With this assumption the first term in the brackets on the right hand side of (4.3) can be neglected in comparison to the other terms and so we obtain

$$(ds^*)^2 - ds^2 = (u_{j,k} + u_{k,j}) dx_j dx_k \tag{4.4}$$

which may be rewritten in the form

$$(ds^*)^2 - ds^2 = 2\varepsilon_{jk} dx_j dx_k \tag{4.5}$$

in which

$$\varepsilon_{jk} = \frac{1}{2}(u_{j,k} + u_{k,j}) \tag{4.6}$$

The two indexed object ε_{jk} is called the **strain tensor**. Its tensor character is a direct consequence of its definition. Also it is a *symmetric* tensor since $\varepsilon_{kj} = \frac{1}{2}(u_{k,j} + u_{j,k}) = \varepsilon_{jk}$.

Obviously if each component of the strain tensor is zero, implying $(ds^*)^2 = ds^2$ then each particle of the body remains a fixed distance apart and so describes a rigid body motion.

Example 4.1
Show that if the displacement vector u_i has the form

$$u_i = t_i + \eta_{ijk} x_k \omega_j$$

where t_i and ω_i are constant vectors then the body undergoes a rigid body motion. Determine t_i and ω_i explicitly in terms of the displacement vector u_i and/or its derivatives.

Solution

Taking derivatives of u_i with respect to x_i we have

$$\begin{aligned} u_{j,k} &= t_{j,k} + \eta_{jam}(x_m \omega_a)_{,k} \\ &= 0 + \eta_{jam} \omega_a x_{m,k} \\ &= \eta_{jam} \omega_a \delta_{mk} \\ &= \eta_{jak} \omega_a \end{aligned}$$

and so, by interchanging indices

$$u_{k,j} = \eta_{kaj} \omega_a = -\eta_{jak} \omega_a$$

therefore

$$(u_{j,k} + u_{k,j}) = 0$$

and so $\varepsilon_{jk} = 0$ implying that this particular displacement field is a rigid body motion. The constant vector t_i is a translation of the body as a whole and $\eta_{jam} x_k \omega_a$ is an infinitessimal rotation of the body through an angle $(\omega_a \omega_a)^{1/2}$. (See (2.24) of Section 2.5). The translation vector is easily expressed in terms of the displacement vector as $t_i = u_i(x_i = 0)$. To isolate the rotation vector we need to take derivatives. We have already developed an expression for $u_{j,k}$ and $u_{k,j}$ above and so

$$u_{j,k} - u_{k,j} = \eta_{jak} \omega_a - \eta_{kaj} \omega_a = 2\eta_{jak} \omega_a$$

from which we easily deduce (by choosing particular values for the indices)

$$\omega_1 = \frac{1}{2}(u_{3,2} - u_{2,3}) \qquad \omega_2 = \frac{1}{2}(u_{1,3} - u_{3,1}) \qquad \omega_3 = \frac{1}{2}(u_{2,1} - u_{1,2})$$

(the more adventurous reader can obtain this result more formally by isolating ω_a through the use of the $\eta - \delta$ identity (see Section 1.6)).

Example 4.2
In the torsion problem of a circular cylinder aligned along the x_3-axis which is rigidly fixed at (0,0,0) it may be shown that the displacement vector is

$$u_i = (-\tau x_2 x_3, \tau x_1 x_3, 0) \qquad \text{in which } \tau \text{ is a constant}$$

Determine the strain tensor components and show that plane sections, originally perpendicular to the axis of the cylinder, suffer no distortion apart from a rotation about the axis.

Solution

Clearly

$$u_{1,1} = 0 \qquad u_{1,2} = -\tau x_3 \qquad u_{1,3} = -\tau x_2$$
$$u_{2,1} = \tau x_3 \qquad u_{2,2} = 0 \qquad u_{2,3} = \tau x_1$$
$$u_{3,1} = 0 \qquad u_{3,2} = 0 \qquad u_{3,3} = 0$$

and so, from (4.6)

$$\varepsilon_{11} = 0 \qquad \varepsilon_{12} = 0 \qquad \varepsilon_{13} = -\frac{1}{2}\tau x_2$$
$$\varepsilon_{21} = 0 \qquad \varepsilon_{22} = 0 \qquad \varepsilon_{23} = \frac{1}{2}\tau x_1$$
$$\varepsilon_{31} = -\frac{1}{2}\tau x_2 \qquad \varepsilon_{32} = \frac{1}{2}\tau x_1 \qquad \varepsilon_{33} = 0$$

If we consider a particle on a plane perpendicular to the x_3-axis with coordinates (x_1, x_2, x_3) in which x_3 is fixed in value, then its change in position is, from (4.1)

$$u_i = x_i^* - x_i = (-\tau x_2 x_3, \tau x_1 x_3, 0)$$

But this vector is of the form $t_i + \eta_{ijk}\omega_j x_k$ in which $t_i = (0,0,0)$ and $\omega_j = (0,0,\tau x_3)$. That is, the plane is not translated but simply rotated through an angle τx_3 (see Example 4.1 and Figure 4.3).

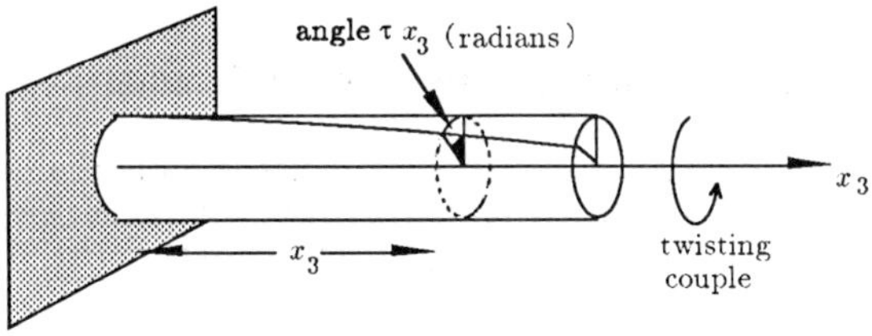

Figure 4.3

We should note here that cylinders with cross-sections other than circular suffer from *warping* – planes perpendicular to the axis of the beam do not remain plane after deformation. This problem will be covered in much greater detail in Chapter 7.

EXERCISES

4.1 Determine the strain tensors for the following displacement vectors:

(a) $$u_1 = -\frac{\nu}{E}\rho g(x_3 x_1 - h x_1) \qquad u_2 = -\frac{\nu}{E}\rho g(x_3 x_2 - h x_2)$$

$$u_3 = \frac{1}{E}\rho g(\frac{1}{2}x_3^2 - h x_3) + \frac{\nu}{E}\rho g(\frac{1}{2}x_1^2 + \frac{1}{2}x_2^2)$$

(The displacement field of a cylinder deforming under its own weight)

(b) $$u_1 = -\frac{\nu M}{EI}x_1 x_2 \quad u_2 = \frac{M}{EI}(-\nu x_2^2 + \frac{1}{2}\nu x_1^2 - \frac{1}{2}x_3^2) \quad u_3 = \frac{M}{EI}x_2 x_3$$

(Beam bending by pure couples gives rise to this displacement field)

(c) $$u_1 = Ax_1 + \frac{Bx_1}{r^2} \qquad u_2 = Ax_2 + \frac{Bx_2}{r^2} \qquad u_3 = 0$$

in which $r^2 = x_1^2 + x_2^2$ and A and B are constants. (This is the displacement field of a cylindrical pipe under pressure).
In each case verify that the displacement field is rotation free (that is, check $u_{i,j} - u_{j,i} = 0$). In case (b) show that the centre line of the beam ($x_1 = x_2 = 0$) deforms into a parabola with radius of curvature $R = \dfrac{EI}{M}$.

4.2 Physical Interpretation of the Strain Tensor

Consider first the interpretation of the diagonal components of the strain tensor ε_{ii}, i not summed. We take an elementary line of particles aligned along the direction l_i choosing ($l_i l_i = 1$) so that $dx_i = l_i ds$. Then, from (4.5),

$$\begin{aligned}(ds^*)^2 - ds^2 &= 2\varepsilon_{ij} dx_i dx_j \\ &= 2\varepsilon_{ij} l_i l_j ds^2\end{aligned}$$

That is,

$$\frac{(ds^*)^2 - ds^2}{ds^2} = 2\varepsilon_{ij} l_i l_j$$

or

$$\frac{(ds^* - ds)}{ds}\frac{(ds^* + ds)}{ds} = 2\varepsilon_{ij} l_i l_j \tag{4.7}$$

Now, to a very good approximation, $ds \approx ds^*$ and so the sum $(ds^* + ds)$ may be replaced by $2ds$. Only the difference between these two small quantities is of significance. Using this in (4.7) we obtain

$$\frac{(ds^* - ds)}{ds} = \varepsilon_{ij} l_i l_j \tag{4.8}$$

By choosing the direction vector l_i appropriately we can isolate the diagonal components of the strain tensor. For example if the filament of particles is, before deformation, aligned along the x_1-axis then $l_i = (1, 0, 0)$ and so (4.8) implies:

$$\varepsilon_{11} = \frac{ds^* - ds}{ds} \tag{4.9}$$

We can interpret ε_{11} as the change in length per unit length of a line of particles originally along the x_1-direction. Note that $\varepsilon_{11} > 0$ at a point indicates that straight line segments in the x_1-direction in the close neighbourhood of the point will be stretched and if $\varepsilon_{11} < 0$ at the point then contraction of these elements will occur.

Similar interpretations follow for ε_{22} and ε_{33} by choosing, in turn, $l_i = (0, 1, 0)$ and $l_i = (0, 0, 1)$. The diagonal components of strain ε_{11}, ε_{22} and ε_{33} are often called the **longitudinal components** of strain.

The formula (4.8) cannot be used to isolate the off-diagonal terms of ε_{ij} (no matter what the choice for l_i). The physical interpretation of these components ε_{12}, ε_{13} and ε_{23}; the so-called **shear components** of strain is less straightforward. As the terminology has anticipated they are connected with shearing effects. To see this we consider two elementary lines of particles of lengths ds and δs which, before deformation, are perpendicular and aligned along the x_1- and x_2-axes respectively so that their vector representations are:

$$dx_i = (ds, 0, 0) \equiv ds\delta_{i1} \quad \text{and} \quad \delta x_i = (0, \delta s, 0) \equiv \delta s \delta_{i2} \tag{4.10}$$

These filaments are, before deformation, orthogonal to each other. The angle between these elements initially is $\pi/2$. Let it be θ^* in the deformed state. After deformation the filaments are denoted by dx_i^* and δx_i^* with respective lengths ds^* and δs^*.

Since $\dfrac{dx_i^*}{ds^*}$ and $\dfrac{\delta x_i^*}{\delta s^*}$ are unit vectors the cosine of the angle θ^* is given by (see Section 1.2)

$$\cos\theta^* = \frac{dx_i^*}{ds^*}\frac{\delta x_i^*}{\delta s^*} \tag{4.11}$$

We shall express the right hand side of (4.11) in terms of the displacement vector. We note, taking derivatives with respect to the original coordinates x_i:

$$\frac{dx_i^*}{ds^*} = \frac{\partial x_i^*}{\partial x_j}\frac{dx_j}{ds^*} \quad \text{and} \quad \frac{\delta x_i^*}{\delta s^*} = \frac{\partial x_i^*}{\partial x_k}\frac{\delta x_k}{\delta s^*}$$

which, from (4.1) becomes

$$\frac{dx_i^*}{ds^*} = (\frac{\partial u_i}{\partial x_j} + \delta_{ij})\frac{dx_j}{ds^*} \quad \text{and} \quad \frac{\delta x_i^*}{\delta s^*} = (\frac{\partial u_i}{\partial x_k} + \delta_{ik})\frac{\delta x_k}{\delta s^*}$$

Hence

$$\cos\theta^* = (\frac{\partial u_i}{\partial x_j} + \delta_{ij})(\frac{\partial u_i}{\partial x_k} + \delta_{ik})\frac{dx_j}{ds^*}\frac{\delta x_k}{\delta s^*}$$
$$= (\frac{\partial u_j}{\partial x_k} + \frac{\partial u_k}{\partial x_j} + \delta_{kj})\frac{dx_j}{ds^*}\frac{\delta x_k}{\delta s^*}$$

where we have again (to be consistent with basic assumptions) neglected products of the displacement derivatives in comparison to linear terms. Employing the strain tensor (4.6) and utilising (4.10) we have

$$\begin{aligned}\cos\theta^* &= (2\varepsilon_{jk} + \delta_{jk})(\delta_{j1})(\delta_{k2})\frac{ds}{ds^*}\frac{\delta s}{\delta s^*} \\ &= (2\varepsilon_{12})\frac{ds}{ds^*}\frac{\delta s}{\delta s^*}\end{aligned} \qquad \textbf{(4.12)}$$

Finally from the interpretations of ε_{11} and ε_{22} we have

$$\frac{ds^*}{ds} - 1 \approx \varepsilon_{11} \qquad \frac{\delta s^*}{\delta s} - 1 \approx \varepsilon_{22}$$

these being the change in length per unit length of filaments along the x_1- and x_2-axes respectively. Using these relations (4.12) becomes

$$\cos\theta^* \approx 2\varepsilon_{12}(1 - \varepsilon_{11})(1 - \varepsilon_{22}) \approx 2\varepsilon_{12}$$

(neglecting products of the strain components, since, being displacement derivatives, are small compared to unity).

As a further approximation if we put $\theta^* = \frac{\pi}{2} - \phi^*$ and note that for small deformations θ^* will be very close to the original value of $\pi/2$ then $\cos\theta^* = \cos(\frac{\pi}{2} - \phi^*) = \sin\phi^* \approx \phi^*$ leading to

$$\varepsilon_{12} = \frac{1}{2}\phi^*$$

So that ε_{12} directly represents the 'angle of shear' at a point (see Figure 4.4). If $\varepsilon_{12} = 0$ at a point then elements, in the close neighbourhood of the point, along the x_1- and x_2-axes which were originally perpendicular will remain perpendicular after deformation.

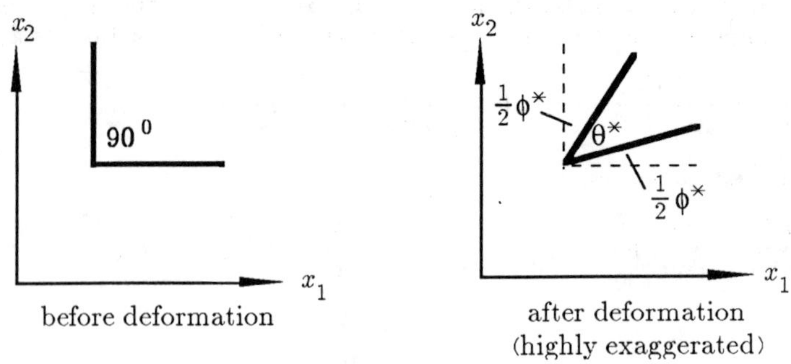

Figure 4.4

Example 4.3
A circular cylinder of length b and radius a has its axis aligned along the x_3-axis. One end is held fixed whilst the other end has a torsional couple applied to it producing a rotation (of that end) of θ^0. (See Figure 4.5).

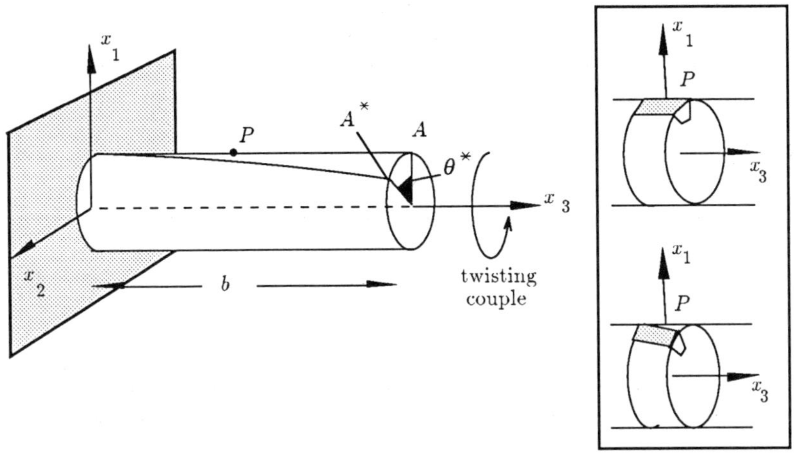

Figure 4.5

Determine the strain tensor and give interpretations for its components.

Solution

From Example 4.2 the combination τx_3 measures the angle of rotation of each plane segment orthogonal to the axis of the cylinder. However, at $x_3 = b$ we know that the angle of rotation is θ^0. Hence

$$\tau(b) = \theta(\frac{2\pi}{360}) \qquad \text{converting to radians}$$

Also, from that same example,

$$\varepsilon_{13} = \varepsilon_{31} = -\frac{1}{2}\tau x_2 = -\frac{\pi}{360}(\frac{\theta}{b})x_2$$

and

$$\varepsilon_{23} = \varepsilon_{32} = \frac{1}{2}\tau x_1 = \frac{\pi}{360}(\frac{\theta}{b})x_1$$

To interpret these strain components we consider two small area elements at P are perpendicular to the x_1-axis and the other perpendicular to the x_2-axis respectively (see Figure 4.5 inset). At P, $x_2 = 0$ and $x_1 = a$ so that

$$\varepsilon_{13} = 0 \qquad \text{and} \qquad \varepsilon_{23} = \frac{\pi}{360}(\frac{\theta}{b})a$$

implying that filaments on the shaded area are sheared due to differential rotations on planes orthogonal to the x_3-axis and that filaments on the unshaded area suffer no distortion but simply undergo a rigid body rotation.
As a specific example if $\theta = 10^0$, $b = 3$m and $a = 0.5$m then

$$\varepsilon_{23} = \frac{\pi}{360}(\frac{10}{3})\frac{1}{2} = \frac{\pi}{216} = 0.0145$$

so that a square region (as at P in Figure 4.5 inset) will, after deformation, appear diamond shaped with smaller angle of 88.34^0.

EXERCISES

4.2 A wire has an undeformed length of 1m. Determine the longitudinal strain in the wire if it is stretched to a length of 1.05m.

4.3 A 10cm cube is deformed as shown in Figure 4.6. Determine the strain tensor under the assumption that the strain tensor components are the same at each point (and hence the displacement vector must be linear in the coordinates).

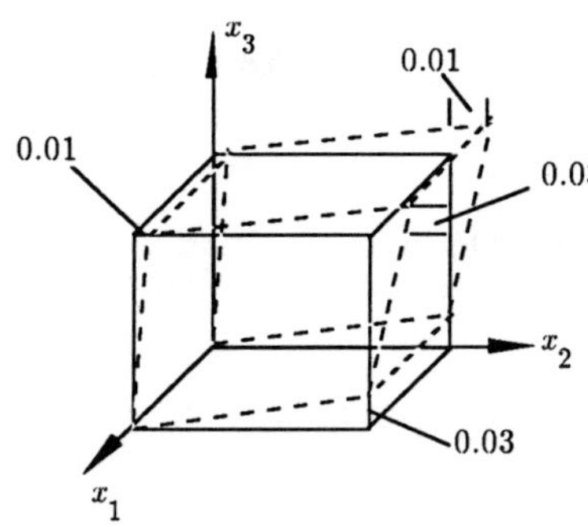

Figure 4.6

4.4 A plate bounded by $x_1 = 0$, $x_1 = 10$, $x_2 = 0$, $x_2 = 20$ is deformed so that at each point the displacement is $u_i = (0.0001x_1x_2,\ -0.00015x_1x_2,\ 0)$. Show that lines parallel to the x_1- and x_2-axes before deformation remain straight after deformation and sketch the deformed shape of the plate.

4.5 By considering an infinitessimal cube with sides dx_1, dx_2, dx_3 before deformation show that the change in volume per unit volume is given by ε_{kk}: this is called the **dilitation**.

4.3 Principal Axes, Principal Strains

The strain tensor ε_{ij} is a second order symmetric tensor and we may, at a point of interest within the material, (as we might for any such tensor), rotate axes so that with respect to primed coordinates, the strain tensor components satisfy $\varepsilon'_{ij} = 0 \; i \neq j$. The primed axes are then **principal axes** for the strain tensor and the diagonal terms of ε'_{ij} (denoted by $\varepsilon^{(1)}$, $\varepsilon^{(2)}$ and $\varepsilon^{(3)}$) are called the **principal strains**. More precisely, the principal axes lie along the principal directions $n_i^{(1)}$, $n_i^{(2)}$ and $n_i^{(3)}$ which are the solutions of the eigenvalue problem:

$$\varepsilon_{ij} n_j = \varepsilon n_i \qquad \text{with } n_i n_j = 1 \tag{4.13}$$

Here, the three possible values of ε : $\varepsilon^{(1)}$, $\varepsilon^{(2)}$ and $\varepsilon^{(3)}$, which are the solutions to $det(\varepsilon_{ij} - \varepsilon\delta_{ij}) = 0$, are the eigenvalues. It immediately follows from (4.13) that $\varepsilon = \varepsilon_{ij} n_i n_j$. (For a more detailed analysis of eigenvalue problems see Sections 2.7 and 3.4).

In this case the interpretation of a principal axis is that for filaments of material which are mutually orthogonal and also perpendicular to the principal axis before deformation *remain* mutually orthogonal after deformation. Also fibres in the direction of the principal axis may be extended or shortened depending on the sign of the corresponding principal strain. (See Figure 4.7). We may say that at a point P, the plane perpendicular to the principal axis through P is free of shearing strains, and a cubical block of material of material with its edges aligned along the principal axes through P would remain cubical after deformation.

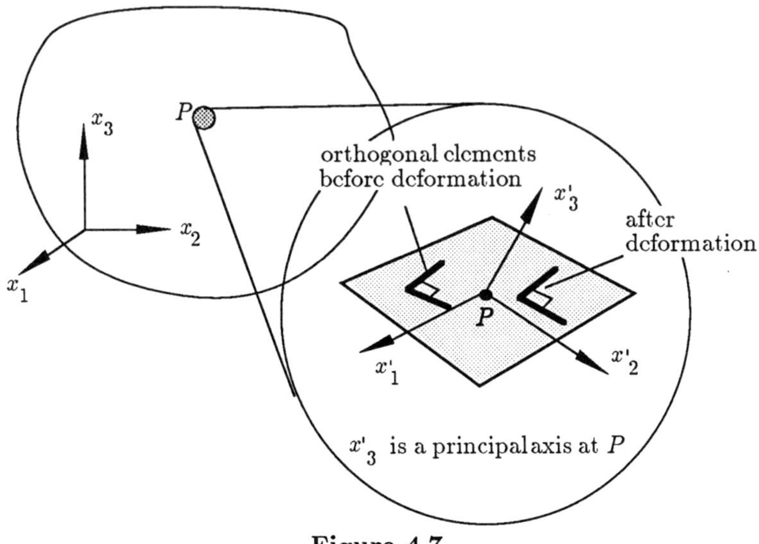

Figure 4.7

We can also show that, at P, the direction l_i along which there occurs the algebraically greatest or least extension of an elementary line of particles is one of the principal axes at P. We do this, as for the stress tensor, by maximising the scalar function $\varepsilon_{ij}l_il_j$ subject to the constraint $l_il_j = 1$ (See (4.8)). This optimisation procedure leads directly to the eigenvalue problem:

$$\varepsilon_{ij}l_i = \lambda l_i \qquad \text{in which} \quad \lambda = \varepsilon_{ij}l_il_j \tag{4.14}$$

But these are precisely the equations (4.13) which are solved by the principal directions (the eigenvectors) and by the principal strains $\varepsilon^{(1)}$, $\varepsilon^{(2)}$ and $\varepsilon^{(3)}$ (the eigenvalues).

Example 4.4
A cube of material of side 10cm is aligned with its edges along the coordinate axes (see Figure 4.8a). The block is deformed by sliding the face $x_3 = 10$ in the direction of the x_2-axis through a length 0.5mm, the opposite face being held fixed as in Figure 4.8b. Determine the principal strains and the principal axes of strain.

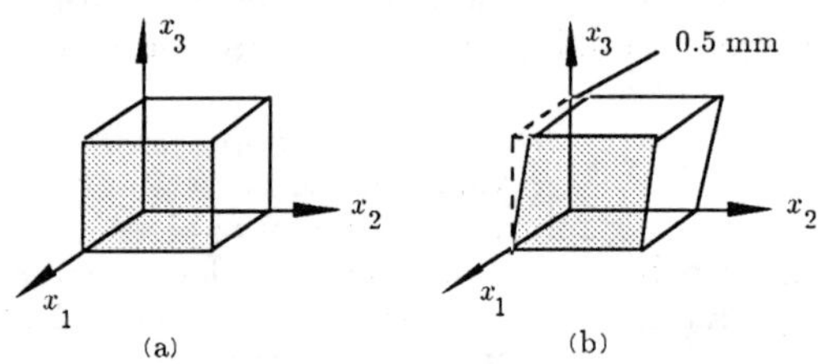

Figure 4.8

Solution

Clearly this deformation is a simple shearing of each plane of the material parallel to the x_2x_3 plane. From the analysis of the shear component of the strain tensor in Section 4.2 we can immediately write $\phi^* = 0.005 = 2\varepsilon_{23}$ with the remaining strain components being zero. Alternatively, it is clear from the deformation that the displacement field u_i is given by

$$u_1 = 0 \qquad u_2 = \frac{x_3}{10}(0.05) \qquad u_3 = 0$$

from which we can readily find, from (4.6), that

$$\varepsilon_{23} = \varepsilon_{32} = 0.0025 \qquad \text{other } \varepsilon_{ij} = 0$$

To determine the principal strains and directions we need to solve the eigenvalue problem of (4.13). In this example we find:

$$\begin{vmatrix} -\varepsilon & 0 & 0 \\ 0 & -\varepsilon & 0.0025 \\ 0 & 0.0025 & -\varepsilon \end{vmatrix} = 0$$

leading to

$$\varepsilon(\varepsilon^2 - 0.0025^2) = 0 \qquad \text{or} \quad \varepsilon^{(1)} = 0,\ \varepsilon^{(2)} = 0.0025,\ \varepsilon^{(3)} = -0.0025$$

These are the principal strains. It is a trivial matter to confirm that the principal directions are $(1, 0, 0)$; $(0, \frac{1}{\sqrt{2}}, \frac{1}{\sqrt{2}})$ and $(0, \frac{1}{\sqrt{2}}, -\frac{1}{\sqrt{2}})$ respectively.

4.4 Principal Strains and the Strain Rosette

We have now defined, at each point within a body, a stress tensor σ_{ij} and a strain tensor ε_{ij}, both of which are symmetric. For each a principal axis system exists. In this text we shall only consider materials for which the principal axes of stress and strain tensors coincide at each point. These materials have the (desirable) property that the greatest (or least) extension of fibres occur in those directions in which the greatest (or least) normal stress is applied. Also area elements which have zero shear stress do not experience any shear strain. As we shall see in the next chapter this *sharing* of principal axes for stress and strain tensors is a property of so-called *isotropic* materials. In Section 3.4 we noted that at a stress free point of the boundary of a material body a principal axis (of σ_{ij}) is normal to the boundary and the corresponding principal stress is zero. The other two principal directions are tangential to the boundary. It follows that at these points in isotropic materials a principal axis of the strain tensor is also normal to the boundary and the other two principal axes of strain are tangential to the bounding surface. These properties are invaluable in the practical measurement of strain. One of the most popular ways of making direct measurements of longitudinal strain is to use a strain gauge.

The strain gauge utilises the linear relationship between the length of a wire and its electrical resistance to measure the strain at a point. The wire, for sensitivity, is very thin (about 0.0005cm in diameter) and is tightly bonded to a thin plate of plastic in such a way as to orient a large length of wire in a single direction (see Figure 4.9). Thus any slight change in the length of the plate will be magnified by the many loops of wire.

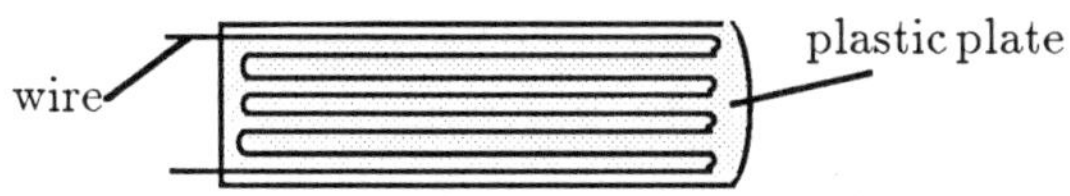

Figure 4.9

The plate is glued to the surface of the body. When the body is strained the gauge is extended (or compressed). The change in resistance of the wire, due to the increase (or decrease) in its length can be measured electrically and the longitudinal strain (being the change in length per unit length) in the direction of the gauge inferred. To obtain the principal strains on the surface of a body a suite of strain gauges called a *rosette* is used, often in the configuration shown in the inset of Figure 4.10.

The longitudinal strains are measured from each of the strain gauges of the rosette at P. We shall refer to these as ε_A, ε_B and ε_C. At P assume the x_1-axis is in the direction of PA and the x_2-axis is in the direction of PC. The x_3-axis is normal to the surface. We can perform a rotation of coordinates in the x_1x_2 plane (the tangent plane), through an angle θ, to obtain new axes x_1', x_2' and x_3'.

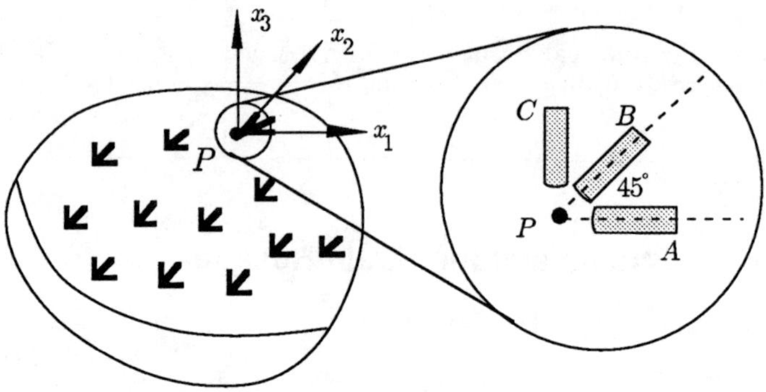

Figure 4.10

Clearly, after such a rotation, the relation between primed and unprimed coordinates is (see Example 2.1):

$$
\begin{aligned}
x_1' &= x_1 \cos\theta + x_2 \sin\theta \\
x_2' &= -x_1 \sin\theta + x_2 \cos\theta \\
x_3' &= x_3
\end{aligned}
\tag{4.15}
$$

so that using $\alpha_{ij} = \dfrac{\partial x_i'}{\partial x_j}$ from Section 2.2 (take partial derivatives of (2.9a)) implies

$$
\begin{aligned}
\alpha_{11} &= \cos\theta & \alpha_{12} &= \sin\theta & \alpha_{13} &= 0 \\
\alpha_{21} &= -\sin\theta & \alpha_{22} &= \cos\theta & \alpha_{23} &= 0 \\
\alpha_{31} &= 0 & \alpha_{32} &= 0 & \alpha_{33} &= 0
\end{aligned}
$$

Hence, since ε_{ij} is a second order tensor its components with respect to the primed system are easily obtained

$$\varepsilon_{ij}' = \alpha_{ik}\alpha_{jm}\varepsilon_{km}$$

which gives, after a short calculation,

$$
\begin{aligned}
\varepsilon_{11}' &= \frac{1}{2}(\varepsilon_{11} + \varepsilon_{22}) + \frac{1}{2}(\varepsilon_{11} - \varepsilon_{22})\cos 2\theta + \varepsilon_{12}\sin 2\theta \\
\varepsilon_{22}' &= \frac{1}{2}(\varepsilon_{11} + \varepsilon_{22}) - \frac{1}{2}(\varepsilon_{11} - \varepsilon_{22})\cos 2\theta - \varepsilon_{12}\sin 2\theta \\
\varepsilon_{12}' &= \frac{1}{2}(\varepsilon_{22} - \varepsilon_{11})\sin 2\theta + \varepsilon_{12}\cos 2\theta \\
\varepsilon_{13}' &= \cos\theta\,\varepsilon_{13} + \sin\theta\,\varepsilon_{23} \\
\varepsilon_{23}' &= -\sin\theta\,\varepsilon_{13} + \cos\theta\,\varepsilon_{23} \\
\varepsilon_{33}' &= \varepsilon_{33}
\end{aligned}
\tag{4.16}
$$

However, if P is a stress free point on the boundary of a body of isotropic material then the normal direction is a principal direction, with principal strain ε_{33}. By choice we have already aligned the x_3'-axis along a principal direction. We can align x_1', x_2' along the other two, mutually orthogonal, principal directions by a suitable choice of θ. The value of θ can be determined since we know that, if referred to a principal axis system, $\varepsilon_{ij}' = 0,\ i \neq j$ and so from the third equation of (4.16)

$$0 = \frac{1}{2}(\varepsilon_{22} - \varepsilon_{11}) \sin 2\theta + \varepsilon_{12} \cos 2\theta$$

giving

$$\tan 2\theta = \frac{2\varepsilon_{12}}{\varepsilon_{11} - \varepsilon_{22}} \tag{4.17}$$

Also, from the fourth and fifth equations of (4.16)

$$0 = \cos\theta\ \varepsilon_{13} + \sin\theta\ \varepsilon_{23} \qquad 0 = -\sin\theta\ \varepsilon_{13} + \cos\theta\ \varepsilon_{23}$$

which implies that ε_{13} and ε_{23} are both zero. Physically, this implies that at a stress-free point on the surface of a body, rectangular plane elements (perpendicular to the surface) remain rectangular after deformation. (As an instance of this see Example 4.3). However, since ε_{12} does not necessarily vanish, rectangular planar elements tangential to the surface may suffer shearing strains. At the value of θ given by (4.17), the corresponding values of ε_{11}', ε_{22}' are the principal strains. The strains ε_{11}', ε_{22}' and ε_{33}' contain the algebraically largest and least longitudinal strains existing at P. In practice it is found that the strain in the normal direction is numerically much smaller than the strains measured tangential to the surface and so we concentrate on the determination of ε_{11}', ε_{22}' alone.

We know, by direct measurement from the strain rosette that

$$\varepsilon_{11} \equiv \varepsilon_A \quad \text{and} \quad \varepsilon_{22} \equiv \varepsilon_C \tag{4.18}$$

We determine ε_{12} (required in (4.17)) by using the strain measurement of strain gauge B. Utilising the first of the equations in (4.16) which apply to *any* planar rotation with a value of θ of 45^0 we obtain for this *specific* rotation

$$\varepsilon_B = \frac{1}{2}(\varepsilon_A + \varepsilon_C) + \varepsilon_{12}$$

giving

$$\varepsilon_{12} = \varepsilon_B - \frac{1}{2}(\varepsilon_A + \varepsilon_C) \tag{4.19}$$

As a specific example let the measurements in gauges A, B, C be 0.002, 0.001 and -0.003 respectively then

$$\varepsilon_A = 0.002 \qquad \varepsilon_B = 0.001 \qquad \varepsilon_C = -0.003$$

and so, from (4.19) $\varepsilon_{12} = 0.001 - \frac{1}{2}(-0.001) = 0.0015$. The angle θ is found from (4.17)

$$\theta = \frac{1}{2}\tan^{-1}\left\{\frac{2(0.0015)}{0.001 + 0.003}\right\} = \frac{1}{2}\tan^{-1}(\frac{3}{4}) = 18.435^0$$

The principal strains are, from (4.16)

$$\varepsilon'_{11} = \frac{1}{2}(-0.001) + \frac{1}{2}(0.004)\cos 2\theta + (0.0015)\sin 2\theta = 0.002$$

$$\varepsilon'_{22} = \frac{1}{2}(-0.001) - \frac{1}{2}(0.004)\cos 2\theta - (0.0015)\sin 2\theta = -0.003$$

Small fibres at an angle of 18.44^0 to the direction of the strain gauge A would be elongated by 0.15% whilst fibres at an angle of 108.44^0 to the direction of the strain gauge A would be shortened by 0.25%.

Coincidently the optimum values of longitudinal strain also occur in the direction of gauges at A and C. However, the analysis which preceeded this question shows that there is no algebraically smaller longitudinal strain than -0.003 and no algebraically larger longitudinal strain than 0.002.

This calculation can be repeated at other points on the surface of a body, enabling a reasonably accurate estimation of the position on the surface which sustains the greatest extensional and compressional strains. This information on surface strains turns out to be more useful than one might imagine since the greatest strains throughout the body often occur on the bounding surface.

Example 4.5

Coordinate axes are drawn on a sheet of material which is subjected to a state of strain given by $\varepsilon_{11} = 0.005$, $\varepsilon_{22} = 0.003$, $\varepsilon_{12} = \varepsilon_{21} = -0.002$. Assuming that a particle at the origin is held fixed against rigid body motion calculate:

(a) the coordinates after deformation of a particle initially at the point (1,3)
(b) the strains on axes at an angle of 30^0 anticlockwise to the given axes.
(c) the directions of the principal axes and the magnitudes of the principal strains.

Solution

(a) We assume the problem is two dimensional in which $u_i = (u_1, u_2, 0)$ with each of the two non-zero components being functions of x_1 and x_2 alone. Then, from (4.6)

$$\frac{\partial u_1}{\partial x_1} = 0.005 \qquad \frac{\partial u_2}{\partial x_2} = 0.003 \qquad \frac{1}{2}(\frac{\partial u_1}{\partial x_2} + \frac{\partial u_2}{\partial x_1}) = -0.002 \tag{4.20}$$

Integrating the first and second of (4.20) implies

$$u_1 = 0.005x_1 + h(x_2) \quad \text{and} \quad u_2 = 0.003x_2 + g(x_1)$$

where h is an arbitrary function of x_2 and g is an arbitrary function of x_1. Substituting these into the last of (4.20) gives

$$\frac{1}{2}(\frac{dh}{dx_2} + \frac{dg}{dx_1}) = -0.002 \tag{4.21}$$

For this to be mathematically consistent (since h and g are functions of different *independent* variables) we must have

$$h = \alpha x_2 + \beta \qquad g = \gamma x_1 + \delta$$

where α, β, γ and δ are constants. Hence

$$u_1 = 0.005x_1 + \alpha x_2 + \beta \qquad u_2 = 0.003x_2 + \gamma x_1 + \delta$$

in which

$$\frac{1}{2}(\alpha + \gamma) = -0.002 \tag{4.22}$$

The conditions under which a displacement vector u_i does not include a rigid body motion have been give in Example 4.1. Since the plate is held fixed at the origin $u_i(0,0) = 0$ implying $\beta = \gamma = 0$, which eliminates the rigid body translation. Also to eliminate the rigid body rotation we require the components of the rotation vector (see Example 4.1) to vanish which, in this two dimensional example, implies

$$\frac{\partial u_2}{\partial x_1} - \frac{\partial u_1}{\partial x_2} = 0$$

That is

$$\gamma - \alpha = 0 \tag{4.23}$$

The relations (4.22) and (4.23) give $\alpha = \gamma = -0.002$ and so, finally:

$$u_1 = 0.005x_1 - 0.002x_2 \qquad u_2 = 0.003x_2 - 0.002x_1$$

A particle originally at $x_i = (1,3)$ is displaced to a position $x_i^* = (x_1^*, x_2^*)$ where $x_i^* = x_i + u_i$ and so

$$x_i^* = (1 + 0.005 - (0.002)3,\ 3 + 0.003(3) - 0.002) = (0.999, 3.007)$$

(b) The strains are obtained through the relations in (4.16)

$$\varepsilon'_{11} = \frac{1}{2}(0.008) + \frac{1}{2}(0.002)\cos 60^0 - 0.002\sin 60^0 = 0.00277$$

$$\varepsilon'_{22} = 0.004 - 0.0005 + 0.00173 = 0.00523$$

$$\varepsilon'_{12} = \varepsilon'_{21} = -0.00186$$

(c) The directions of the principal axes are given by (4.17):

$$\tan 2\theta = -\frac{2(0.002)}{0.002} = -2$$

leading to $\theta = 58.28^0,\ 148.28^0$. The principal strains, again found from (4.16) are found to be

$$\varepsilon_A = 0.00176 \qquad \varepsilon_B = 0.00624$$

EXERCISES

4.6 (a) The longitudinal strains measured in the elements A, B, C of a strain rosette are 5.9×10^{-4}, 4.4×10^{-4} and 2.0×10^{-4} respectively. Elements A, C are aligned orthogonally to each other whilst B is positioned between A and C at an angle of 30^0 to A. Show that the principal strains are approximately 6×10^{-4} and 2×10^{-4} at an angle -8.63^0 and 81.36^0 to the direction of A respectively.
(b) Determine the longitudinal strain along a line inclined at 45^0 to the direction of A.

4.7 Suppose the strain state at a point is:

$$\varepsilon_{ij} = \begin{bmatrix} 0.003 & -0.006 & -0.004 \\ -0.006 & 0.004 & 0.002 \\ -0.004 & 0.002 & -0.001 \end{bmatrix}$$

What are the maximum and minimum values of longitudinal strain at this point? What are the maximum and minimum longitudinal strains in the $x_1 x_2$ plane?

4.5 The Compatibility Equations for Strain

We have seen that the strain tensor ε_{ij} is intimately related to the displacement vector u_i through the equation

$$\varepsilon_{ij} = \frac{1}{2}(u_{i,j} + u_{j,i}) \qquad \textbf{(4.24)}$$

If we are *given* the displacement vector u_i (as a function of the coordinates x_i) then it is a simple matter, by taking appropriate derivatives, to determine each component of the strain tensor.

However, in the more realistic situation in which the strain components are known and it is desired to find the displacement vector then (4.24) constitutes a system of first order linear partial differential equations in (u_1, u_2, u_3). A solution to this system of equations may not exist. That is, we cannot specify ε_{ij} *arbitrarily* and hope to find a displacement vector u_i which is a single-valued and continuous function of position. In this text we *demand* these characteristics in the displacement field. If we accepted a discontinuous displacement field then we would be allowing a situation in which material points of the body *infinitessimally* close before deformation would be a *finite* distance apart after deformation. This would not be consistent with linear infinitessimal elasticity theory. Also, if multi-valued displacements were allowed then material points might, mathematically, be considered to be in two different positions after deformation – that is a crack or tear has developed within the body. Such a situation can be considered, quite properly, when developing a theory of cracks and crack propagation but these topics are well outside the limited scope of this text and so such displacement fields will be considered invalid.

The problem of compatibility arises each time the existence of a solution of a system of partial differential equations is questioned. For example, consider the simple problem of finding a single function $\phi(x,y)$ satisfying

$$\frac{\partial \phi}{\partial x} = h(x,y) \qquad \frac{\partial \phi}{\partial y} = g(x,y) \tag{4.25}$$

If $\phi(x,y)$ is sufficiently smooth so that second partial derivatives exist then

$$\frac{\partial^2 \phi}{\partial y \partial x} = \frac{\partial h}{\partial y} \qquad \text{and} \qquad \frac{\partial^2 \phi}{\partial x \partial y} = \frac{\partial g}{\partial x}$$

But, for such smooth functions, mixed partial derivatives are always equal and so h and g *cannot* be specified arbitrarily; they *must* satisfy

$$\frac{\partial h}{\partial y} = \frac{\partial g}{\partial x} \tag{4.26}$$

if a smooth solution for $\phi(x,y)$ is to exist.

If h, g satisfy (4.26) then the equations in (4.25) are said to be compatible and the existence of $\phi(x,y)$ is guaranteed. (The guarantee applies to the *local* region around the point for which (4.25) applies. Extra complications can arise if the region within which (4.25) is true is not simply connected. See Section 7.2).

The system of equations in (4.24) is similar in structure, though with far greater complication, to those in (4.25). It is relatively easy to derive *necessary* conditions for the existence of a single-valued continuous u_i given ε_{ij} from (4.24). Taking second partial derivatives of ε_{ij} with respect to x_k and x_m and then rearranging indices we find

$$\begin{aligned}
\varepsilon_{ij,km} &= \frac{1}{2}(u_{i,jkm} + u_{j,ikm}) \\
\varepsilon_{km,ij} &= \frac{1}{2}(u_{k,mij} + u_{m,kij}) \\
\varepsilon_{mj,ki} &= \frac{1}{2}(u_{j,mik} + u_{m,jik}) \\
\varepsilon_{ki,mj} &= \frac{1}{2}(u_{k,ijm} + u_{i,kjm})
\end{aligned} \tag{4.27}$$

Assuming the equality of mixed derivatives:

$$u_{i,jkm} = u_{i,kjm} \qquad u_{j,ikm} = u_{j,mik} \qquad u_{k,mij} = u_{k,ijm} \qquad u_{m,kij} = u_{m,jik}$$

we obtain, by adding the first two and subtracting the second two of the relations in (4.27), the set of relations involving the strains

$$\varepsilon_{ij,km} + \varepsilon_{km,ij} - \varepsilon_{mj,ki} - \varepsilon_{ki,mj} = 0 \tag{4.28}$$

Although there are eighty one equations here only six are independent due to the symmetries inherent in the strain tensor and in the derivatives. The six independent equations are (conveniently) obtained by contracting the k and m indices:

$$\varepsilon_{ij},_{kk} + \varepsilon_{kk},_{ij} - \varepsilon_{kj},_{ki} - \varepsilon_{ki},_{kj} = 0 \tag{4.29}$$

It can be shown that these conditions (4.29) known as the *strain compatibility conditions* are also sufficient, in a simply connected region, to uniquely determine a smooth, single-valued displacement field u_i.

EXERCISES

4.8 (a) Show that the strain components which are *linear* functions of the coordinates are always physically plausible.

(b) Is it possible to find values of the constants a, b, c, d so that the strain tensor:

$$\varepsilon_{11} = x_1^2 + ax_2^2 \quad \varepsilon_{22} = bx_2^2 + cx_3^2 \quad \varepsilon_{12} = dx_1x_2x_3 \quad \text{other } \varepsilon_{ij} = 0$$

is physically plausible?

4.9 Given the following state of strain at a point on the surface of a body

$$\varepsilon_{11} = 0.002 \quad \varepsilon_{22} = -0.001 \quad \varepsilon_{12} = 0.003$$

show that the principal strains and their directions are 0.00385, −0.00285, and $31.7^0, 121.7^0$ respectively.

4.10 The displacement components in a rod subjected to pure bending are:

$$u_1 = -\frac{x_1x_3}{\rho} \quad u_2 = k\frac{x_2x_3}{\rho} \quad u_3 = \frac{x_1^2 + k(x_3^2 - x_2^2)}{2\rho}$$

Find the components of strain for this displacement field and verify that they satisfy the compatibility relations.

4.11 A strain gauge rosette has its elements aligned at angles $-\theta$, 0, θ with its centre line. The longitudinal strains measured in each are $\varepsilon_A, \varepsilon_B$ and ε_C respectively. Show that the principal axes of strain makes angles ϕ and $\phi + \pi/2$ with the centre line where

$$\tan 2\phi = \frac{\varepsilon_A - \varepsilon_C}{\varepsilon_A + \varepsilon_C - 2\varepsilon_B} \tan\theta$$

In the case in which $\theta = \pi/4$, prove that the principal strains are

$$\frac{1}{2}(\varepsilon_A + \varepsilon_B) \pm [\varepsilon_B^2 + \frac{1}{2}(\varepsilon_A^2 + \varepsilon_C^2) - \varepsilon_A\varepsilon_B - \varepsilon_B\varepsilon_C]^{1/2}$$

Chapter 5

LINEAR ELASTICITY

5.1 Hooke's Law and the Simple Tension Experiment

Perhaps the simplest experiment one can consider in structural mechanics is the stretching of a thin wire by the application of a force of magnitude F (see Figure 5.1a).

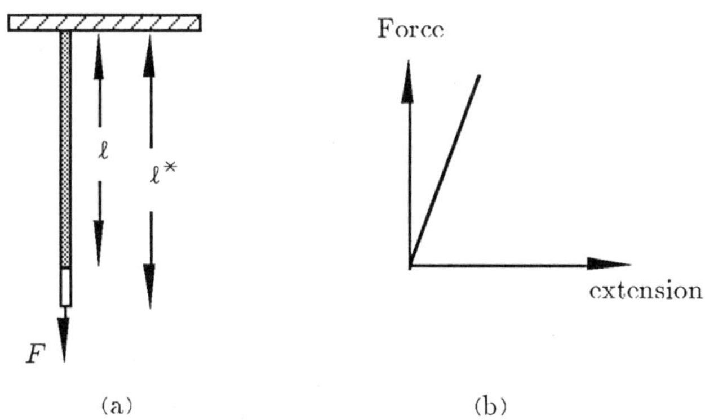

Figure 5.1

For each value of F the length of the wire can be measured and so its extension from its undeformed length can be determined. It is found (if F is not too large) that, for structural materials like steel, for example, the relation between F and the extension is linear (see Figure 5.1b). Mathematically, we have:

$$F = c(\ell^* - \ell) \qquad \textbf{(5.1)}$$

where c, called the stiffness of the wire, is the constant of proportionality. This simple result is Hooke's law, first reported in 1676.

If A_0 is the original cross-sectional area and ℓ the original length then

$$\frac{F}{A_0} = E(\frac{\ell^* - \ell}{\ell}) \qquad \text{where} \quad E = \frac{c\ell}{A_0} \qquad \textbf{(5.2)}$$

E is known as the Young's modulus of the material from which the wire has been formed. For engineering materials E is very large, ranging from $7 \times 10^{10}\text{Nm}^{-2}$ for aluminium to $2 \times 10^{11}\text{Nm}^{-2}$ for structural steel. (Most metals have the same Young's modulus in compression as in tension).

In (5.2) F/A_o is the longitudinal stress and $(\ell^* - \ell)/\ell$ is the longitudinal strain. For this simple configuration the relation between stress and strain is *linear*.

As well as an obvious longitudinal extension in the simple tension experiment the wire is also observed to 'thin'. The lateral dimensions of the wire contract if F is a tensile force. The lateral 'strain' can be measured and it is found that it is proportional to the longitudinal strain (and hence to longitudinal stress):

$$\text{lateral strain} = -\nu(\text{longitudinal strain}) \qquad \textbf{(5.3)}$$

The constant of proportionality ν is known as Poisson's ratio. For steel $\nu \approx 0.3$ and for most structural materials varies between ≈ 0.2 to the (theoretically) maximum value of 0.5. The Young's modulus E and the Poisson ratio ν are mechanical characteristics of the material.

The simple tension experiment is essentially one dimensional. To proceed, we must exhibit a relationship between the stress tensor and the strain tensor for generally shaped bodies in two and three dimensions. Rather than attempt to describe *all* materials in *all* possible situations we shall restrict our attention in this text to linear homogeneous isotropic elastic materials. Fortunately, most structural materials may be so described.

A material is said to be **elastic** if the state of stress at a point is dependent only on the deformation (strain) at that point and not on the history of the deformation. Such a material when unloaded will always return to its initial configuration. The typical example is rubber. Although the relation between longitudinal stress and longitudinal strain, in the simple tension experiment with a rubber 'wire' is not linear the material is nevertheless elastic. The force/extension curve is identical whether the wire is being loaded or unloaded. (See Figure 5.2a)

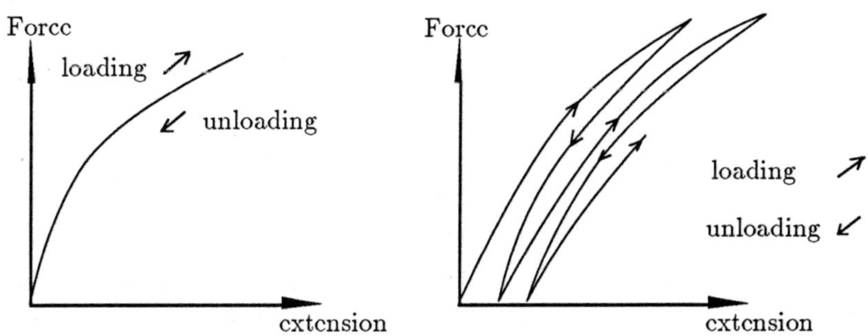

elastic material | inelastic material

(a) | (b)

Figure 5.2

If the force/extension curve has the appearance of Figure 5.2b in which the deformation is dependent upon the history of the loading the material is said to be inelastic (or plastic).

For elastic materials we can write the relation between the stress and strain tensors as

$$\sigma_{ij} = F_{ij}(\varepsilon_{11}, \varepsilon_{12}, \varepsilon_{13}, \varepsilon_{22}, \varepsilon_{23}, \varepsilon_{33}) \qquad \textbf{(5.4)}$$

That is, each component of the stress tensor is a function of the six components of the strain tensor.

A **linear elastic** material is one for which the functions F_{ij} in (5.4) are linear. For this class of materials we may write

$$\sigma_{ij} = C_{ijkm}\varepsilon_{km} \tag{5.5}$$

The coefficients C_{ijkm} are called *stiffnesses*. Each elastic material will be characterised by a particular set of stiffnesses. The right hand side of (5.5) is written without additive constants implying that the stresses vanish when the strains are zero. This relation, generalising the linear stress/strain relation (5.2) in the simple tension experiment is known as the **generalised Hooke's law**.

We shall further assume that the coefficients C_{ijkm} do not vary from point to point. A material with this property is called **homogeneous**.

An **isotropic** material is one which has the same physical properties in *all* directions. If one removes a rectangular block (say) of the material and determine its elastic constants (Young's modulus etc) one would find that the values obtained would be the same irrespective of the *orientation* of the block within the material. Thus a uniformly grained plank of wood is homogeneous (as the elastic properties do not change from point to point) though it is not isotropic as the grain makes the wood stronger in certain directions. Its elastic constants do not depend on position but do depend on direction.

Most structural metals, being crystalline in structure, cannot be considered isotropic at the microscopic level. However since the dimensions of the crystals are small compared with the dimensions of the material body and are (generally) randomly distributed, it may be assumed that, on the macroscopic level, the metallic body behaves isotropically.

Mathematically, an isotropic medium is one for which the stiffnesses C_{ijkm} take the same values in all coordinate systems. That is, C_{ijkm} is a fourth order isotropic tensor having general form (see Section 2.6):

$$C_{ijkm} = a\delta_{ij}\delta_{km} + b\delta_{ik}\delta_{jm} + c\delta_{im}\delta_{jk} \tag{5.6}$$

in which a, b and c are constants. Substituting into (5.5) we obtain

$$\begin{aligned}\sigma_{ij} &= a\delta_{ij}\varepsilon_{kk} + b\varepsilon_{ij} + c\varepsilon_{ji} \\ &= a\delta_{ij}\varepsilon_{kk} + (b+c)\varepsilon_{ij}\end{aligned}$$

where the symmetry of ε_{ij} has been used. Thus, for a linear homogeneous isotropic elastic material the stress-strain relation (known as the constitutive relations) can be written:

$$\sigma_{ij} = 2\mu\varepsilon_{ij} + \lambda\delta_{ij}\varepsilon_{kk} \tag{5.7}$$

These equations were first introduced (though not in terms of tensors) by Cauchy in 1828. The constants λ and μ are known as the Lamé constants. They can be directly related to the Young's modulus and the Poisson ratio of the material. Whilst it can be shown that a general crystalline material is characterised by 21 independent stiffnesses, most structural materials can be fully characterised by just two stiffnesses.

An important characteristic of materials whose stresses and strains are related according to (5.7) is that, at each point in the material, the stress and strain tensors have a principal axis system in common. (This property has already been used to advantage in Sections 4.3 and 4.4). This is easily proved. All we need do is, at each point of interest, rotate coordinates so that they are aligned with the principal axes of the strain tensor. Since (5.7) is a tensor equation it is valid in all cooordinate systems:

$$\sigma'_{ij} = 2\mu\varepsilon'_{ij} + \lambda\delta_{ij}\varepsilon'_{kk}$$

But, with respect to this new system of coordinates $\varepsilon'_{ij} = 0\ i \neq j$. Also since

$$\sigma'_{ij} = 2\mu\varepsilon'_{ij} \quad i \neq j$$

we can conclude $\sigma'_{ij} = 0\ i \neq j$ and so the principal axis system for the strain tensor is also a principal axis system for the stress tensor.

As (5.7) is a linear relation for the stresses in terms of the strains we can easily 'invert' the relation to express the strains in terms of the stresses. First we contract the i and j indices in (5.7) to give

$$\sigma_{ii} = 2\mu\varepsilon_{ii} + 3\lambda\varepsilon_{kk}$$

leading to

$$\varepsilon_{kk} = \frac{\sigma_{kk}}{2\mu + 3\lambda} \qquad \text{as } i \text{ and } k \text{ are dummy indices}$$

Hence replacing ε_{kk} in (5.7) and transposing all stress terms to one side we have

$$\varepsilon_{ij} = \frac{1}{2\mu}\{\sigma_{ij} - \frac{\lambda}{2\mu + 3\lambda}\delta_{ij}\sigma_{kk}\} \tag{5.8}$$

which completes the 'inversion'. The relations (5.8) and (5.7) are completely equivalent.

5.2 The Governing Equations of Linear Elasticity

We are now in a position to group together the governing equations of linear homogeneous isotropic elasticity (which we shall henceforth shorten to linear elasticity). They are

$$\sigma_{ij} = \sigma_{ji} \qquad \varepsilon_{ij} = \varepsilon_{ji} \qquad \text{Symmetry conditions} \tag{5.9}$$

$$T_i^{(n)} = \sigma_{ij}n_j \qquad \text{Cauchy relation} \tag{5.10}$$

$$\sigma_{ij},_j + \rho B_i = 0 \qquad \text{Equations of equilibrium} \tag{5.11}$$

$$\varepsilon_{ij} = \frac{1}{2}(u_i,_j + u_j,_i) \qquad \text{Strain – displacement} \tag{5.12}$$

$$\sigma_{ij} = 2\mu\varepsilon_{ij} + \lambda\delta_{ij}\varepsilon_{kk} \qquad \text{Constitutive equations} \tag{5.13}$$

$$\varepsilon_{ij},_{kk} + \varepsilon_{kk},_{ij} - \varepsilon_{kj},_{ki} - \varepsilon_{ki},_{kj} = 0 \qquad \text{Compatibility} \tag{5.14}$$

In this rather formidable looking set of equations there are 15 basic unknown quantities (being 6 stress components, 6 strain components and 3 displacement components). There are 15 basic equations, the equations of equilibrium (3), the constitutive equations (6) and the strain-displacement relations (6). The set is a mixture of linear partial differential equations and linear algebraic equations. They need to be supplemented by appropriate boundary conditions if a unique solution is to be obtained.

The equations may be rearranged into different forms, for convenience, depending on the type of problem under consideration. Consider, for example, a situation in which a linearly elastic body V, bounded by a surface S is deformed by the application of a known body force B_i and by the application of known surface forces (or stresses) $T_i^{(n)}$ applied at each point of S. Such a problem is called a *boundary value problem of the first kind* of linear elasticity. Since the boundary conditions (those applied on the bounding surface S) are expressed explicitly in terms of forces, and will require the use of the Cauchy relation (5.10), it is likely to prove advantageous to express, as far as possible, the basic equations of linear elasticity in terms of the stress tensor alone.

Since a valid solution for the displacement vector u_i is not possible unless the compatibility conditions (5.14) are satisfied we must express these relations in terms of the stress tensor.

Substituting (5.8) into (5.14) we obtain

$$\frac{1}{2\mu}[\sigma_{ij},_{kk} - \frac{\lambda}{2\mu+3\lambda}\delta_{ij}\sigma_{mm},_{kk}] + \frac{1}{2\mu}[\sigma_{kk},_{ij} - \frac{3\lambda}{2\mu+3\lambda}\sigma_{kk},_{ij}]$$
$$-\frac{1}{2\mu}[\sigma_{kj},_{ki} - \frac{\lambda}{2\mu+3\lambda}\sigma_{mm},_{ji}] - \frac{1}{2\mu}[\sigma_{ki},_{kj} - \frac{\lambda}{2\mu+3\lambda}\sigma_{mm},_{ij}] = 0$$

Use of (5.11) then implies

$$\sigma_{ij},_{kk} - \frac{\lambda}{2\mu+3\lambda}\delta_{ij}\sigma_{mm},_{kk} + \frac{2\mu+2\lambda}{2\mu+3\lambda}\sigma_{kk},_{ij} + \rho(B_j,_i + B_i,_j) = 0 \qquad \textbf{(5.15)}$$

The six equations in (5.15) are known as the Beltrami-Mitchell equations of linear elasticity. The solution of these equations, subject to the satisfaction of the Cauchy relation $T_i^{(n)} = \sigma_{ij}n_j$ (with *given* $T_i^{(n)}$) at each point of the boundary will lead to a set of stress components that both satisfy the equilibrium equations and from which (through (5.13) and (5.12)) a single valued displacement vector can be obtained.

EXERCISES

5.1 Show that, for constant body forces, a stress tensor which is a **linear** function of the coordinates is always a possible solution to the equations of linear elasticity.

5.2 Show that for constant body forces the dilitation, ε_{kk}, is an harmonic function (i.e. satisfies the two dimensional Laplace equation).

5.3 By expressing the equilibrium equations in terms of the displacement vector derive the Navier equation of linear elasticity in standard vector notation:

$$\mu\nabla^2\underline{u} + (\lambda+\mu)\underline{\nabla}(\underline{\nabla}.\underline{u}) + \rho\underline{B} = 0$$

Why are the compatibility equations not required?

5.4 Show by contracting (5.15) that, in the absence of body forces, $\nabla^2\sigma_{mm} = 0$ and so in this case (5.15) can be written in the form

$$\nabla^2\sigma_{ij} + \frac{1}{1+\nu}\sigma_{kk,ij} = 0$$

5.3 Simple Solutions

At the early stages in any theoretical development it is always worthwhile, if possible, to obtain exact solutions to 'ideal' problems. By this process one can get a *feel* not only of the usefulness (or otherwise) of the theory being developed but also one can begin to understand the mathematical nature of the subject. A knowledge of exact solutions, albeit to very simple problems, will prove to be an invaluable aid even though, eventually, we shall have to resort to numerical techniques or to approximations in order to obtain solutions to realistic problems.

In this section we shall consider a circular cylinder deformed in three distinct ways; firstly by the action of tensile forces applied along its axis, secondly by the action of couples applied at either end producing bending and lastly by the application of torsional couples producing twisting.

- **The Simple Tension Experiment**

Using the theory developed we can analyse, theoretically, the simple tension experiment and hence (as we shall see) determine certain relationships between the elastic constants.

Consider a circular cylinder (a wire) of radius a, length ℓ and cross-sectional area A deformed by the action of equal and opposite forces applied at either end (see Figure 5.3).

If we imagine the cylinder aligned along the x_3-direction with the lower face on the plane $x_3 = 0$, and further assume that the force F is uniformly distributed over each end (so that the stress vector has magnitude F/A) then the stress boundary conditions are that

$$T_i^{(n)} = (0,0,-\frac{F}{A}) \quad \text{at } x_3 = 0 \qquad (\text{where } n_i = (0,0,-1))$$

$$T_i^{(n)} = (0,0,\frac{F}{A}) \quad \text{at } x_3 = \ell \qquad (\text{where } n_i = (0,0,1))$$

and

$$T_i^{(n)} = (0,0,0) \quad \text{on the curved boundary of the cylinder.}$$

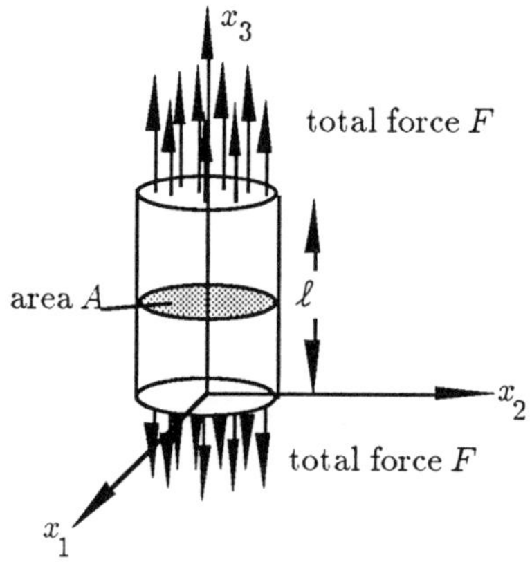

Figure 5.3

We shall also assume that the effects of gravity (a body force) are negligible in comparison to the effects of the force F. As we shall see in Example 5.1, below, this turns out to be an excellent assumption.

An 'obvious' solution to this problem is

$$\sigma_{33} = \frac{F}{A} \qquad \text{other } \sigma_{ij} = 0 \tag{5.16}$$

This stress tensor, with each component being constant, clearly satisfies the Beltrami-Mitchell equations as each term in (5.15) (in the absence of body forces) involves second derivatives. All we need check is that, with the choice of stress tensor in (5.16), the boundary conditions are satisfied.

On the face $x_3 = \ell$, $n_i = (0, 0, 1)$ and so

$$\sigma_{ij} n_j = \sigma_{i3} n_3 = (0, 0, \frac{F}{A}) \quad \text{as required}$$

On the face $x_3 = 0$, $n_i = (0, 0, -1)$ and so

$$\sigma_{ij} n_j = \sigma_{i3} n_3 = (0, 0, -\frac{F}{A}) \quad \text{as required}$$

Finally, on the curved face $n_i = (\frac{x_1}{a}, \frac{x_2}{a}, 0)$, and so

$$\sigma_{ij} n_j = \sigma_{i3} n_3 = (0, 0, 0) \quad \text{as required}$$

Having confirmed the solution for the stress tensor, the strain tensor components follow from (5.8)

$$\varepsilon_{33} = \frac{1}{2\mu}[\sigma_{33} - \frac{\lambda}{2\mu + 3\lambda}\sigma_{33}] = \frac{\mu + \lambda}{\mu(2\mu + 3\lambda)}\sigma_{33}$$

However, Young's modulus E is defined as the quotient of the longitudinal stress (σ_{33}) to the longitudinal strain (ε_{33}). Hence, for consistency, we must conclude

$$E = \frac{\mu(2\mu + 3\lambda)}{(\mu + \lambda)} \tag{5.17}$$

Also from (5.8)

$$\varepsilon_{11} = \varepsilon_{22} = -\frac{\lambda}{2\mu(2\mu + 3\lambda)}\sigma_{33} = -\frac{\lambda}{2\mu + 2\lambda}\varepsilon_{33}$$

Since the Poisson ratio ν is defined in terms of the longitudinal strain (ε_{33}) and lateral strain (ε_{11} or ε_{22}) through (5.3) we must have the relation

$$\nu = \frac{\lambda}{2\mu + 2\lambda} \tag{5.18}$$

As we have already remarked the materials that we are considering in this text are characterised by just two elastic constants. Which two we use is irrelevant. In fact, depending on the application, we shall use any two of E, ν, μ, λ. The relations (5.17) and (5.18) can be solved for λ and μ to give

$$\lambda = \frac{E\nu}{(1+\nu)(1-2\nu)} \tag{5.17a}$$

$$\mu = \frac{E}{2(1+\nu)} \tag{5.18a}$$

As a result of the deformation the volume changes. We can obtain an expression for the change in volume by considering an elementary cube of material with sides ds_1, ds_2, ds_3 before deformation and ds_1^*, ds_2^*, ds_3^* after deformation. Then utilising the interpretation of the longitudinal components of strain derived in Section 4.2 (equation (4.9)) we find:

$$ds_1^* = (1+\varepsilon_{11})ds_1 \qquad ds_2^* = (1+\varepsilon_{22})ds_2 \qquad ds_3^* = (1+\varepsilon_{33})ds_3$$

and so the change in volume, δV, is

$$\begin{aligned}\delta V = V^* - V &= ds_1^* ds_2^* ds_3^* - ds_1 ds_2 ds_3 \\ &= [(1+\varepsilon_{11})(1+\varepsilon_{22})(1+\varepsilon_{33}) - 1] ds_1 ds_2 ds_3\end{aligned}$$

that is, ignoring products of the strains:

$$\delta V = (\varepsilon_{11} + \varepsilon_{22} + \varepsilon_{33}) ds_1 ds_2 ds_3$$

The material is incompressible (when $\delta V = 0$) only if the dilitation, ε_{kk}, vanishes.

An expression (obtained quite generally) for the diliatation in terms of the stress derives from (5.8):

$$\varepsilon_{kk} = \frac{1}{2\mu + 3\lambda}\sigma_{kk}$$

But, from (5.18) $1 - 2\nu = \dfrac{\mu}{\mu + \lambda}$ and so, from (5.17), $\dfrac{1}{2\mu + 3\lambda} = \dfrac{(1-2\nu)}{E}$ leading to

$$\varepsilon_{kk} = \frac{(1-2\nu)}{E}\sigma_{kk}$$

If $\nu = 0.5$ the dilitation vanishes and the material is incompressible. Rubber is an example of a material which is very nearly incompressible with a Poisson ratio of approximately 0.49.

For a wire in the simple tension experiment the percentage relative change in volume is $\frac{\delta V}{V}.100\%$ or

$$\varepsilon_{kk}.100\% = (\frac{1-2\nu}{E})(\frac{F}{A})100\%$$

so that for structural steel, ($\nu \approx 0.3,\ E \approx 2 \times 10^{11}\mathrm{Nm}^{-2}$) with $A \approx 0.05\mathrm{m}^2$ loaded to near its elastic limit ($F \approx 2 \times 10^9\mathrm{N}$) then

$$\varepsilon_{kk}.100\% \approx \frac{(1-0.6)}{2 \times 10^{11}}.\frac{(2 \times 10^9)}{0.005}100\% \approx 8\%$$

In the exact solution of the simple tension experiment, outlined above, we have not obtained an expression for the displacement vector. The techniques for carrying out this calculation are described in the following interesting example. After studying this example the reader is encouraged to return to the simple tension experiment to verify that the solution for the displacement vector is (in the absence of rigid body motions):

$$u_1 = -\frac{\nu F}{EA}x_1 \qquad u_2 = -\frac{\nu F}{EA}x_2 \qquad u_3 = \frac{F}{EA}x_3$$

Example 5.1
Determine the deformation of a circular cylinder of uniform density ρ under its own weight.

Solution

Consider the cylinder of radius r and height h with its axis aligned along the x_3-axis as shown in Figure 5.4. We assume that gravity is acting in the negative (downward) x_3-direction.

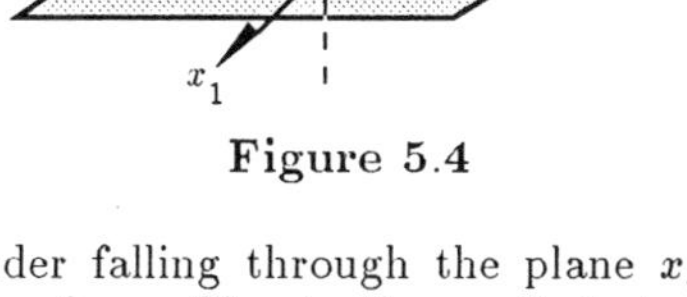

Figure 5.4

To prevent the cylinder falling through the plane $x_3 = 0$ a force per unit area is applied to the lower face. Clearly the applied stress vector, at each point, has the form $T_i^{(n)} = (0, 0, T_3^{(n)})$. The total force acting upwards is $T_3^{(n)}(\pi r^2)$. This is balanced by the force on the cylinder due to gravity, being $(\pi r^2 h)\rho g$ (g is the acceleration due to gravity). Hence, for equilibrium the forces must balance:

$$T_3^{(n)}(\pi r^2) = (\pi r^2 h)\rho g$$

therefore

$$T_3^{(n)} = \rho g h$$

Every other point on the surface of the cylinder is stress free. Since the body force vector $B_i = (0, 0, -g)$ is a *constant* vector the Beltrami-Mitchell equations (5.15) reduce to

$$\sigma_{ij},_{kk} - \frac{\lambda}{2\mu + 3\lambda}\delta_{ij}\sigma_{mm},_{kk} + \sigma_{kk},_{ij}\left(\frac{2\mu + 2\lambda}{2\mu + 3\lambda}\right) = 0$$

We shall propose a tentative solution for the stress tensor to this problem:

$$\sigma_{33} = \rho g x_3 + c \qquad \text{other } \sigma_{ij} = 0$$

where c is a constant. The Beltrami-Mitchell equations are clearly satisfied since each term involves a second derivative of the stress tensor components. We must check that the boundary conditions are satisfied.

$$\text{on } x_3 = 0 \quad T_i^{(n)} = (0, 0, \rho g h) \quad n_i = (0, 0, -1)$$

and so the Cauchy relation $T_i^{(n)} = \sigma_{ij} n_j$ becomes:

$$\rho g h = -\sigma_{33} = -c \quad \text{on } x_3 = 0$$

and so

$$\sigma_{33} = \rho g x_3 - \rho g h$$

$$\text{On } x_3 = h \quad T_i^{(n)} = (0, 0, 0) \quad n_i = (0, 0, 1)$$

and so the Cauchy relation implies:

$$0 = \sigma_{33} = \rho g h - \rho g h = 0 \quad \text{on } x_3 = h$$

that is, the boundary condition on this surface is satisfied. On the curved surface $T_i^{(n)} = (0, 0, 0)$ and the normal is $n_i = (\frac{x_1}{r}, \frac{x_2}{r}, 0)$. On this surface the Cauchy relation is satisfied identically:

$$0 = \sigma_{ij} n_j = \sigma_{i3} n_3 = 0$$

Hence *the* solution to the boundary value problem of a cylinder deforming under its own weight is

$$\sigma_{33} = \rho g x_3 - \rho g h \qquad \text{other } \sigma_{ij} = 0$$

From (5.13) (or more usefully (5.8)) we find (using (5.17),(5.18))

$$\varepsilon_{33} = \frac{1}{2\mu}(\sigma_{33} - \frac{\lambda}{2\mu + 3\lambda}\sigma_{33}) = \frac{(2\mu + 2\lambda)}{2\mu(2\mu + 3\lambda)}\sigma_{33} \equiv \frac{1}{E}\sigma_{33}$$

$$\varepsilon_{11} = \varepsilon_{22} = -\frac{\lambda}{2\mu(2\mu + 3\lambda)}\sigma_{33} \equiv -\frac{\nu}{E}\sigma_{33} \qquad \text{other } \varepsilon_{ij} = 0$$

and so, from (5.12):

$$\frac{\partial u_1}{\partial x_1} = -\frac{\nu}{E}\rho g(x_3 - h) \quad \frac{\partial u_2}{\partial x_2} = -\frac{\nu}{E}\rho g(x_3 - h) \quad \frac{\partial u_3}{\partial x_3} = -\frac{1}{E}\rho g(x_3 - h) \qquad \textbf{(5.19a)}$$

$$\frac{\partial u_1}{\partial x_2} + \frac{\partial u_2}{\partial x_1} = 0 \quad \frac{\partial u_1}{\partial x_3} + \frac{\partial u_3}{\partial x_1} = 0 \quad \frac{\partial u_2}{\partial x_3} + \frac{\partial u_3}{\partial x_2} = 0 \qquad \textbf{(5.19b)}$$

We may determine a *particular* solution to these equations knowing that the full general solution is obtained by adding the *complimentary function* (obtained by solving $\varepsilon_{ij} = 0$). The complimentary function takes the form of a rigid body motion $u_i^{(cf)} = t_i + \eta_{ijk}x_j\omega_k$ (see Example 4.1). In this problem we shall assume that the base of the cylinder is held fixed so as to exclude a rigid body motion. This implies that (in this problem) we may take $t_i = 0$ and $\omega_i = 0$. Integrating (the first two equations of (5.19a) implies

$$u_1 = -\frac{\nu}{E}\rho g(x_3 x_1 - h x_1) + f_1(x_2, x_3)$$

$$u_2 = -\frac{\nu}{E}\rho g(x_3 x_2 - h x_2) + f_2(x_1, x_3)$$

whilst the last equation of (5.19a) implies

$$u_3 = \frac{1}{E}\rho g(\frac{1}{2}x_3^2 - h x_3) + f_3(x_1, x_2)$$

in which f_1, f_2 and f_3 are arbitrary functions of their arguments. Substituting these three into the equations in (5.19b) implies

$$\frac{\partial f_1}{\partial x_2} + \frac{\partial f_2}{\partial x_1} = 0$$

$$-\frac{\nu}{E}\rho g x_1 + \frac{\partial f_1}{\partial x_3} + \frac{\partial f_3}{\partial x_1} = 0$$

$$-\frac{\nu}{E}\rho g x_2 + \frac{\partial f_2}{\partial x_3} + \frac{\partial f_3}{\partial x_2} = 0$$

To satisfy these equations we can *choose* as a particular solution:

$$f_1 = f_2 = 0 \qquad \text{and} \qquad f_3 = \frac{\nu}{E}\rho g(\frac{x_1^2 + x_2^2}{2})$$

giving, finally

$$u_1 = -\frac{\nu}{E}\rho g(x_3 x_1 - h x_1) \qquad u_2 = -\frac{\nu}{E}\rho g(x_3 x_2 - h x_2)$$

$$u_3 = \frac{1}{E}\rho g(\frac{1}{2}x_3^2 - h x_3) + \frac{\nu}{E}\rho g(\frac{x_1^2 + x_2^2}{2}) \qquad (5.20)$$

We can now use the solution to sketch the shape of the cylinder after deformation. For points on the $x_1 = 0$ cross-section

$$u_1 = 0, \quad u_2 = -\frac{\nu}{E}\rho g(x_3 - h)x_2 \quad u_3 = \frac{\rho g}{E}[\frac{x_3^2 + \nu x_2^2}{2} - h x_3)] \qquad (5.21)$$

The deformed shapes of the $x_3 = h$ and $x_3 = 0$ surfaces now follow from the displacements in (5.21). We obtain the two dotted curves shown in Figure 5.5a.

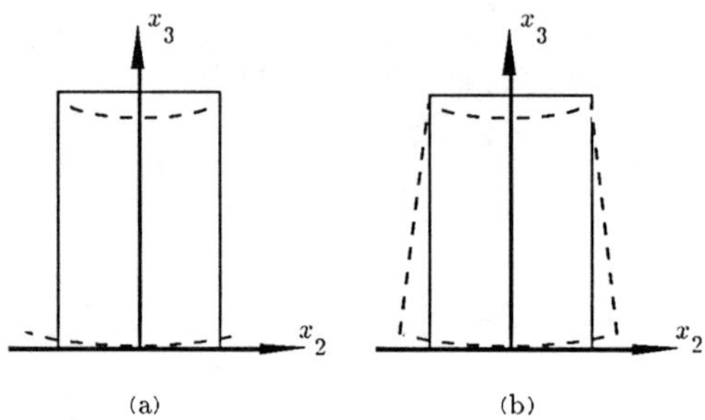

Figure 5.5

Again using (5.21) (and also (4.1)) we obtain the coordinates of the line $x_2 = r$ after deformation as

$$x_1^* = 0 \quad x_2^* = -\frac{\nu}{E}\rho g(x_3 - h)r + r \quad x_3^* = \frac{\rho g}{E}(\frac{x_3^2}{2} - h x_3) + \frac{\nu}{E}\rho g\frac{r^2}{2} + x_3$$

Realising that, for structural materials, E is extremely large we can solve the last of these equations to obtain

$$x_3 \approx x_3^* + O(\frac{1}{E})$$

and so

$$x_2^* = -\frac{\nu\rho g}{E}(x_3^* - h)r + r + O(\frac{1}{E^2})$$

so that the straight line $x_2 = r$ is deformed into a straight line (with a very steep gradient) completing the deformed shape of the cross-section shown in Figure 5.5b. This deformation is highly exaggerated. For example, for a steel rod ($\nu \approx 0.3$, $\rho \approx 10^4$ kg m^{-3}, $g \approx 10$ m s^{-2}, $E \approx 2 \times 10^{11}$ N m^{-2}) with length $h = 10$m and $r = 1$m then the maximum radial displacement occurs at the base of the column (where $x_3 = 0$) of magnitude

$$u_2 = \frac{\nu\rho g h r}{E} \approx 2 \times 10^{-7}\text{m}$$

In the application of steel and similar materials to structures such deformations are entirely negligible. However, we should not be over confident about neglecting the effects due to gravity. In a very sensitive optical experiment small deformations of this magnitude may be of significance.

• Beam Bending by Pure Couples

A circular cylinder of radius a is deformed by the action of equal and opposite couples M applied at each end. The axis of the beam is along the x_3-axis, with the origin of coordinates chosen to be the right hand end face of the beam shown in Figure 5.6. For simplicity we choose the x_2-axis so that the couple M_i (a vector) has a magnitude M in the x_1-direction at the origin:

$$M_i = (M, 0, 0) \tag{5.22}$$

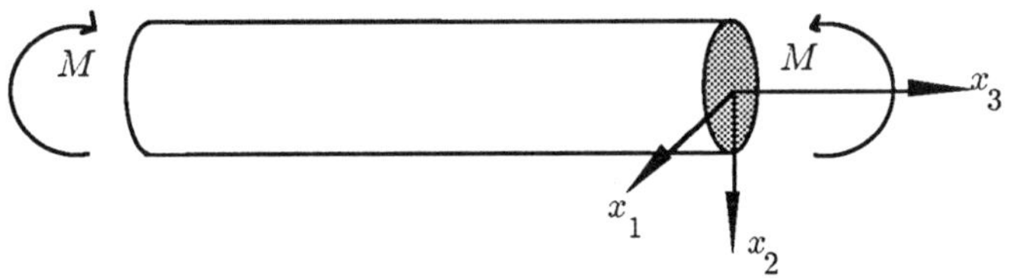

Figure 5.6

As with the previous two examples we shall assume the form taken by the stress tensor and then proceed to justify our 'guess'. In this case take

$$\sigma_{33} = kx_2 \qquad \text{other } \sigma_{ij} = 0 \tag{5.23}$$

in which k is a constant, as yet unknown.

Sketching the stress vectors, which result from this assumption as to the form taken by the stress tensor, at each point of the cross-section (on $x_1 = 0$ only, for clarity) we obtain the force field described in Figure 5.7

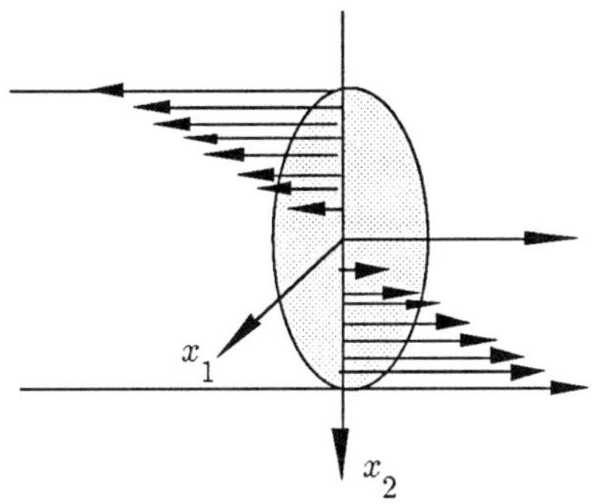

Figure 5.7

Clearly (by symmetry) the resultant force on the face is zero though there is a resultant couple (as required) about the x_1-axis. Our 'guess' is, at least, physically reasonable.

In the absence of body forces the assumed solution to the problem, (5.23), satisfies the Beltrami-Mitchell equations and all we need do is to satisfy the boundary conditions. On the lateral (curved) stress free surface of the cylinder a unit normal has the form $n_i = (\frac{x_1}{a}, \frac{x_2}{a}, 0)$ and so, on this surface

$$T_i^{(n)} = \sigma_{ij} n_j = \sigma_{i3} n_3 = 0$$

as required.

We have already noted that on the end face ($x_3 = 0$) the resultant force is zero and so the resultant couple is independent of position and its value may be determined about any convenient point. We choose the origin.

Before we proceed with this calculation we should note that if S is a general surface at each point of which a stress vector $T_i^{(n)}$ is defined then the resultant couple due to all the forces on S about the origin is given by

$$M_i = \iint_S \eta_{ijk} x_j T_k^{(n)} dS$$

(This follows since $T_k^{(n)} dS$ is the force acting over the elementary area dS). Writing this in terms of the stress tensor we find:

$$M_i = \iint_S \eta_{ijk} x_j \sigma_{km} n_m dS$$

In our particular example $n_m = (0, 0, 1)$ and so

$$\begin{aligned} M_1 \equiv M &= \iint_A \eta_{1jk} x_j \sigma_{k3} dx_1 dx_2 \\ &= \iint_A \eta_{123} x_2 \sigma_{33} dx_1 dx_2 = \iint_A k x_2^2 dx_1 dx_2 \end{aligned}$$

in which A is the cross-sectional area of the cylinder. (The reader should verify that $M_2 = M_3 = 0$. This can be done by direct calculation or by realising that M_2 and M_3 are, for constant density, directly related to the products of inertia. Since *any* pair of orthogonal axes is a principal set for a circle this result follows immediately).

If we denote the second moment of area of the cross-section by I:

$$I = \iint_A x_2^2 dx_1 dx_2$$

and so $M = kI$. Thus the solution to this problem, for the stress tensor is

$$\sigma_{33} = \frac{M}{I} x_2 \qquad \text{other } \sigma_{ij} = 0$$

What are the consequences of this solution? The non-zero strain components, from (5.8) are (and using (5.17), (5.18))

$$\varepsilon_{33} = \frac{1}{E}\sigma_{33} = \frac{M}{EI} x_2 \qquad \varepsilon_{11} = \varepsilon_{22} = -\frac{\nu}{E}\sigma_{33} = -\frac{\nu M}{EI} x_2$$

Utilising the strain-displacement equations (5.12) and ignoring rigid body motions (assuming the beam is held fixed at the origin) we find

$$u_1 = -\frac{\nu M}{EI}x_1x_2 \quad u_2 = \frac{M}{EI}(-\nu x_2^2 + \frac{\nu x_1^2}{2} - \frac{x_3^2}{2}) \quad u_3 = \frac{M}{EI}x_2x_3 \qquad \textbf{(5.24)}$$

(Although we have only considered the case of a beam with a circular cross-section one may show, by a suitable choice of axes x_i that in fact the above results apply to a beam with an arbitrarily shaped cross-section. In the more general case we would rotate axes x_1, x_2 so that they are aligned with the principal axes of inertia of the cross-section. M_2, M_3 would then vanish and I would still be defined as above).

Obviously, as a result of applying equal and opposite couples to a beam of this kind the beam will bend and the line of centroids, initially straight before deformation, will be curved after deformation. (Imagine flexing a wooden ruler between finger and thumb of each hand). We easily ascertain the shape of the deformed line of centroids. A point initially on the axis of the beam, with coordinates $x_i = (0, 0, x_3)$ is deformed to a position with coordinates x_i^* such that (from (5.24) and (4.1))

$$\begin{aligned} x_1^* &= u_1 + x_1 = -\frac{\nu M}{EI}x_1x_2 + x_1 = 0 \\ x_2^* &= u_2 + x_2 = \frac{M}{EI}(-\nu x_2^2 + \frac{\nu x_1^2}{2} - \frac{x_3^2}{2}) + x_2 = -\frac{M}{2EI}x_3^2 \\ x_3^* &= u_3 + x_3 = \frac{M}{EI}x_3x_2 + x_3 = x_3 \end{aligned} \qquad \textbf{(5.25)}$$

where, in each of the equations of (5.25) we have used $x_1 = x_2 = 0$.

Hence from the second two of equations of (5.25)

$$x_2^* = -\frac{M}{2EI}(x_3^*)^2 \qquad \textbf{(5.26)}$$

This is the equation of the deformed line of centroids. This is a parabola as shown in Figure 5.8

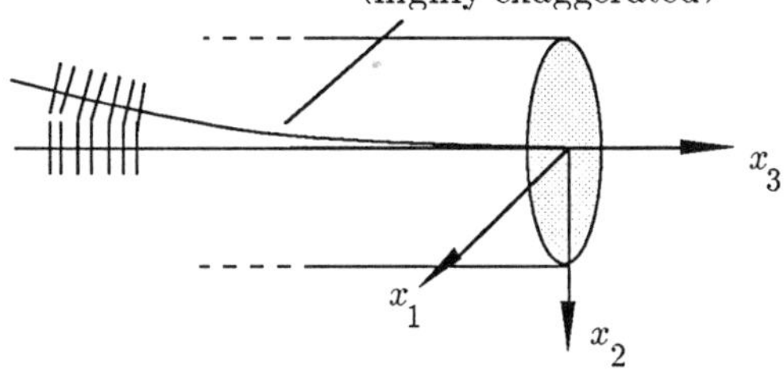

Figure 5.8

The curvature of the parabola is

$$\kappa = \frac{\left|\frac{d^2x_2^*}{d(x_3^*)^2}\right|}{\left\{1+\left[\frac{dx_2^*}{dx_3^*}\right]^2\right\}^{\frac{3}{2}}} = \frac{M}{EI}$$

where we have ignored the squared derivative term in the denominator. (Remember *all* displacement derivatives are small in linear elasticity theory and, in particular, $\left|\frac{\partial u_3}{\partial x_2}\right| = \frac{M}{EI}x_3 = \left|\frac{dx_2^*}{dx_3^*}\right|$).

That is, to the order of approximation used in linear elasticity theory the curvature of the deformed line of centroids is a constant with a very large radius of curvature R:

$$R = \frac{EI}{M} \tag{5.27}$$

(Technically, although the deformed line of centroids is a parabola it cannot be distinguished from a circle. As an example, if a moment M of magnitude 5000 Nm is applied to a circular steel cylinder of radius 2cm with $E \approx 2 \times 10^{11}\text{Nm}^{-2}$ then $\text{I} = \frac{1}{4}\pi a^4 \approx 12 \times 10^{-8}\text{m}$ and so $R \approx 4.8\text{m}$) We should also note that points on the plane $x_3 = c$, that is, normal to the centre line are also normal to the deformed centre line. To see this put $x_3 = c$ in the second two equations of (5.25). We have

$$x_2^* = x_2 + O(\frac{1}{R}) \quad \text{and} \quad x_3^* = \frac{c}{R}(x_2 + 1)$$

Hence

$$x_3^* = \frac{c}{R}(x_2^* + 1) + O(\frac{1}{R^2})$$

which, to the order of approximation considered in linear elasticity is the equation of a plane with normal $n_i = (0, \frac{c}{R}, -1)$. The tangent to the deformed line of centroids given in (5.26) is

$$t_i = (0, \frac{dx_2^*}{dx_3^*}, 1) = (0, -\frac{c}{R}, 1)$$

and so the deformed plane is normal to the deformed line as indicated in Figure 5.8.

Equation (5.27) is a mathematical statement of the Euler-Bernoulli Law:

To bend an initially straight beam made from a material with a Young's Modulus E and with uniform cross-section having second moment of area I into a circle of radius R using only pure couples applied at either end requires a couple of moment $M = EI/R$.

The magnitude of the couple required will increase as EI increases. This combination, EI, is a measure of the rigidity (in bending) of the beam. To increase the rigidity in bending we must either use a 'stiffer' material (thereby increasing E) or else design the cross-section so that I is increased. This is one of the reasons why beams are often designed with cross-sections in the shape of cetain letters; the I-beam, the T-beam etc. Shapes such as these have a large second moment of area. The two beams shown in Figure 5.9 have the same area. For the solid circular cross-section $I \approx 1.3$ whilst for the I-beam $I \approx 6$ which represents a substantial increase in bending stiffness without any increase in material quantity.

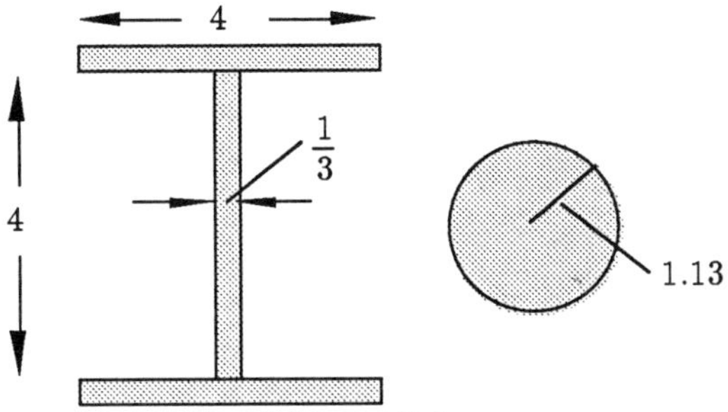

Figure 5.9

- **Torsion of a Beam with a Uniform cross-section**

Aspects of this problem have been treated earlier in Examples 3.2 and 4.2. However, we shall examine the problem once again and verify that the stress tensor components given in Example 3.2 and the displacement components given in Example 4.2 are compatible and are valid solutions of the equations of linear elasticity.

A circular cylinder of radius a has its axis aligned along the x_3-axis and is deformed by the action of equal and opposite torsional couples M applied at either end (see Figure 5.10). The origin of coordinates is chosen on the right hand face of the cylinder.

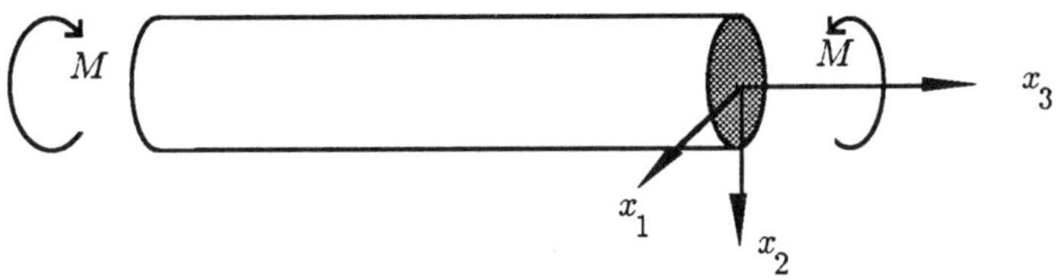

Figure 5.10

It has already been validated, in Example 3.2, that the stress field

$$\sigma_{13} = \sigma_{31} = -kx_2 \quad \sigma_{23} = \sigma_{32} = kx_1 \quad \text{other } \sigma_{ij} = 0 \tag{5.28}$$

in which k is a constant, adequately accounts for the physical arrangement of forces shown in Figure 5.10 with $M = \frac{1}{2}k\pi a^4$. That is, the stress field (5.28) satisfies the boundary conditions (5.10). Since the stress components are, at worst, linear in the coordinates, they satisfy the Beltrami-Mitchell equations (5.15). (We are again ignoring the effects of body forces). We proceed to determine the displacement field. From (5.8) the only non-zero strain components are:

$$\varepsilon_{13} = \varepsilon_{31} = \frac{1}{2\mu}\sigma_{13} = -\frac{k}{2\mu}x_2 \qquad \varepsilon_{23} = \varepsilon_{32} = \frac{1}{2\mu}\sigma_{23} = \frac{k}{2\mu}x_1 \tag{5.29}$$

Thus the displacement vector is the solution to the system of equations in (5.12):

$$\begin{gathered} \frac{\partial u_1}{\partial x_1} = 0 \qquad \frac{\partial u_2}{\partial x_2} = 0 \qquad \frac{\partial u_3}{\partial x_3} = 0 \\ \frac{\partial u_1}{\partial x_2} + \frac{\partial u_2}{\partial x_1} = 0 \qquad \frac{\partial u_1}{\partial x_3} + \frac{\partial u_3}{\partial x_1} = -\frac{k}{2\mu}x_2 \qquad \frac{\partial u_2}{\partial x_3} + \frac{\partial u_3}{\partial x_2} = \frac{k}{2\mu}x_1 \end{gathered} \tag{5.30}$$

We shall fix the point of the beam at the origin so as to exclude rigid body motions. The first three equations of (5.30) integrate to

$$u_1 = f_1(x_2, x_3) \qquad u_2 = f_2(x_1, x_3) \qquad u_3 = f_3(x_1, x_2) \tag{5.31}$$

respectively. The final three equations of (5.30) then imply:

$$\frac{\partial f_1}{\partial x_2} + \frac{\partial f_2}{\partial x_1} = 0 \qquad \frac{\partial f_1}{\partial x_3} + \frac{\partial f_3}{\partial x_1} = -\frac{k}{\mu}x_2 \qquad \frac{\partial f_2}{\partial x_3} + \frac{\partial f_3}{\partial x_2} = \frac{k}{\mu}x_1$$

A *particular* solution to this set is

$$f_1 = -\frac{kx_2x_3}{\mu} \qquad f_2 = \frac{kx_1x_3}{\mu} \qquad \text{and} \quad f_3 = 0$$

Hence the solution for u_i is

$$u_1 = -\frac{k}{\mu}x_2x_3 \qquad u_2 = \frac{k}{\mu}x_1x_3 \qquad u_3 = 0$$

It is usual, in the torsion problem to rewrite the constant k/μ as τ so that (as in Example 4.1)

$$u_1 = -\tau x_2x_3 \qquad u_2 = \tau x_1x_3 \qquad u_3 = 0 \tag{5.32}$$

The constant τ is called the **twist** and is directly related to the angle through which each plane, $x_3 = c$, normal to the axis of the cylinder, is rotated. The reader is referred to Examples 3.2, 4.2 and 4.3 for more detail on the torsion problem.

EXERCISES

5.5 An arbitrarily shaped body is subjected to so-called hydrostatic pressure (as if submerged in liquid). At each point on the surface the stress vector is in the normal direction at that point. That is $T_i^{(n)} = -pn_i$ where p is called the **pressure** (n_i is the outward pointing normal).

Make an inspired guess as to the form taken by the stress tensor in this problem. Validate your guess by showing that the equations and boundary conditions of linear elasticity are satisfied. Finally, determine the displacement vector assuming the body is fixed in such a way as to exclude rigid body motions

5.6 The stress distribution in a beam of length ℓ of circular cross-section and radius a with its axis aligned along the x_1-axis is given by:

$$\sigma_{11} = \frac{4W}{\pi a^4}(x_1 - \ell)x_3 \qquad \sigma_{12} = -\frac{W(1+2\nu)}{\pi a^4(1+\nu)}x_2x_3$$

$$\sigma_{13} = \frac{2W}{\pi a^4(1+\nu)}[(\frac{3}{4}+\frac{\nu}{2})(a^2 - x_3^2) - (\frac{1}{4}-\frac{\nu}{2})x_2^2] \qquad \text{other } \sigma_{ij} = 0$$

Is this a possible solution to the equations of linear elasticity? Verify that the lateral surface of the cylinder is stress free and that the resultant of the forces on the end $x = \ell$ is $(0, 0, W)$. Show that the resultant couple about the origin is $(0, -\ell W, 0)$. Interpret the physical situation described by this solution.

5.4 The Navier Equation in Linear Elasticity

In the previous section the governing equations of linear elasticity were, effectively, expressed entirely in terms of the stress tensor alone so as to ease the solution of boundary value problems of the first kind.

Although this type of boundary value problem often arises in practice other forms can also arise. The problem in which an exact solution to the equations of linear elasticity is required given that at each point of the boundary of a body the displacement vector u_i is known is called a *boundary value problem of the second kind.* It proves advantageous in problems of this kind to express, as far as possible, the equations of linear elasticity soley in terms of the displacement vector. Obviously, since we will only use a physically meaningful form for the displacement vector the compatibility equations become redundant.

We begin by expressing the equilibrium equations (5.11) in terms of strain, using (5.13), to give:

$$2\mu\varepsilon_{ij},_j + \lambda\varepsilon_{kk},_i + \rho B_i = 0 \tag{5.33}$$

Re-expressing this in terms of the displacement vector through (5.12) implies

$$\mu(u_i,_{jj} + u_j,_{ij}) + \lambda u_k,_{ki} + \rho B_i = 0 \tag{5.34}$$

that is,

$$\mu u_i,_{jj} + (\lambda + \mu)u_k,_{ki} + \rho B_i = 0 \tag{5.35}$$

It is convenient to rewrite this in standard vector notation

$$\mu\nabla^2\underline{u} + (\lambda+\mu)\underline{\nabla}(\underline{\nabla}.\underline{u}) + \rho\underline{B} = 0 \tag{5.36a}$$

Using the vector identity $\nabla^2\underline{u} = -\underline{\nabla}\wedge(\underline{\nabla}\wedge\underline{u}) + \underline{\nabla}(\underline{\nabla}.\underline{u})$ (see Section 1.7) we can write (5.36a) in the alternative form:

$$-\mu\underline{\nabla}\wedge(\underline{\nabla}\wedge\underline{u}) + (\lambda+2\mu)\underline{\nabla}(\underline{\nabla}.\underline{u}) + \rho\underline{B} = 0 \tag{5.36b}$$

These equations (either (5.35) or (5.36)) are the Navier equations of linear elasticity.

Although they were derived with the aim of solving boundary value problems of the second kind we shall in fact use them to solve a boundary value problem of the first kind. The application that we have in mind, the analysis of a cylindrical pressure vessel, enjoys a high degree of symmetry which we shall exploit in forming possible expressions for $\underline{u}$ in (5.36).

- **Cylindrical Pressure Vessel**

To simplify the analysis we shall consider an infinitely long cylinder (in practice this means that the longitudinal dimension should be very much greater than the lateral dimension). The cylinder sustains an internal pressure P_1 (on the inner circle, radius a) and an external pressure P_2 (on the outer circle, radius b). See Figure 5.11.

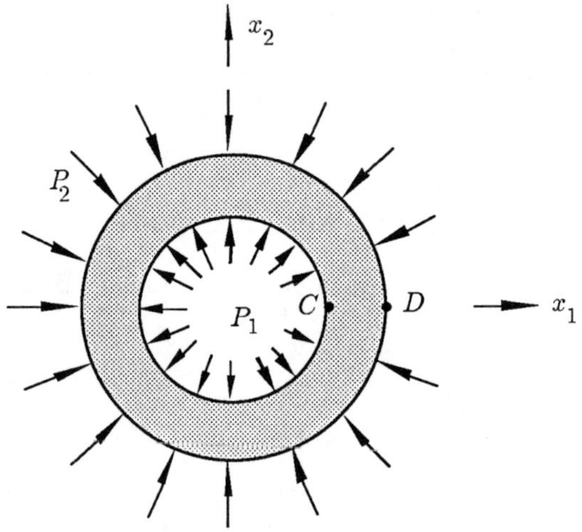

Figure 5.11

Due to the symmetry in the problem the displacement vector must be radially directed away from the axis of the cylinder and can only be a function of r, the cylindrical radial coordinate ($r = \sqrt{x_1^2 + x_2^2}$). That is, in terms of standard vector notation

$$\underline{u} = u_r(r)\hat{e}_r \tag{5.37}$$

(the 'r' subscript is a *label* not a tensor index) where $\hat{e}_r$ is a unit vector in the radial direction. Using the formula for curl in cylindrical polar coordinates it is a simple matter to verify, in this example, that $\underline{\nabla} \wedge \underline{u} = 0$ and so the Navier equation, in the absence of body forces, reduces to

$$(\lambda + 2\mu)\underline{\nabla}(\underline{\nabla}.\underline{u}) = 0$$

which clearly implies that

$$\underline{\nabla}.\underline{u} = \text{constant} \tag{5.38}$$

Now, in cylindrical polar coordinates:

$$\underline{\nabla}.\underline{u} = \frac{1}{r}u_r + \frac{du_r}{dr}$$

and so

$$\frac{1}{r}u_r + \frac{du_r}{dr} = A' \qquad \mathbf{(5.39)}$$

where A' is a constant. This is easily rearranged to the form $\frac{d}{dr}(ru_r) = A'r$ and integrated to give $u_r = Ar + \frac{B}{r}$ (where $A = A'/2$). Hence the displacement is $\underline{u} = (Ar + \frac{B}{r})\hat{e}_r$. Reverting to Cartesian coordinates we find

$$\hat{e}_r = \cos\theta\hat{i} + \sin\theta\hat{j} = \frac{x_1}{r}\hat{i} + \frac{x_2}{r}\hat{j}$$

where $\hat{i}, \hat{j}$ are the usual unit vectors in the x_1- and x_2-directions. Thus the Cartesian components of $\underline{u}$ are

$$u_1 = Ax_1 + \frac{Bx_1}{r^2} \qquad u_2 = Ax_2 + \frac{Bx_2}{r^2} \qquad u_3 = 0 \qquad \mathbf{(5.40)}$$

From these we determine the strain and stress components in the usual way and find the non-zero strain components to be

$$\varepsilon_{11} = A + B(\frac{x_2^2 - x_1^2}{r^4}) \quad \varepsilon_{22} = A + B(\frac{x_1^2 - x_2^2}{r^4}) \quad \varepsilon_{12} = -\frac{2Bx_1x_2}{r^4} \qquad \mathbf{(5.41)}$$

whilst the non-zero stress components are

$$\begin{aligned} \sigma_{11} &= 2\mu[A + B(\frac{x_2^2 - x_1^2}{r^4})] + 2\lambda A \\ \sigma_{22} &= 2\mu[A + B(\frac{x_1^2 - x_2^2}{r^4})] + 2\lambda A \\ \sigma_{12} &= -4\mu B\frac{x_1x_2}{r^4} \qquad \sigma_{33} = 2\lambda A \end{aligned} \qquad \mathbf{(5.42)}$$

The inherent symmetry in the problem implies that if we satisfy the stress boundary conditions at one point on the interior surface and at one point on the exterior surface then the boundary conditions will be satisfied at *every* boundary point. We choose these points at $C(x_1 = a, x_2 = 0)$ and at $D(x_1 = b, x_2 = 0)$.

At C (where $n_i = (-1,0,0)$) the boundary condition is that $T_i^{(n)} = (P_1, 0, 0)$ and at D (where $n_i = (1,0,0)$) the boundary condition is that $T_i^{(n)} = (-P_2, 0, 0)$. Hence using (5.10) at C and D we find:

$$P_1 = -2\mu(A - \frac{B}{a^2}) - 2\lambda A \qquad -P_2 = 2\mu(A - \frac{B}{b^2}) + 2\lambda A$$

On solving these equations for A and B we obtain

$$A = \frac{1}{(2\mu + 2\lambda)}[\frac{a^2P_1 - b^2P_2}{b^2 - a^2}] \qquad B = \frac{1}{2\mu}\frac{a^2b^2}{(b^2 - a^2)}(P_1 - P_2)$$

and so

$$\sigma_{11} = \frac{a^2P_1 - b^2P_2}{b^2 - a^2} + \frac{a^2b^2}{(b^2 - a^2)}(P_1 - P_2)\frac{(x_2^2 - x_1^2)}{r^4}$$
$$\sigma_{22} = \frac{a^2P_1 - b^2P_2}{b^2 - a^2} + \frac{a^2b^2}{(b^2 - a^2)}(P_1 - P_2)\frac{(x_1^2 - x_2^2)}{r^4} \qquad (5.43)$$
$$\sigma_{12} = -\frac{2a^2b^2}{(b^2 - a^2)}(P_1 - P_2)\frac{x_1x_2}{r} \qquad \sigma_{33} = 2\nu[\frac{a^2P_1 - b^2P_2}{b^2 - a^2}]$$

It is perhaps a little inappropriate in a problem with such a high degree of cylindrical symmetry to continue to express the stresses and strains in terms of Cartesian coordinates. If, at a point Q on the cylinder we rotate axes through an angle θ from (x_1, x_2) to (x_1', x_2') so that x_1' is pointing in the outward normal direction then in terms of the new coordinates (see Section 4.4)

$$\sigma_{11}' = \cos^2\theta\sigma_{11} + 2\cos\theta\sin\theta\sigma_{12} + \sin^2\theta\sigma_{22}$$
$$= \frac{a^2P_1 - b^2P_2}{b^2 - a^2} - \frac{a^2b^2}{(b^2 - a^2)}\frac{(P_1 - P_2)}{r^2} \qquad (5.44)$$

where, of course, we have used the transformation equations connecting Cartesian and cylindrical coordinates: $x_1 = r\cos\theta,\ x_2 = r\sin\theta$. In an exactly similar way we find

$$\sigma_{22}' = \frac{a^2P_1 - b^2P_2}{b^2 - a^2} + \frac{a^2b^2}{(b^2 - a^2)}\frac{(P_1 - P_2)}{r^2} \qquad (5.45)$$

$$\sigma_{12}' = 0 \qquad \text{and} \qquad \sigma_{33}' = \sigma_{33} = \frac{2\nu(a^2P_1 - b^2P_2)}{b^2 - a^2} \qquad (5.46)$$

Clearly, since off-diagonal components of σ_{ij}' are zero, the new coordinate system is a principal axis system for the stress tensor at Q.

(In fact we could have 'guessed' these results without explicitly performing the rotation. At ($x_1 = r,\ x_2 = 0$) the stress components given in (5.43) are the principal stresses at this point, since $\sigma_{ij} = 0\ i \neq j$. But the direction of x_1 has been chosen arbitrarily and so we should expect these stress components no matter in which radial direction x_1 is pointing. This is precisely what we have shown in equations (5.44),(5.45) and (5.46)).

It is usual (in problems with cylindrical symmetry) to rename $\sigma_{11}', \sigma_{22}'$ and σ_{33}' as $\sigma_{rr}, \sigma_{\theta\theta}$ and σ_{zz} respectively so that

$$\sigma_{rr} = \frac{a^2P_1 - b^2P_2}{b^2 - a^2} - \frac{a^2b^2}{b^2 - a^2}\frac{(P_1 - P_2)}{r^2} \qquad (5.47)$$

$$\sigma_{\theta\theta}' = \frac{a^2P_1 - b^2P_2}{b^2 - a^2} + \frac{a^2b^2}{b^2 - a^2}\frac{(P_1 - P_2)}{r^2} \qquad (5.48)$$

$$\sigma_{zz} = \frac{2\nu(a^2P_1 - b^2P_2)}{b^2 - a^2} \qquad (5.49)$$

σ_{rr} is called the **radial stress** and $\sigma_{\theta\theta}$ is called the **hoop stress.**

The practical case occurs when $P_1 >> P_2$ and a and b are comparable so that

$$\sigma_{rr} \approx \frac{a^2}{b^2-a^2}(1-\frac{b^2}{r^2})P_1 \quad \sigma_{\theta\theta} \approx \frac{a^2}{b^2-a^2}(1+\frac{b^2}{r^2})P_1 \quad \sigma_{zz} \approx \frac{2\nu a^2}{b^2-a^2}P_1$$

In this case, since $r \leq b$ then $\sigma_{rr} < 0$ and $\sigma_{\theta\theta} > 0$. The principal stress with the largest magnitude is clearly $\sigma_{\theta\theta}$ having a maximum value of $\dfrac{(b^2+a^2)}{(b^2-a^2)}P_1$ (at $r = a$). The hoop stress is *always* tensile and *always* greater than P_1 whilst the radial stress is *always* compressive.

It is interesting to note that using a 'stiffer' material does not affect the hoop stress in any way. The only way to decrease the magnitude of the hoop stress (and so prevent the pressure pipe from tearing apart under extreme pressure) is to increase the ratio of b to a implying the use of more material.

Also note that when the vessel is thin-walled ($b = a + t$ where t is small) then

$$\sigma_{\theta\theta} = \frac{a^2}{(b-a)(b+a)}(1+\frac{b^2}{r^2})P_1 \approx \frac{b}{2t}(1+\frac{b^2}{r^2})P_1$$

but $r \approx b$ throughout the thickness and so

$$\sigma_{\theta\theta} \approx \frac{b}{t}P_1$$

That is, for thin-walled pressure vessels one may assume that the hoop stress is constant throughout the thickness of the cylinder.

EXERCISES

5.7 A cylindrical shell $a \leq r \leq b$ is subject to an internal pressure P_1 whilst the outer surface is rigidly restrained from expansion. Determine the displacement vector in this case and also find the principal stress components.

5.8 (a) By making an appropriate choice for the form taken by the displacement vector determine a solution to the equations of linear elasticity which describes a spherical pressure vessel subjected to an internal pressure P_1 and an external pressure P_2.

(b) By transforming to x'_i coordinates with x'_1 pointing along the outward radial direction and x'_2, x'_3 in the $\hat{\theta}, \hat{\phi}$ directions of spherical polar coordinates show that the principal stresses occur in these directions. In particular in the absence of external pressure show that (introducing a similar notation for stress components to that above)

$$\sigma_{rr} = \frac{P_1 a^3}{(b^3 - a^3)}(1 - \frac{b^3}{r^3}) \qquad \sigma_{\theta\theta} = \sigma_{\phi\phi} = \frac{P_1 a^3}{(b^3 - a^3)}(1 + \frac{b^3}{2r^3})$$

in which a and b are the internal and external radii respectively.
(c) Derive appropriate thin-shell formulae.

5.9 Show that the deformation of a solid sphere of radius a due to its own gravitational field (variable body force, in polar coordinates $B_r = -g\frac{r}{a}$) is such that

$$u_r = -g\rho a \frac{(1-2\nu)(1+\nu)}{10E(1-\nu)}\left[\left(\frac{3-\nu}{1+\nu}\right)r - \frac{r^3}{a^2}\right]$$

(Utilise the Beltrami-Mitchell equations in spherical polar coordinate form). Confirm that the solid is compressed ($\varepsilon_{rr} < 0$) within a spherical surface of radius $a\sqrt{(3-\nu)/3(1+\nu)}$ and stretched outside it.

5.10 A condition of plane strain is said to exist in an elastic solid if the displacement components take the form:

$$u_1 = f(x_1, x_2) \qquad u_2 = g(x_1, x_2) \qquad u_3 = 0$$

Show that in this special case that the compatibility equations reduce to a single constraint:

$$\varepsilon_{11},_{22} + \varepsilon_{22},_{11} - 2\varepsilon_{21},_{21} = 0$$

and that the Beltrami-Mitchell equations reduce to (in the absence of body forces)

$$\nabla^2(\sigma_{11} + \sigma_{22}) = 0$$

5.11 (i) From the equations of equilibrium, in the absence of body forces, show that

$$\varepsilon_{ii},_{jj} = 0$$

(ii) Under the same conditions show that the stress tensor satisfies the equations

$$\sigma_{kk},_{ij} + (1+\nu)\sigma_{ij},_{kk} = 0$$

Hence show that, although the expressions

$$\sigma_{11} = a[x_2^2 + \eta(x_1^2 - x_2^2)] \qquad \sigma_{22} = 2[x_1^2 + \eta(x_2^2 - x_1^2)]$$

$$\sigma_{33} = a\eta(x_1^2 + x_2^2) \qquad \sigma_{12} = -2a\eta x_1 x_2 \qquad \sigma_{13}\sigma_{23} = 0$$

may satisfy the equilibrium equations for the case of zero body force, they are nevertheless not physically possible stresses in an isotropic elastic material.

Chapter 6

ENERGY

6.1 Strain Energy and Work

When a force F_i moves its point of application along a vector u_i it does work of amount $F_i u_i$. When an elastic body is deformed by the application of surface forces or by body forces the work that is done by these forces is stored in the body and used to ensure the body recovers its original configuration when the deforming forces are removed.

A simple example is a wire of stiffness c. If the wire is deformed a certain amount by applying a force then on releasing the force the wire recovers to its original length.

Work stored within a body in this way is called **strain energy**. It is exactly equal to the work done by the external forces in deforming the body.

For the case of a simple wire it is a relatively straightforward matter to determine its strain energy. Consider the wire of unstressed length ℓ_0 deformed to a length ℓ. From Hooke's law (see Section 5.1), $F = c(\ell - \ell_0)$. The force required to extend the spring by a further amount $\delta\ell$ is $F + \delta F$ in which

$$F + \delta F = c(\ell + \delta\ell - \ell_0)$$

that is,

$$\delta F = c\delta\ell$$

The work done δW_w by this force is $(F + \delta F)\delta\ell$ and so, to a good approximation:

$$\delta W_w \approx c(\ell - \ell_0)\delta\ell$$

Hence, the total work done to extend the spring from ℓ_0 to ℓ^* is

$$\begin{aligned} W_w &= \int_{\ell_0}^{\ell^*} c(\ell - \ell_0)d\ell = \frac{1}{2}c\Big[(\ell - \ell_0)^2\Big]_{\ell_0}^{\ell^*} \\ &= \frac{1}{2}c(\ell^* - \ell_0)^2 \end{aligned}$$

We can rewrite this in the form:

$$W_w = \frac{1}{2}F^*(\ell^* - \ell_0)$$

where F^* is the magnitude of the final force applied to the wire. Thus, for this simple system, the strain energy is equal to half of the final force times the final displacement at the point of application of F^*. We also note that if A is the cross-sectional area of the wire then, in terms of stress and strain

$$W_w = \frac{1}{2}\frac{F^*}{A}A\ell_0\frac{(\ell^* - \ell_0)}{\ell_0} = \frac{1}{2}\sigma_{33}\varepsilon_{33}V$$

We can experiment *mathematically* with the form of the strain energy obtained for the wire. An obvious result is that, if W_w is expressed entirely in terms of the displacement of the endpoint, $u = (\ell^* - \ell_0)$, then $W_w = \frac{1}{2}c(\ell^* - \ell_0)^2 = \frac{1}{2}cu^2$ and

$$\frac{\partial W_w}{\partial u} = cu = c(\ell^* - \ell_0) = F^* \quad \textbf{(6.1a)}$$

Alternatively, if W_w is expressed solely in terms of the final force applied then $W_w = \dfrac{1}{2c}(F^*)^2$ and

$$\frac{\partial W_w}{\partial F^*} = \frac{1}{c}F^* = u \quad \textbf{(6.1b)}$$

These relations (6.1a,b) are special cases of the Castigliano relations (to be derived below) connecting strain energy to the applied forces and measured displacements.

Another purely mathematical result is that if $(F^*)'$ and u' are a second set of applied force and measured displacement of the wire then the work done by $(F^*)'$ acting through u is equal to the work done by (F^*) acting through u'. That is

$$(F^*)'u = F^*(u')$$

This is easy to show since $(F^*)' = cu'$ and $F^* = cu$ and so $cu'u = cuu'$ which is obviously true. This result is a special case of Rayleigh's reciprocal theorem to be derived in the next section.

It will be interesting to discover what form the strain energy takes for a more general elasic body.

- **The Strain Energy of an Elastic Body**

Consider an element δS of surface on an arbitrarily shaped elastic body which is deformed by the action of surface stresses. We imagine that the final displacement u_i at each point to be the result of building up the applied stresses from a zero value to the final value $T_i^{(n)}$. The final force on δS is $T_i^{(n)}\delta S$. Overall, the displacement of the element is u_i and so we can argue that the work done by the surface stress on this element is $\frac{1}{2}(T_i^{(n)}\delta S)u_i$. (The $\frac{1}{2}$ arises because, as for the wire in the simple tension experiment, the final value of the force $T_i^{(n)}\delta S$does not move its point of application throughout the full magnitude of u_i). Repeating this argument over all the bounding surface we find that the total work done by the surface stresses is $\dfrac{1}{2}\iint_S T_i^{(n)}u_i dS$ whilst, proceeding similarly, the total work done by the body force B_i is $\dfrac{1}{2}\iiint_V \rho B_i u_i dV$. Hence the strain energy W of the body V bounded by S is:

$$W = \frac{1}{2}\iint_S T_i^{(n)}u_i dS + \frac{1}{2}\iiint_V \rho B_i u_i dV \quad \textbf{(6.2)}$$

The argument, above, that W is the work done by the external forces is plausible though perhaps not very convincing. This does not matter. We *define* the strain energy of an elastic body as in (6.2) irrespective of its physical interpretation.

Obviously if the body is deformed then at each point there will be defined a stress tensor σ_{ij} and a strain tensor ε_{ij}. It will be useful to rework (6.2) in terms of the stresses and the strains. Using the Cauchy relation (5.10) and the equations of equilibrium (5.11) we have

$$W = \frac{1}{2}\iint_S \sigma_{ij} n_j u_i dS - \frac{1}{2}\iiint_V \sigma_{ij},_j\, u_i dV \qquad \textbf{(6.3)}$$

Converting the surface integral in (6.3) into a volume integral using the divergence theorem (see Section 1.8) implies:

$$\begin{aligned} W &= \frac{1}{2}\iiint_V (\sigma_{ij}u_i),_j\, dV - \frac{1}{2}\iiint_V \sigma_{ij},_j\, u_i dV \\ &= \frac{1}{2}\iiint_V [\sigma_{ij},_j\, u_i + \sigma_{ij}u_i,_j\,]dV - \frac{1}{2}\iiint_V \sigma_{ij},_j\, u_i dV \\ &= \frac{1}{2}\iiint_V \sigma_{ij}u_i,_j\, dV \end{aligned}$$

However, since σ_{ij} is symmetric, $\sigma_{ij}u_i,_j = \frac{1}{2}\sigma_{ij}(u_i,_j + u_j,_i\,) = \sigma_{ij}\varepsilon_{ij}$ and so, finally:

$$W = \frac{1}{2}\iiint_V \sigma_{ij}\varepsilon_{ij} dV \qquad \textbf{(6.4a)}$$

or equivalently:

$$\frac{1}{2}\iiint_V \sigma_{ij}\varepsilon_{ij} dV = \frac{1}{2}\iint_S T_i^{(n)} u_i dS + \frac{1}{2}\iiint_V \rho B_i u_i dV \qquad \textbf{(6.4b)}$$

(Note that for a simple wire deformed as in the simple tension experiment then, according to (5.16) and (6.4a) $W_w = \frac{1}{2}\iiint_V \sigma_{33}\varepsilon_{33} dV = \frac{1}{2}\sigma_{33}\varepsilon_{33} V$ as in this case the stresses and strains are constant. This confirms a result obtained earlier).

An important property of the strain energy is that it is always positive, only vanishing when $\varepsilon_{ij} = 0$. The proof is a little inelegant. First we express W in terms of the stress tensor alone using (5.8):

$$\begin{aligned} W = \frac{1}{2}\iiint_V \sigma_{ij}\varepsilon_{ij} dV &= \frac{1}{2}\iiint_V \sigma_{ij}\Big[\frac{1}{2\mu}(\sigma_{ij} - \frac{\lambda}{2\mu + 3\lambda}\delta_{ij}\sigma_{kk})\Big] dv \\ &= \frac{1}{2}\iiint_V \frac{1}{E}\big[(1+\nu)\sigma_{ij}\sigma_{ij} - \nu\sigma_{kk}^2\big] dV \end{aligned}$$

where we have used (5.17) and (5.18) to replace the Lamé constants by the Young's modulus and the Poisson ratio. We find that when expanded, the integrand is (up to a factor of $1/2E$)

$$\sigma_{11}^2 + \sigma_{22}^2 + \sigma_{33}^2 + 2(1+\nu)(\sigma_{12}^2 + \sigma_{13}^2 + \sigma_{23}^2) - 2\nu(\sigma_{11}\sigma_{22} + \sigma_{11}\sigma_{33} + \sigma_{22}\sigma_{33})$$

This is a quadratic form in six 'variables' $\sigma_{11}, \sigma_{12}, \sigma_{13}, \sigma_{22}, \sigma_{23}, \sigma_{33}$ with coefficient matrix:

$$J = \begin{bmatrix} 1 & -\nu & -\nu & 0 & 0 & 0 \\ -\nu & 1 & -\nu & 0 & 0 & 0 \\ -\nu & -\nu & 1 & 0 & 0 & 0 \\ 0 & 0 & 0 & 2(1+\nu) & 0 & 0 \\ 0 & 0 & 0 & 0 & 2(1+\nu) & 0 \\ 0 & 0 & 0 & 0 & 0 & 2(1+\nu) \end{bmatrix}$$

Moreover this matrix J is positive definite since $det J$, and each sub-determinant of J is positive depending only on the assumption that the value of the Poisson ratio is less than 1/2. (The reader should verify this calculation). But all elastic materials satisfy this physical constraint and so W is a positive definite scalar.

As we shall see in the following sections in this chapter the strain energy concept occupies a central role in linear elasticity. The first application that we shall study which uses strain energy in a fundamental way is Kirchoff's uniqueness theorem.

6.2 Kirchoff's Uniqueness Theorem

What we wish to show in this section is that the solution to the governing equations (5.9) – (5.14) of Section 5.2, which satisfy certain classes of boundary conditions admit (up to a rigid body motion) a unique solution for the displacement vector u_i.

There are three classes of problem that we shall consider (though uniqueness can in fact be proved for many other classes of boundary value problems). These are the boundary value problems of first, second and third kinds.

A boundary value problem of the third kind in linear elasticity is one for which the boundary conditions take the form, on a bounding surface S:

$$\sigma_{ij} n_j = T_i^{(n)} \quad \text{on } S_0 \qquad \text{and} \qquad u_i = h_i \quad \text{on } S - S_0 \tag{6.5}$$

That is, on part of the surface S_0, the stress vector $T_i^{(n)}$ is given, whilst on the remainder $S - S_0$, the displacement vector is given as h_i. Boundary value problems of the first and second kinds arise as special cases of boundary value problems of the third kind when $S_0 \equiv S$ or when $S_0 \equiv 0$ (empty) respectively.

To prove the theorem, we consider two possible solutions $(\overline{u}_i, \overline{\varepsilon}_{ij}, \overline{\sigma}_{ij})$ and $(u_i, \varepsilon_{ij}, \sigma_{ij})$ to the *same* boundary value problem of the third kind which, as well as satisfying the equations of equilibrium (5.11) also satisfy the boundary conditions:

$$\overline{\sigma}_{ij} n_j = T_i^{(n)} \quad \text{on } S_0 \qquad \overline{u}_i = h_i \quad \text{on } S - S_0 \tag{6.6a}$$

$$\sigma_{ij} n_j = T_i^{(n)} \quad \text{on } S_0 \qquad u_i = h_i \quad \text{on } S - S_0 \tag{6.6b}$$

We form v_i the *difference* in the displacement vectors between the solutions:

$$v_i = \overline{u}_i - u_i \tag{6.7}$$

and attempt to show that v_i describes a rigid body motion. It follows from (6.7) that if e_{ij} and s_{ij} are strain and stress tensors for v_i then

$$e_{ij} = \overline{\varepsilon}_{ij} - \varepsilon_{ij} \qquad s_{ij} = \overline{\sigma}_{ij} - \sigma_{ij} \tag{6.8}$$

Clearly, from the boundary conditions on the displacement vectors in (6.6)

$$v_i = 0 \quad \text{on } S - S_0 \tag{6.9}$$

Also, from (6.8) and the stress boundary conditions of (6.6)

$$s_{ij}n_j = 0 \quad \text{on } S_0 \tag{6.10}$$

Now since both $\overline{\sigma}_{ij}$ and σ_{ij} satisfy the equations of equilibrium it follows that

$$s_{ij,j} = 0 \tag{6.11}$$

Consider the strain energy W_d corresponding to v_i:

$$\begin{aligned} W_d &= \frac{1}{2}\iiint_V s_{ij}e_{ij}dV \\ &= \frac{1}{2}\iiint_V s_{ij}v_{i,j}\,dV \\ &= \frac{1}{2}\iint_S s_{ij}n_jv_idS - \frac{1}{2}\iiint_V s_{ij,j}\,v_idV \end{aligned} \tag{6.12}$$

(where, essentially, we have reversed the calculation carried through in Section 6.1 which transformed (6.3) into (6.4a)).

Now, because of (6.11) the volume integral in (6.12) vanishes and so

$$\begin{aligned} W_d &= \frac{1}{2}\iint_S s_{ij}n_jv_idS \\ &= \frac{1}{2}\iint_{S_0} s_{ij}n_jv_idS + \frac{1}{2}\iint_{S-S_0} s_{ij}n_jv_idS \end{aligned} \tag{6.13}$$

However, both of these integrals are zero since, from (6.9) v_i vanishes on $S - S_0$ and, from (6.10), $s_{ij}n_j$ vanishes on S_0. Hence,

$$W_d = \frac{1}{2}\iint_S s_{ij}e_{ij}dV = 0 \tag{6.14}$$

But we have already deduced from the previous section that the strain energy tensor only vanishes if $e_{ij} = 0$; that is, if v_i is a rigid body motion. Thus the two proposed solutions u_i and $\overline{u}_i$ are identical up to a rigid body motion. This completes the proof of Kirchoff's theorem.

6.3 The Reciprocal Theorem

A set of surface stresses $T_i^{(n)}$ and body forces B_i produces, in a body of volume V bounded by a surface S, a displacement field u_i, a strain tensor ε_{ij} and a stress tensor σ_{ij}. We refer to the collection $(T_i^{(n)}, B_i, u_i, \varepsilon_{ij}, \sigma_{ij})$ as an *elastic force system* on V. We can imagine a second elastic force system $(\overline{T}_i^{(n)}, \overline{B}_i, \overline{u}_i, \overline{\varepsilon}_{ij}, \overline{\sigma}_{ij})$ also acting on V and can define a scalar K:

$$K = \iiint_V \overline{\sigma}_{ij}\varepsilon_{ij}dV - \iiint_V \sigma_{ij}\overline{\varepsilon}_{ij}dV \tag{6.15}$$

It is relatively straightforward to show that K is zero. Now

$$\overline{\sigma}_{ij} = 2\mu\overline{\varepsilon}_{ij} + \lambda\delta_{ij}\overline{\varepsilon}_{kk} \quad \text{and} \quad \sigma_{ij} = 2\mu\varepsilon_{ij} + \lambda\delta_{ij}\varepsilon_{kk}$$

Hence (6.15) is

$$\begin{aligned} K &= \iiint_V [2\mu\overline{\varepsilon}_{ij} + \lambda\delta_{ij}\overline{\varepsilon}_{kk}]\varepsilon_{ij}dV - \iiint_V [2\mu\varepsilon_{ij} + \lambda\delta_{ij}\varepsilon_{kk}]\overline{\varepsilon}_{ij}dV \\ &= \iiint_V [2\mu\overline{\varepsilon}_{ij}\varepsilon_{ij} + \lambda\varepsilon_{mm}\overline{\varepsilon}_{kk}]dV - \iiint_V [2\mu\varepsilon_{ij}\overline{\varepsilon}_{ij} + \lambda\overline{\varepsilon}_{mm}\varepsilon_{kk}]dV \\ &= 0 \end{aligned}$$

so that

$$\iiint_V \overline{\sigma}_{ij}\varepsilon_{ij}dV = \iiint_V \sigma_{ij}\overline{\varepsilon}_{ij}dV \tag{6.16}$$

The relation (6.16) can be re-expressed in terms of the surface stresses and body forces. If we work with the volume integral on the left hand side of (6.16)

$$\begin{aligned} \iiint_V \overline{\sigma}_{ij}\varepsilon_{ij}dV &= \frac{1}{2}\iiint_V \overline{\sigma}_{ij}(u_{i,j} + u_{j,i})dV \\ &= \iiint_V \overline{\sigma}_{ij}u_{i,j}\, dV \qquad \text{(due to the symmetry in } \overline{\sigma}_{ij}) \end{aligned}$$

Now, by the divergence theorem:

$$\iint_S \overline{\sigma}_{ij}u_i n_j dS = \iiint_V (\overline{\sigma}_{ij}u_i)_{,j}\, dV = \iiint_V [\overline{\sigma}_{ij,j}\, u_i + \overline{\sigma}_{ij}u_{i,j}]dV$$

and so

$$\begin{aligned} \iiint_V \overline{\sigma}_{ij}u_{i,j}\, dV &= \iint_S \overline{\sigma}_{ij}u_i n_j dS - \iiint_V \overline{\sigma}_{ij,j}\, u_i dV \\ &= \iint_S \overline{T}_i^{(n)} u_i dS + \iiint_V \rho\overline{B}_i u_i dV \end{aligned}$$

Repeating this for the right hand side of (6.16) we have

$$\iint_S \overline{T}_i^{(n)} u_i dS + \iiint_V \rho\overline{B}_i u_i dV = \iint_S T_i^{(n)}\overline{u}_i dS + \iiint_V \rho B_i \overline{u}_i dV \tag{6.17}$$

This is the mathematical statement of Rayleigh's reciprocal theorem. In words it states that the work done by $(\overline{T}_i^{(n)}, \overline{B}_i)$ acting through u_i is equal to the work done by $(T_i^{(n)}, B_i)$ acting through $\overline{u}_i$. It has far reaching theoretical consequences in linear elasticity.

As a practical application of the result we shall consider two elastic force systems acting on a cylinder of arbitrary, though uniform, cross-sectional area A

aligned along the x_3-axis. The first elastic force system corresponds to the simple tension experiment described in Section 5.3. See Figure 6.1a. (In that application the cross-section was circular but in fact, as the reader should verify the solution given there for the stress tensor applies to any cylinder of arbitrary though uniform cross-section). In the second elastic force system we will also assume a zero body force and take the surface stresses to be zero on the curved surfaces. On the plane end for which $n_i = (0, 0, -1)$ we take $\overline{T}_i^{(n)} = (0, 0, -\frac{M}{I}x_2)$ whilst on the end $x_3 = \ell$ where $n_i = (0, 0, 1)$ we take $\overline{T}_i^{(n)} = (0, 0, \frac{M}{I}x_2)$

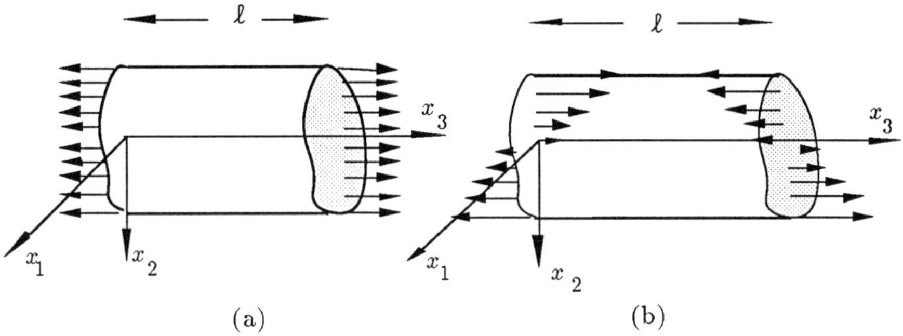

Figure 6.1

The solution for the displacement u_i in the first elastic force system has already been noted in Section 5.3:

$$u_i = \frac{F}{EA}(-\nu x_1, -\nu x_2, x_3)$$

In the absence of body forces Rayleigh's reciprocal theorem states:

$$\iint_S \overline{T}_i^{(n)} u_i dS = \iint_S T_i^{(n)} \overline{u}_i dS \tag{6.18}$$

Both surface integrals (due to the applied stress distributions) reduce to integrals taken over each end face. Now since $T_i^{(n)} = (0, 0, -\frac{F}{A})$ on $x_3 = 0$ and $T_i^{(n)} = (0, 0, \frac{F}{A})$ on $x_3 = \ell$ the surface integral on the right hand side of (6.18) reduces to

$$\frac{F}{A}\iint_A [\overline{u}_3(x_1, x_2, \ell) - \overline{u}_3(x_1, x_2, 0)]dA$$

Also, since $\overline{T}_i^{(n)} u_i = 0$ when $x_3 = 0$ and $\overline{T}_i^{(n)} u_i = \frac{M}{I}x_2(\frac{F}{EA}\ell)$ when $x_3 = \ell$, then the integral on the left hand side of (6.18) reduces to

$$\frac{MF}{IEA}\ell \iint_A x_2 dA$$

Hence the reciprocal theorem implies

$$\iint_A [\overline{u}_3(x_1, x_2, \ell) - \overline{u}_3(x_1, x_2, 0) - \frac{M}{IE}\ell x_2] dA = 0$$

from which it follows (since A is arbitrary) that

$$\overline{u}_3(x_1, x_2, \ell) - \overline{u}_3(x_1, x_2, 0) = \frac{M}{IE}\ell x_2$$

If we eliminate the rigid body translation by taking $\overline{u}_3(x_1, x_2, 0) = 0$ we obtain, finally

$$\overline{u}_3(x_1, x_2, \ell) = \frac{M}{IE}\ell x_2 \tag{6.19}$$

We have obtained the displacement in the direction of the x_3-axis for particles on the end face $x_3 = \ell$ for a cylinder (of arbitrary cross-sectional area) which is deformed by the action of pure couples applied at either end. The reader should verify that the result in (6.19) is consistent with the exact solution for this problem obtained in Section 5.3 (see equation (5.24)).

6.4 The Castigliano Theorem

Consider an elastic body V, bounded by S, which is rigidly supported and acted upon by a system of surface and body forces which include a point force F_i at a point Q on S. If the strain energy of the body is W and u_i is the displacement measured at Q then Castigliano's theorem states that

$$u_i = \frac{\partial W}{\partial F_i} \quad i = 1, 2, 3$$

The proof of this extremely useful result depends upon an application of the reciprocal theorem. We shall, for simplicity, ignore body forces though the theorem is easily extended to include them.

Proof

We consider two elastic force systems which are identical except for an 'extra' surface stress $\delta T_i^{(n)}$ acting over a small area element δS, which contains the point Q. See Figure 6.2.

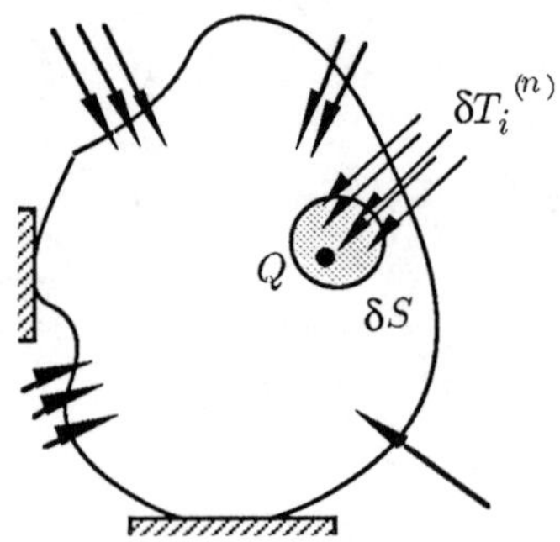

Figure 6.2

Therefore

$$\overline{T}_i^{(n)} = \begin{cases} T_i^{(n)} & \text{on } S - \delta S \\ T_i^{(n)} + \delta T_i^{(n)} & \text{on } \delta S \end{cases} \tag{6.20}$$

(Eventually we shall take the limit as $\delta S \to 0$ so that $\delta T_i^{(n)} \to 0$ and $\overline{u}_i \to u_i$).

Now, in the absence of body forces, the reciprocal theorem (6.18) states,

$$\begin{aligned} \iint_S T_i^{(n)}\overline{u}_i dS &= \iint_S \overline{T}_i^{(n)} u_i dS \\ &= \iint_{S-\delta S} T_i^{(n)} u_i dS + \iint_{\delta S} (T_i^{(n)} + \delta T_i^{(n)}) u_i dS \\ &= \iint_S T_i^{(n)} u_i dS - \iint_{\delta S} T_i^{(n)} u_i dS + \iint_{\delta S} (T_i^{(n)} + \delta T_i^{(n)}) u_i dS \\ &= \iint_S T_i^{(n)} u_i dS + \iint_{\delta S} \delta T_i^{(n)} u_i dS \end{aligned}$$

That is

$$\iint_S T_i^{(n)}(\overline{u}_i - u_i) dS = \iint_{\delta S} \delta T_i^{(n)} u_i dS \tag{6.21}$$

To maintain equilibrium the extra force $\delta T_i^{(n)} \delta S$ needs to be balanced by increases in the supporting (reaction) forces. However, if the supports are rigid, as we are assuming, the work done by the reaction forces will be zero. Now the strain energies W and $\overline{W}$ of the original and amended elastic force systems are

$$W = \frac{1}{2} \iint_S T_i^{(n)} u_i dS \qquad \text{and} \qquad \overline{W} = \frac{1}{2} \iint_S \overline{T}_i^{(n)} \overline{u}_i dS$$

If, as seems reasonable, $\overline{W}$ is not too different from W we can write $\delta W = \overline{W} - W$ in which case

$$\begin{aligned} \delta W &= \frac{1}{2} \iint_S (\overline{T}_i^{(n)} \overline{u}_i - T_i^{(n)} u_i) dS \\ &= \frac{1}{2} \iint_{S-\delta S} T_i^{(n)} \overline{u}_i dS + \frac{1}{2} \iint_{\delta S} (T_i^{(n)} + \delta T_i^{(n)}) \overline{u}_i dS - \frac{1}{2} \iint_S T_i^{(n)} u_i dS \\ &= \frac{1}{2} \iint_S T_i^{(n)} \overline{u}_i dS - \frac{1}{2} \iint_{\delta S} T_i^{(n)} \overline{u}_i dS \\ &\qquad + \frac{1}{2} \iint_{\delta S} (T_i^{(n)} + \delta T_i^{(n)}) \overline{u}_i dS - \frac{1}{2} \iint_S T_i^{(n)} u_i dS \end{aligned}$$

That is, on using (6.21) the expression for the change in strain energy due to the change in surface stress is:

$$\delta W = \frac{1}{2} \iint_{\delta S} \delta T_i^{(n)} (u_i + \overline{u}_i) dS$$

Now if δS is sufficiently small then the integrand in this surface integral can, to a good approximation, be taken as constant over δS and so

$$\delta W = \frac{1}{2}\delta T_i^{(n)}(u_i + \overline{u}_i)\delta S$$

The force acting at Q is $\delta F_i = \delta T_i^{(n)}\delta S$ and as we have already noted as $\delta S \to 0$, then $\overline{u}_i \to u_i$ and so

$$\delta W \to \delta F_i u_i$$

In the above calculation we have considered $\delta T_i^{(n)}$ to be unrestrained in direction. If we had instead considered the change in the surface stress at Q to have a component only in the x_1-direction: $\delta T_i^{(n)} = (\delta T_1^{(n)}, 0, 0)$ then the above relation implies

$$\frac{\partial W}{\partial F_1} = \lim_{\delta F_1 \to 0} \frac{\delta W}{\delta F_1} = u_1$$

Similarly, the rates of change of strain energy with respect to F_2 and F_3 would be

$$\frac{\partial W}{\partial F_2} = u_2 \qquad \text{and} \qquad \frac{\partial W}{\partial F_3} = u_3$$

or, more succinctly

$$\frac{\partial W}{\partial F_i} = u_i \quad i = 1, 2, 3 \tag{6.22}$$

which is the mathematical statement of Castigliano's theorem. We must emphasise that u_i is the displacement measured at the point of application of F_i. In an exactly similar way, by defining

$$\overline{u}_i = \begin{cases} u_i & \text{on } S - \delta S \\ u_i + \delta u_i & \text{on } \delta S \end{cases}$$

one can easily show that

$$\frac{\partial W}{\partial u_i} = F_i \quad i = 1, 2, 3 \tag{6.23}$$

The castigliano relations have particular application to the analysis of structures.

Example 6.1
Use the Castigliano theorem to determine the vertical deflection of node B (a pin-joint or hinge) subject to a force P as shown in Figure 6.3. Both elements are made of the same material and have the same, uniform, cross-sectional area. The elements are fixed at A, C with pin-joints and connected together at B with another pin-joint. Such joints do not allow couples of any kind to be transmitted from one element to another.

Solution

In problems of this kind, which are outside the scope of *exact* analysis using the equations of linear elasticity, it is assumed that the elements AB, BC sustain only longitudinal stresses, as in the simple tension experiment. We will refer to AB as element [1] and BC as element [2]. To use the Castigliano relation (6.22) we need to express the total strain energy of the system in terms of P alone.

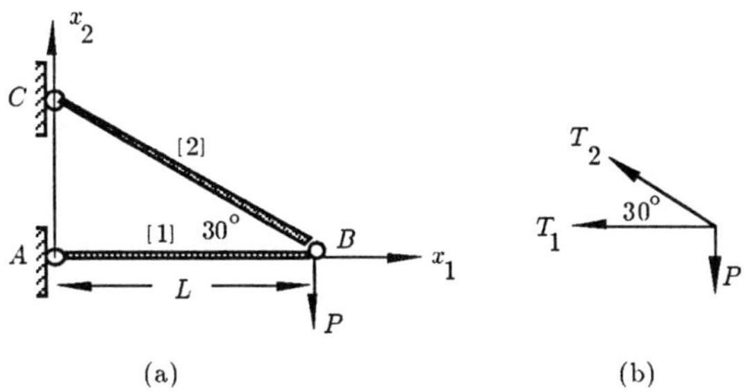

Figure 6.3

The force P is distributed (vectorially) between the elements meeting at B. If the forces acting through elements [1] and [2] are denoted by T_1 and T_2 respectively (in the directions shown in Figure 6.3b) then

$$T_2 \cos 60 = -P \qquad \text{and} \qquad -T_1 - T_2 \cos 30 = 0$$

therefore

$$T_2 = -2P \qquad T_1 = \sqrt{3}P$$

so that element [1] is in compression whilst element [2] is in tension. Hence the longitudinal stresses measured in each element are (using an obvious notation)

$$\sigma_{[1]} = \frac{T_1}{A} = \frac{\sqrt{3}P}{A} \qquad \text{and} \qquad \sigma_{[2]} = \frac{T_2}{A} = -\frac{2P}{A}$$

The longitudinal strains are $\varepsilon_{[1]} = \dfrac{\sigma_{[1]}}{E}$ and $\varepsilon_{[2]} = \dfrac{\sigma_{[2]}}{E}$ (See Section 5.3 on the simple tension experiment). From (6.4a) we can calculate the strain energy of the system:

$$\begin{aligned} W &= \frac{1}{2}\iiint_{V_{[1]}} \sigma_{ij}\varepsilon_{ij}\,dV + \frac{1}{2}\iiint_{V_{[2]}} \sigma_{ij}\varepsilon_{ij}\,dV \\ &= \frac{1}{2}\sigma_{[1]}^2\frac{AL}{E} + \frac{1}{2}\sigma_{[2]}^2\frac{2AL}{\sqrt{3}E} \\ &= \frac{1}{2}\frac{LP^2}{AE}[3 + \frac{8}{\sqrt{3}}] \end{aligned}$$

Now using the Castigliano relation (6.22) in which $F_i = (0, -P, 0)$ we have

$$u_2 = \frac{\partial W}{\partial F_2} = -\frac{\partial W}{\partial P} = -\frac{PL}{AE}[3 + \frac{8}{\sqrt{3}}]$$

This is the displacement of joint B in the direction of F_i.
Using the present formulation there is no way of finding the deflection in the x_1-direction. However, if we add a fictious force Q at joint B in the direction of AB then, repeating the above analysis we find

$$T_2 = -2P \qquad T_1 = -Q + \sqrt{3}P$$

Hence

$$W = \frac{1}{2}\frac{AL}{E}[\frac{(\sqrt{3}P - Q)^2}{A^2} + \frac{2}{\sqrt{3}}(\frac{4P^2}{A^2})]$$

Now $F_i = (Q, -P, 0)$ and so by the Castigliano relation (6.22)

$$u_1 = \frac{\partial W}{\partial F_1} = \frac{\partial W}{\partial Q} = \frac{1}{2}\frac{AL}{E}[-2\frac{(\sqrt{3}P - Q)}{A^2}]$$

If we take the limit as $Q \to 0$ we obtain the original system and, in this limit,

$$u_1 = -\frac{\sqrt{3}LP}{AE}$$

being the displacement in the x_1-direction of the system described in Figure 6.3a.

EXERCISES

6.1 Shown in Figure 6.4 is a 4-element pin-jointed structure. Use an appropriate Castigliano relation to determine the deflection of joint 5. All members have the same length, cross-sectional area and Young's modulus. Determine also the forces developed in each member.

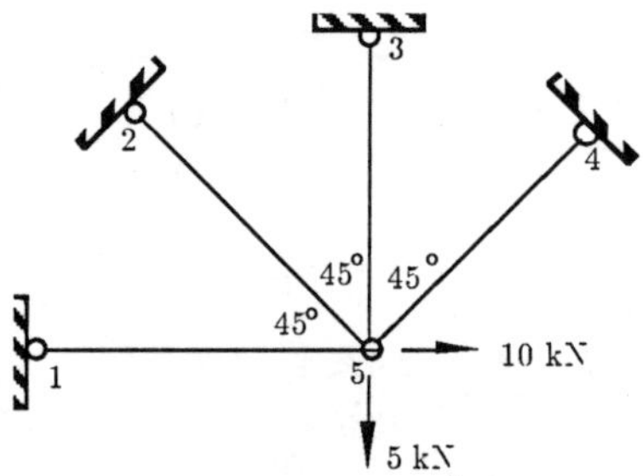

Figure 6.4

6.2 It may be shown that the strain energy of a slender beam of length L (with, possibly, a variable cross-section) is

$$W = \frac{1}{2}\int_0^L \frac{M^2}{EI} dx_3$$

where $M(x_3)$ is the moment measured at a cross-section x_3 =constant with normal $(0, 0, 1)$ due to the applied or reaction forces to the *left* of this plane. As in the pure bending deformation of a beam, I is the second moment of area. A simply supported beam is shown in Figure 6.5. At the left hand end the support (being a rolling suport) is such that the reaction force has only a vertical component. Of course, simply supported means that at either end there is a hinge allowing the beam to bend at these points as no couple can be applied to prevent it. Show, using considerations of overall equilibrium, that

$$M(\xi) = \begin{cases} -\frac{1}{2}P\xi & \xi < \frac{1}{2}L \\ \frac{1}{2}P(\xi - L) & \frac{1}{2}L < \xi < L \end{cases}$$

Determine the deflection at $x_3 = L/2$ and also (by use of a fictious force Q) at $x_3 = L/4$.

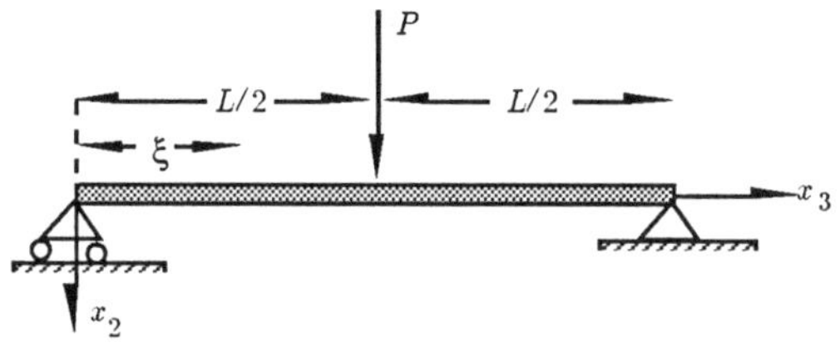

Figure 6.5

6.5 Potential Energy

Whereas the strain energy has a direct physical interpretation in terms of the work done on a body in order to deform it, the potential energy associated with a strained body, is a pure mathematical invention. However, this concept has far reaching applications in elasticity theory particularly, as we shall see later, to the development of approximate formulations in solid mechanics and to the development of approximate solutions to realistic elasticity problems.

Consider an elastic body V, bounded by a surface S, and let $(T_i^{(n)}, B_i, u_i, \varepsilon_{ij}, \sigma_{ij})$ be an elastic force system associated with the body. This system satisfies the equations of linear elasticity (5.9) – (5.14) together with the boundary conditions

$$\sigma_{ij}n_j = T_i^{(n)} \quad \text{on } S_0$$

and

$$u_i = h_i \quad \text{on } S - S_0$$

in which $T_i^{(n)}$ and h_i are *given* functions.

We *define* the potential energy, denoted by U, to be:

$$U = \frac{1}{2}\iiint_V \theta_{ij}(2\mu\theta_{ij} + \lambda\delta_{ij}\theta_{kk})dV - \iint_{S_0} T_i^{(n)}\gamma_i dS - \iiint_V \rho B_i\gamma_i dV \qquad \textbf{(6.24)}$$

In this 'form', γ_i is an *arbitrary* vector field, except that it satisfies the **same** boundary conditions as u_i, (that is $\gamma_i = h_i$ on $S - S_0$) and that θ_{ij} is related to γ_i through:

$$\theta_{ij} = \frac{1}{2}(\gamma_{i,j} + \gamma_{j,i}) \qquad \textbf{(6.25)}$$

that is, θ_{ij} is the 'strain tensor' corresponding to γ_i.

The reader may be tempted to think of γ_i as being the same as u_i (the true displacement) and θ_{ij} as being the same as ε_{ij} (the true strain tensor). However, don't be tempted. In the potential energy there is not necessarily any connection between $(\sigma_{ij}, T_i^{(n)}, B_i)$ and (γ_i, θ_{ij}).

We note that U is a scalar, and for a given elastic problem, (that is, a given elastic force system), the value of the scalar U changes as different choices are made for γ_i. In mathematical terminology U is a **functional**.

An obvious mathematical question to ask about the character of the scalar U is 'what vector field γ_i, if any, makes U a minimum?' This question is answered in the following important theorem known as the Theorem of Minimum Potential Energy.

Theorem

Of all the values of U obtained by using different vector fields γ_i the minimum value is attained when γ_i is the true displacement field u_i (and hence θ_{ij} is the true strain tensor).

Proof

Let u_i be the true displacement field and γ_i be any other. So as to ease the calculation that is to follow let us take γ_i as being equal to $(u_i + e_i)$. As γ_i and u_i satisfy the same boundary conditions (on $S - S_0$) so $e_i = 0$ on $S - S_0$. Calculating the 'strains' θ_{ij}:

$$\theta_{ij} = \frac{1}{2}[(u_i + e_i)_{,j} + (u_j + e_j)_{,i}] = \varepsilon_{ij} + h_{ij}$$

in which $h_{ij} = \frac{1}{2}(e_{i,j} + e_{j,i})$. We shall evaluate the difference between the potential energy obtained using u_i and that obtained by using any other γ_i and show that the difference is always positive.

$$\begin{aligned} U(\gamma_i) - U(u_i) =& \frac{1}{2}\iiint_V (\varepsilon_{ij} + h_{ij})\Big[2\mu(\varepsilon_{ij} + h_{ij}) + \lambda\delta_{ij}(\varepsilon_{kk} + h_{kk})\Big]dV \\ &- \frac{1}{2}\iiint_V \varepsilon_{ij}\Big[2\mu\varepsilon_{ij} + \lambda\delta_{ij}\varepsilon_{kk}\Big]dV - \iint_{S_0} T_i^{(n)}(u_i + e_i)dS \\ &+ \iint_{S_0} T_i^{(n)}u_i dS - \iiint_V \rho B_i(u_i + e_i)dV + \iiint_V \rho B_i u_i dV \end{aligned}$$

that is

$$\begin{aligned} U(\gamma_i) - U(u_i) =& \frac{1}{2}\iiint_V h_{ij}[2\mu\varepsilon_{ij} + \lambda\delta_{ij}\varepsilon_{kk}]dV \\ &+ \frac{1}{2}\iiint_V \varepsilon_{ij}[2\mu h_{ij} + \lambda\delta_{ij}h_{kk}]dV \\ &+ \frac{1}{2}\iiint_V h_{ij}[2\mu h_{ij} + \lambda\delta_{ij}h_{kk}]dV - \iint_{S_0} T_i^{(n)}e_i dS \\ &- \iiint_V \rho B_i e_i dV \end{aligned} \tag{6.26}$$

Now, due to the symmetry of ε_{ij} and h_{ij}, the first two integrals are identical, and both may be expressed in terms of the stress tensor σ_{ij}

$$\frac{1}{2}\iiint_V h_{ij}\left[2\mu\varepsilon_{ij} + \lambda\delta_{ij}\varepsilon_{kk}\right]dV \equiv \frac{1}{2}\iiint_V h_{ij}\sigma_{ij}dV \tag{6.27}$$

However, using the integral identity (6.4b) (which was proved entirely using the properties of the stress tensor alone) we have

$$\begin{aligned} \frac{1}{2}\iiint_V h_{ij}\sigma_{ij}dV &= \frac{1}{2}\iint_S T_i^{(n)}e_i dS + \frac{1}{2}\iiint_V \rho B_i e_i dV \\ &= \frac{1}{2}\iint_{S_0} T_i^{(n)}e_i dS + \frac{1}{2}\iiint_V \rho B_i e_i dV \end{aligned} \tag{6.28}$$

since e_i vanishes on $S - S_0$. Using this, (6.26) becomes (the first two terms cancel with the last two):

$$U(\gamma_i) - U(u_i) = \frac{1}{2}\iiint_V h_{ij}(2\mu h_{ij} + \lambda\delta_{ij}h_{kk})dV \tag{6.29}$$

As we have already noted in Section 6.1, the right hand side, being the 'strain energy' for the vector field e_i is a positive definite quantity and so for all γ_i

$$U(\gamma_i) - U(u_i) \geq 0 \tag{6.30}$$

vanishing only when $h_{ij} = 0$ (that is, when u_i and γ_i differ by a rigid body motion) and so the minimum value of U occurs when γ_i is the true displacement field u_i. This completes the proof.

Example 6.2

Repeat Example 6.1 but use the theorem of minimum potential energy to determine the deflection of joint B.

Solution

We shall assume joint B is displaced to (γ_1, γ_2) from its unstressed position and (consistent with the assumptions of linear elasticity and to our advantage) take γ_1 and γ_2 as being small. See Figure 6.6. Note that the proposed configuration satisfies the displacement boundary conditions.

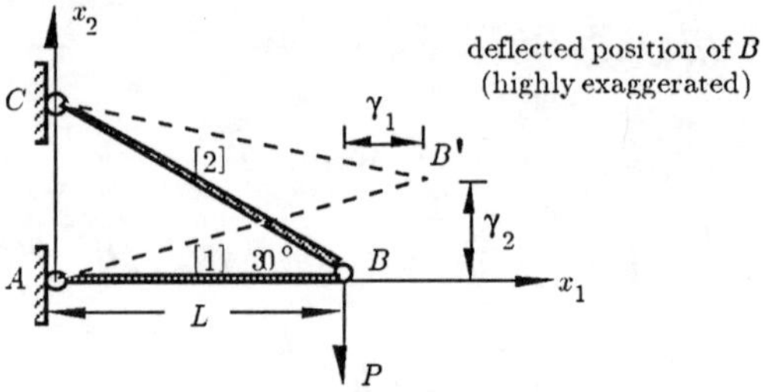

Figure 6.6

The new length of AB is

$$AB' = \sqrt{(L+\gamma_1)^2 + \gamma_2^2} \approx L + \gamma_1$$

whilst

$$CB' = \sqrt{(L+\gamma_1)^2 + (\frac{L}{\sqrt{3}} - \gamma_2)^2} \approx \frac{2}{\sqrt{3}}L + \frac{6}{8\sqrt{3}}(2\gamma_1 - \frac{2}{\sqrt{3}}\gamma_2)$$

The longitudinal strains in each element are (change in length/original length):

$$\varepsilon_{[1]} = \frac{\gamma_1}{L} \qquad \varepsilon_{[2]} = \frac{3}{8L}(2\gamma_1 - \frac{2}{\sqrt{3}}\gamma_2)$$

whilst the longitudinal stresses are

$$\sigma_{[1]} = E\varepsilon_{[1]} \qquad \sigma_{[2]} = E\varepsilon_{[2]}$$

Thus the 'strain energy' of the system for the proposed displacement γ_i is:

$$\frac{1}{2}\iiint_{V_{[1]}} \sigma_{[1]}\varepsilon_{[1]}dV + \frac{1}{2}\iiint_{V_{[2]}} \sigma_{[2]}\varepsilon_{[2]}dV = \frac{1}{2}\frac{EA}{L}[\gamma_1^2 + \frac{3\sqrt{3}}{32}(2\gamma_1 - \frac{2}{\sqrt{3}}\gamma_2)^2]$$

Also the work done by the force P acting through the displacement γ_i is $-P\gamma_2$ (that is, the scalar product of $(0,-P,0)$ with $(\gamma_1,\gamma_2,0)$). Hence, in the absence of body forces, the potential energy for this proposed displacement field is

$$U = \frac{1}{2}\frac{EA}{L}[\gamma_1^2 + \frac{3\sqrt{3}}{32}(2\gamma_1 - \frac{2}{\sqrt{3}}\gamma_2)^2] + P\gamma_2 \tag{6.31}$$

We now choose $(\gamma_1,\gamma_2,0)$ so that U is minimised. The values of γ_i so obtained will, by the theorem of minimum potential energy, be the required displacement of joint B.

Minimising:

$$\frac{\partial U}{\partial \gamma_1} = \frac{1}{2}\frac{EA}{L}[2\gamma_1 + \frac{3\sqrt{3}}{8}(2\gamma_1 - \frac{2}{\sqrt{3}}\gamma_2)] = 0 \tag{6.32}$$

$$\frac{\partial U}{\partial \gamma_2} = \frac{1}{2}\frac{EA}{L}[-\frac{3}{8}(2\gamma_1 - \frac{2}{\sqrt{3}}\gamma_2)] + P = 0 \tag{6.33}$$

The solution to the equations (6.32), (6.33) is

$$\gamma_1 = -\frac{\sqrt{3}LP}{AE} \quad \gamma_2 = -\frac{LP}{AE}(3 + \frac{8}{\sqrt{3}})$$

as already determined in Example 6.1

In the last example the optimisation of the potential energy reduced to a standard max/min problem. This is typical when the application is in the field of structural analysis.

We now examine an important application of the theorem, to the development of an approximate theory of transverse deflections in uniform beams; the so-called Euler-Bernoulli theory in which the optimisation does not reduce to standard form but one for which we need to invoke the techniques of the variational calculus.

The Euler-Bernoulli theory of beam bending is the result of making a number of 'reasonable' assumptions on how a beam deforms under the action of transverse loads and then utilising the theorem of minimum potential energy.

Consider a beam aligned along the x_3-axis with cross-sectional area $A(x_3)$ which, as we have anticipated in the notation, may be variable. (See Figure 6.7).

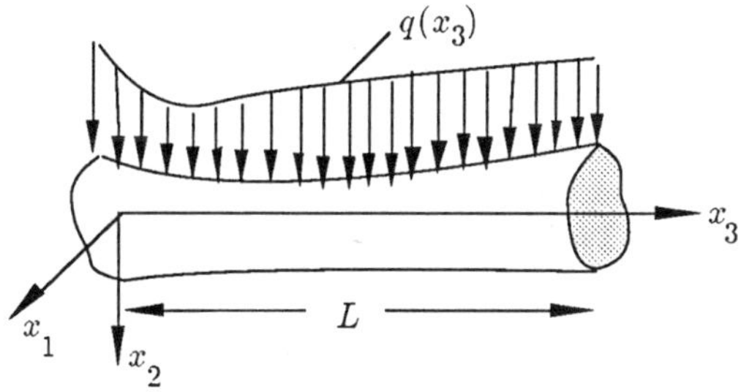

Figure 6.7

The beam is held fixed, against rigid body motions, in some way at one or both ends $x_3 = 0$, $x_3 = L$. It is deformed by the action of a transverse force $q(x_3)$, also possibly variable (units are force per unit length). The beam is assumed long in comparison to lateral dimensions. The deformation is characterised by the deformation of the centre line (initially on the x_3-axis) which as a result of the transverse force is deformed into a curve in the x_2x_3 plane with equation: $x_2 = w(x_3)$. Our purpose is to determine the function $w(x_3)$.

Rather like the pure bending problem of Section 5.3 each plane section is assumed to have the same x_3-coordinate after deformation, though the section may have been rotated (see Figure 6.8).

As the deformation is small $\alpha \approx \tan\alpha = \dfrac{dw}{dx_3}$. The displacement components u_i of a typical element in the beam, a distance x_2 from the centre line can now be deduced. Clearly $u_1 = 0$. Also

$$u_2 = w(x_3) - x_2(1 - \cos\alpha) \approx w(x_3) \qquad \text{and} \qquad u_3 \approx -x_2 \frac{dw}{dx_3}$$

Hence

$$u_i = (0, w(x_3), -x_2 \frac{dw}{dx_3})$$

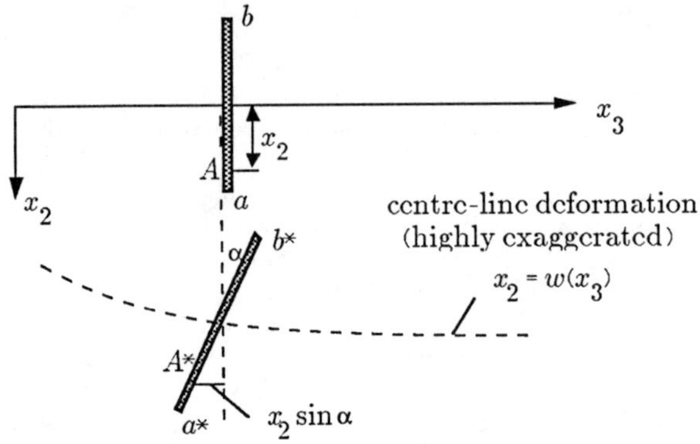

Figure 6.8

From this displacement field the corresponding 'strains' are found to be

$$\varepsilon_{33} = -x_2 \frac{d^2w}{dx_3^2} \qquad \text{other } \varepsilon_{ij} = 0$$

The strain energy for this configuration is therefore

$$\frac{1}{2} \iiint_V \varepsilon_{33}^2 (2\mu + \lambda) dV = \frac{E'}{2} \int_0^L \left[\iint_A x_2^2 (\frac{d^2w}{dx_3^2})^2 dx_1 dx_2 \right] dx_3 \tag{6.34}$$

in which $E' = 2\mu + \lambda = \dfrac{E(1-\nu)}{(1+\nu)(1-2\nu)}$. It is common to make a further approximation by replacing E' by E. Physically this implies that the Poisson ratio is small $\nu \approx 0$ or, more importantly, that the longitudinal stress σ_{33} is very much larger than σ_{11} or σ_{22} (that is, $\varepsilon_{33} = \sigma_{33}/E$ only if $\sigma_{11} = \sigma_{22} = 0$; see (5.8)). It turns out that this is a reasonable assumption for long thin beams and so we shall make this approximation also.

Now, in (6.34) define, for each plane section x_3=constant, the second moment of area I(x_3):

$$I(x_3) = \iint_A x_2^2 dx_1 dx_2 \tag{6.35}$$

(it is a function of x_3 since A is a function of x_3), then

$$\frac{1}{2} \iiint_V \sigma_{33}\varepsilon_{33} dV = \frac{1}{2} E \int_0^L I(x_3) (\frac{d^2w}{dx_3^2})^2 dx_3 \tag{6.36}$$

The work done by the transverse forces $q(x_3)$ is clearly

$$\int_0^L q(x_3)w(x_3)dx_3 \tag{6.37}$$

Depending on the boundary conditions, other terms may need to be considered in the expression for the potential energy. To be definite, consider a beam (see Figure 6.9) which is clamped at the left hand end and has, in addition to the linear force $q(x_3)$ a couple M_0 and a shearing force F_0 applied at $x_3 = L$.

The work done by F_0 is $F_0 w(L)$ whilst the work done by the moment M_0 is $-M_0 w'(L)$ (see Figure 6.9 inset) and so the potential energy of this elastic system is

$$U = \frac{1}{2}\int_0^L \left[EI(\frac{d^2w}{dx_3^2})^2 - 2qw\right]dx_3 - F_0 w(L) + M_0 w'(L) \tag{6.38}$$

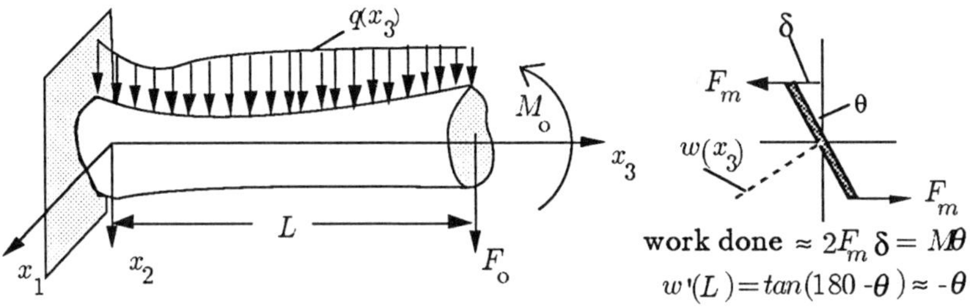

Figure 6.9

Now, given $E, q(x_3), F_0, M_0$ and L we have to find that function $w(x_3)$ satisfying only the displacement boundary conditions:

$$w(0) = 0 \qquad \frac{dw}{dx_3}(0) = 0$$

which makes the potential energy U a minimum. This is a classical variational calculus problem. The solution for $w(x_3)$ is (see Appendix 1) that function which satisfies the boundary value problem:

$$\begin{aligned} &\frac{d^2}{dx_3^2}(EI\frac{d^2w}{dx_3^2}) = q \qquad 0 < x_3 < L \\ &w(0) = 0 \qquad \frac{dw}{dx_3}(0) = 0 \\ &-EI(L)\frac{d^2w}{dx_3^2}(L) = M_0 \qquad -\frac{d}{dx_3}(EI(\frac{d^2w}{dx_3^2})(L) = F_0 \end{aligned} \tag{6.39}$$

EXERCISES

6.3 Show that the strain energy of an 'Euler-Bernoulli' beam of length L, Young's modulus E and second moment of area $I(x_3)$ is

$$W = \frac{1}{2}\int_0^L \frac{M^2}{EI} dx_3$$

where $M(x_3)$ is the moment developed at cross-section x_3.

6.4 Shown in Figure 6.10 is a 6-element pin-jointed structure. It is supporting concentrated loads at pins B and D, (in the directions shown). Use the theorem of minimum potential energy to determine the resulting deflection of pin C. All members have the same cross-sectional area A and are linear elastic with Young's modulus E.

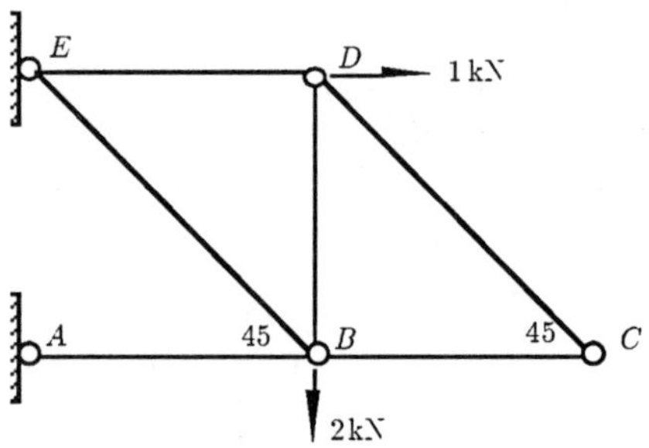

Figure 6.10

6.5 The Castigliano relations (6.22) and (6.33) can be extended to include point couples and *their* corresponding 'displacements' which are rotations, so that if M_i is a point couple acting at a point P on the surface of a rigidly supported elastic body then the rotation θ_i measured at P is related to M_i through the strain energy via:

$$\frac{\partial W}{\partial M_i} = \theta_i \quad \text{and} \quad \frac{\partial W}{\partial \theta_i} = M_i \quad i = 1, 2 \text{ or } 3$$

Using these generalised relations (with the help of a fictious couple M) determine the rotation at $x_3 = L/2$ in Exercise 6.2

6.6 (a) Using the Euler-Bernoulli equation (6.39) determine the shape of a beam, clamped at one end and free at the other which is deformed by its own weight.

(b) A light (that is, ignore gravity) rod is simply supported at both ends and deformed by a transverse force of magnitude P. Use an appropriate Euler-Bernoulli boundary value problem to determine the deflection of the rod. Perform a partial check on your answer by using the Castigliano relations (6.22), (6.23) to deduce the deflection and the rotation (the value of $\frac{dw}{dx_3}$) at the centre point.

6.7 A uniform cantilever beam of length L with its centre line along the x_3-axis is loaded by a transverse point force $-P$ at one end ($x_3 = 0$) producing a deflection curve $u_2 = -\frac{P}{6EI}(2L^3 - 3L^2x_1 + x_1^3)$. Using this result and the reciprocal theorem show that the end deflection resulting from a transverse point force $-Q$ applied distance a from the free end is $u_Q = -\frac{Q}{6EI}(2L^3 - 3L^2a + a^3)$.

6.8 A uniform rod of cross sectional area A of length ℓ deforms under its own weight. Show that if both ends are clamped that the deflection of the centre line is

$$w = \frac{\rho g A}{24EI}x_3^2(x_3 - \ell)^2$$

whereas if one end ($x_3 = \ell$) is clamped and the other simply supported that

$$w = \frac{\rho g A}{48EI}x_3(2x_3^3 - 3\ell x_3^2 + \ell^3)$$

Chapter 7

THE GENERAL TORSION PROBLEM

7.1 Introduction

The requirement to transmit power from an engine to various forms of machinery through a rotating shaft is commonplace. Often, in this dynamic transmission, the shaft is solid with a circular cross-section. In other instances it may be desired to prevent, as far as possible, a torsional rotation being transmitted from one point to another. That is, we may wish to design a cylinder (perhaps of least weight) of some cross section; not necessarily circular nor solid, which is 'stiff' with respect to torsion.

In yet other instances it may be that we are constrained geometrically to consider a cylinder with a given cross-section and wish to analyse its torsional characteristics. Such an analysis is often required in order that the magnitude of any applied torsional couple is kept strictly within certain limits.

It is only by analysing the general torsion problem that we can attempt this process of *designing for torsion.* We shall see below that the mathematical treatment of the torsion problem for uniform cylinders with arbitrarily shaped cross sections is, in almost all respects, complete. However we shall find that for cross-sections not displaying a high degree of symmetry a numerical analysis of the torsion problem may be the only practical approach to use. Nonetheless the analytical work provides many dividends – perhaps the greatest being the proof that, in the torsion problem, the shearing stress magnitude always takes its maximum value on the boundary of the cylinder. In other ways the analysis of certain analytical solutions suggests ways in which the torsion problem could be treated in an approximate manner.

The torsion of a circular cylinder has already been discussed in Section 5.3. We found that the displacement components of a cylinder, with the x_3-axis aligned along the axis of the cylinder, are:

$$u_1 = -\tau x_2 x_3 \qquad u_2 = \tau x_1 x_3 \qquad u_3 = 0 \tag{7.1}$$

and an examination of this solution deduced that plane sections orthogonal to the x_3-axis remained plane after deformation. The constant τ, called the **twist** is directly related to the angle through which each transverse plane is rotated. (See Examples 3.2, 4.2 and 4.3).

We now wish to examine the more general problem of a cylinder of arbitrary cross-sectional shape, though we shall still demand that the cylinder is uniform from one end to the other. The cross-section is not restricted to be simply connected so that cylinders with holes (parallel to the axis) may be considered. See Figure 7.1a in which there are k holes (though only three are displayed in the figure).

The cross-section D is bounded by an outer boundary C_0, described in the positive (anti-clockwise) direction, and by internal boundaries $C_1, C_2, ..., C_k$

each being described in a clockwise direction. The direction in which the boundary is described is intimately related to the direction of the outward pointing normal. The vector product of the outward pointing normal with the tangent to C_0 (in the direction of the arrow) points along the positive x_3-axis. Similarly, at an internal boundary the cross product of the outward pointing normal (pointing *away* from the material) and the tangent vector to the internal boundary also points along the positive x_3-axis.

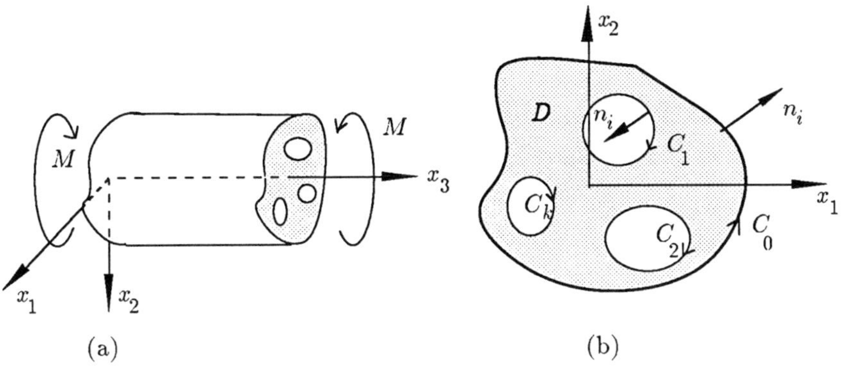

Figure 7.1

In the solution to this more general torsion problem we make an educated 'guess' as to the solution for the displacements:

$$u_1 = -\tau x_2 x_3 \qquad u_2 = \tau x_1 x_3 \qquad u_3 = \tau\phi(x_1, x_2) \tag{7.2}$$

in which $\phi(x_1, x_2)$ is a function (called the **warping** function) of the variables x_1, x_2 to be determined.

We are proposing that as a result of applying a torsional (twisting) couple at each end of the cylinder a particle on a plane x_3 =constant rotates through an angle τx_3 about the x_3-axis (which we expect) and displaces through a distance $\tau\phi(x_1, x_2)$ in the x_3-direction. To check that this is a possible solution to the torsion problem we must check that the equilibrium equations and boundary conditions of linear elasticity are satisfied.

From (7.2) the non-zero components of the strain tensor are

$$\varepsilon_{13} = \varepsilon_{31} = \frac{1}{2}\tau(\frac{\partial\phi}{\partial x_1} - x_2) \qquad \varepsilon_{23} = \varepsilon_{32} = \frac{1}{2}\tau(\frac{\partial\phi}{\partial x_2} + x_1) \tag{7.3}$$

whilst the non-zero stress components are:

$$\sigma_{13} = \sigma_{31} = \mu\tau(\frac{\partial\phi}{\partial x_1} - x_2) \qquad \sigma_{23} = \sigma_{32} = \mu\tau(\frac{\partial\phi}{\partial x_2} + x_1) \tag{7.4}$$

In the absence of body forces the first two ($i = 1, 2$) equilibrium equations (5.11) are satisfied identically. The third ($i = 3$) gives:

$$\frac{\partial}{\partial x_1}\left[\mu\tau(\frac{\partial\phi}{\partial x_1} - x_2)\right] + \frac{\partial}{\partial x_2}\left[\mu\tau(\frac{\partial\phi}{\partial x_2} + x_1)\right] = 0$$

that is,

$$\frac{\partial^2 \phi}{\partial x_1^2} + \frac{\partial^2 \phi}{\partial x_2^2} = 0 \qquad \text{within } D \tag{7.5}$$

so that ϕ is an harmonic function.

On the lateral stress-free surfaces (outer boundary C_0 and inner boundaries C_i $i = 1, 2, ...k$) the boundary condition is that $T_i^{(n)} = 0$; in which the unit normal n_i takes the form $n_i = (n_1, n_2, 0)$. This implies, from the Cauchy relation (5.10) that

$$0 = \sigma_{ij} n_j \tag{7.6}$$

Of the three constraints in (7.6) only the last ($i = 3$) is not satisfied identically:

$$0 = \sigma_{31} n_1 + \sigma_{32} n_2$$

which, on using (7.4) becomes

$$n_1(\frac{\partial \phi}{\partial x_1} - x_2) + n_2(\frac{\partial \phi}{\partial x_2} + x_1) = 0 \tag{7.7}$$

which is to be satisfied on the boundary C_i $i = 0, 1, 2, ...k$.

7.2 The Torsion Function

Before proceeding to check that the forces on the two ends of the cylinder constitute a torsional couple about the x_3-axis we first simplify the form taken by the boundary conditions (7.7). We do this in two stages. First we introduce a new function ψ as the conjugate function to ϕ satisfying:

$$\frac{\partial \phi}{\partial x_1} = \frac{\partial \psi}{\partial x_2} \qquad \frac{\partial \phi}{\partial x_2} = -\frac{\partial \psi}{\partial x_1} \tag{7.8}$$

The existence of a suitably smooth function ψ is guaranteed since the second order mixed derivatives are equal:

$$\frac{\partial^2 \psi}{\partial x_1 \partial x_2} - \frac{\partial^2 \psi}{\partial x_2 \partial x_1} = 0 \quad \text{or} \quad \frac{\partial^2 \phi}{\partial x_1^2} + \frac{\partial^2 \phi}{\partial x_2^2} = 0$$

which, due to (7.5) of course, is satisfied.

With this change of function (7.7) becomes

$$n_1(\frac{\partial \psi}{\partial x_2} - x_2) - n_2(\frac{\partial \psi}{\partial x_1} - x_1) = 0 \quad \text{on } C_i \;\; i = 0, 1, 2, ...k \tag{7.9}$$

whilst (7.5) remains the same, in form:

$$\frac{\partial^2 \psi}{\partial x_1^2} + \frac{\partial^2 \psi}{\partial x_2^2} = 0 \qquad \text{within } D \tag{7.10}$$

With the introduction of the conjugate function ψ the boundary condition (7.7) takes a more symmetrical form (7.9) without inducing any changes in the differential equation (7.10).

Now, from vector field theory, the rate of change of a scalar function G along a curve with arc-length parameter s is $\dfrac{dG}{ds} = \hat{t}.\underline{\nabla}G$ where $\hat{t}$ is a unit tangent vector to the curve. However, on a curve lying in the x_1x_2 plane with outward normal n_i then the unit tangent vector has components $\hat{t} = (-n_2, n_1, 0)$ and so

$$\frac{dG}{ds} = -n_2\frac{\partial G}{\partial x_1} + n_1\frac{\partial G}{\partial x_2} \quad \textbf{(7.11)}$$

Using (7.11) we can rewrite (7.9) in the form

$$\frac{d}{ds}\left[\psi - \frac{1}{2}(x_1^2 + x_2^2)\right] = 0 \quad \text{on } C_i \quad i = 0, 1, 2, ...k$$

Hence

$$\psi - \frac{1}{2}(x_1^2 + x_2^2) = k_i \quad \text{a constant on each } C_i \quad i = 0, 1, 2, ...k$$

This suggests we introduce yet another function Ψ, called the **torsion function**:

$$\Psi = \psi - \frac{1}{2}(x_1^2 + x_2^2) \quad \textbf{(7.12)}$$

which within the interior D satisfies, from (7.10):

$$\frac{\partial^2\Psi}{\partial x_1^2} + \frac{\partial^2\Psi}{\partial x_2^2} + 2 = 0 \quad \textbf{(7.13)}$$

and on the boundary $C_i \quad i = 0, 1, 2, ...k$

$$\Psi = k_i \quad \text{a constant} \quad \textbf{(7.14)}$$

For non-simply connected regions (cross-sections with holes) this is not the end of the story. A function Ψ determined from (7.13) and satisfying boundary conditions (7.14) need not give rise to a single valued displacement function ϕ. This is due to the fact that ϕ is related to Ψ through its conjugate ψ in the partial differential equations (7.8). The existence of a single valued solution to (7.8) needs to be considered.

Consider the cross-section shown in Figure 7.2a

If the value of ϕ is known at a point M_0 then its value at any other point M within D may be determined by adding up all the changes in ϕ as we move along a path from M_0 to M. That is,

$$\phi(M) = \phi(M_0) + \int_{M_0M} d\phi \quad \textbf{(7.15)}$$

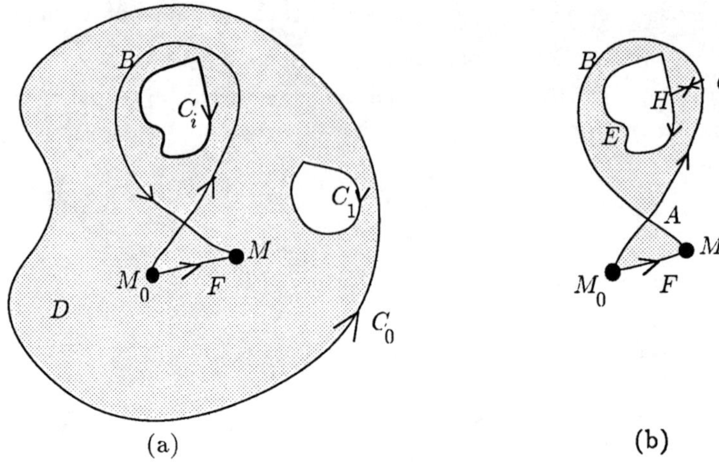

Figure 7.2

However, the path M_0M may include the hole C_i, say, which may be traversed any number of times. (See Figure 7.2b). This would make ϕ multivalued unless

$$\oint_{C_i} d\phi \equiv \oint_{C_i} \left[\frac{\partial \phi}{\partial x_1} dx_1 + \frac{\partial \phi}{\partial x_2} dx_2\right] = \oint_{C_i} \left[\frac{\partial \psi}{\partial x_2} dx_1 - \frac{\partial \psi}{\partial x_1} dx_2\right] = 0$$

for each *interior* boundary $i = 1, 2, ...k$.

To see this we note that the path M_0ABAM may be split into three paths M_0A, ABA and along AM. Now Stokes' theorem can be used to show that (use the path $ACHEHCBA$ of Figure 7.2b and realise that ψ is harmonic within the shaded region):

$$\int_{ABA} \left[\frac{\partial \psi}{\partial x_2} dx_1 - \frac{\partial \psi}{\partial x_1} dx_2\right] = -\oint_{C_i} \left[\frac{\partial \psi}{\partial x_2} dx_1 - \frac{\partial \psi}{\partial x_1} dx_2\right] \qquad \textbf{(7.16)}$$

Also, again appealing to Stokes' theorem,

$$\int_{M_0FM} \left[\frac{\partial \psi}{\partial x_2} dx_1 - \frac{\partial \psi}{\partial x_1} dx_2\right] = \int_{M_0AM} \left[\frac{\partial \psi}{\partial x_2} dx_1 - \frac{\partial \psi}{\partial x_1} dx_2\right] \qquad \textbf{(7.17)}$$

since ψ is harmonic within the region bounded by M_0FMAM_0. Hence ϕ will be single valued only if the contribution to the line integral along a path such as ABA is zero. This will be the case if

$$\oint_{C_i} \left[\frac{\partial \psi}{\partial x_2} dx_1 - \frac{\partial \psi}{\partial x_1} dx_2\right] = 0 \quad i = 1, 2, ...k \qquad \textbf{(7.18)}$$

Now in terms of the torsion function Ψ, (7.18) becomes

$$\oint_{C_i} \left[\frac{\partial \Psi}{\partial x_2} dx_1 - \frac{\partial \Psi}{\partial x_1} dx_2\right] + \oint_{C_i} (x_2 dx_1 - x_1 dx_2) = 0 \quad i = 1, 2, ...k \qquad \textbf{(7.19)}$$

and, on using Stokes' theorem on the second of these integrals (remembering that, in these line integrals $C_i \quad i = 1, 2, ...k$ is traversed in a negative direction)

$$\oint_{C_i} \left[\frac{\partial \Psi}{\partial x_2} dx_1 - \frac{\partial \Psi}{\partial x_1} dx_2 \right] + 2 \iint_{R_i} dx_1 dx_2 = 0 \quad i = 1, 2, ...k$$

in which R_i is the region (with area A_i) enclosed by C_i. that is,

$$\oint_{C_i} \left[\frac{\partial \Psi}{\partial x_2} dx_1 - \frac{\partial \Psi}{\partial x_1} dx_2 \right] + 2A_i = 0 \quad i = 1, 2, ...k \tag{7.20}$$

In the interests of clarity we now pull together what we have discovered about the torsion function:
The function Ψ satisfies the boundary value problem:

$$\frac{\partial^2 \Psi}{\partial x_1^2} + \frac{\partial^2 \Psi}{\partial x_2^2} + 2 = 0 \qquad \text{within } D \tag{7.21}$$

in which, on the boundary $C_i \quad i = 1, 2, ...k$

$$\Psi = k_i \quad \text{a constant} \tag{7.22}$$

However, these constants are not arbitrary. If the displacement u_3 is to be single valued then the constraints:

$$\oint_{C_i} \left[\frac{\partial \Psi}{\partial x_2} dx_1 - \frac{\partial \Psi}{\partial x_1} dx_2 \right] + 2A_i = 0 \quad i = 1, 2, ...k \tag{7.23}$$

must also be satisfied. There are only k equations here and so one of the constants can be chosen arbitrarily. We normally choose $k_0 = 0$.

(An alternative form of (7.23) can be obtained by realising that the rate of change of Ψ in a direction normal to the contour C_i that is, $\dfrac{d\Psi}{dn}$ is given by

$$\frac{d\Psi}{dn} = n_i \Psi_{,i} \quad (\equiv \hat{n}.\underline{\nabla}\Psi)$$

But if the unit tangent vector is $(\dfrac{dx_1}{ds}, \dfrac{dx_2}{ds}, 0)$ in which s is the measure of arc-length on the boundary then $n_i = (-\dfrac{dx_2}{ds}, \dfrac{dx_1}{ds}, 0)$ and so

$$\frac{d\Psi}{dn} ds \equiv -\frac{\partial \Psi}{\partial x_1} dx_2 + \frac{\partial \Psi}{\partial x_2} dx_1$$

Hence (7.23) can be written:

$$\oint_{C_i} \frac{d\Psi}{dn} ds + 2A_i = 0 \quad i = 1, 2, ...k \tag{7.23a}$$

This form will be of some use in Section 7.5).

In terms of the stress function the non-zero strain and stress components are, respectively

$$\varepsilon_{13} = \varepsilon_{31} = \frac{1}{2} \tau \frac{\partial \Psi}{\partial x_2} \qquad \varepsilon_{23} = \varepsilon_{32} = -\frac{1}{2} \tau \frac{\partial \Psi}{\partial x_1} \tag{7.24}$$

$$\sigma_{13} = \sigma_{31} = \mu\tau\frac{\partial\Psi}{\partial x_2} \qquad \sigma_{23} = \sigma_{32} = -\mu\tau\frac{\partial\Psi}{\partial x_1} \tag{7.25}$$

We are now in a better position to check that the forces on each end of the cylinder are equivalent to a torsional couple about the x_3-axis.

Consider the right hand end of the cylinder with unit normal $m_i = (0,0,1)$. The resultant force on this end is

$$F_i = \iint_D T_i^{(m)} dS = \iint_D \sigma_{i3} dx_1 dx_2$$

Hence, from (7.25)

$$F_1 = \mu\tau \iint_D \frac{\partial\Psi}{\partial x_2} dx_1 dx_2 \qquad F_2 = -\mu\tau \iint_D \frac{\partial\Psi}{\partial x_1} dx_1 dx_2 \tag{7.26}$$

But, in the first of (7.26)

$$\iint_D \frac{\partial\Psi}{\partial x_2} dx_1 dx_2 = \iint_{R_0} \frac{\partial\Psi}{\partial x_2} dx_1 dx_2 - \ldots\ldots - \iint_{R_k} \frac{\partial\Psi}{\partial x_2} dx_1 dx_2 \tag{7.27}$$

However, if we integrate by parts (see Figure 7.3), and use the fact, from (7.22), that Ψ is constant on each boundary we find

$$\iint_{R_i} \frac{\partial\Psi}{\partial x_2} dx_1 dx_2 = \int_a^b \left\{ \left[\Psi \right]_{f_-(x_1)}^{f_+(x_1)} \right\} dx_1 = 0$$

Hence $F_1 = 0$. In an exactly similar manner $F_2 = 0$ also.

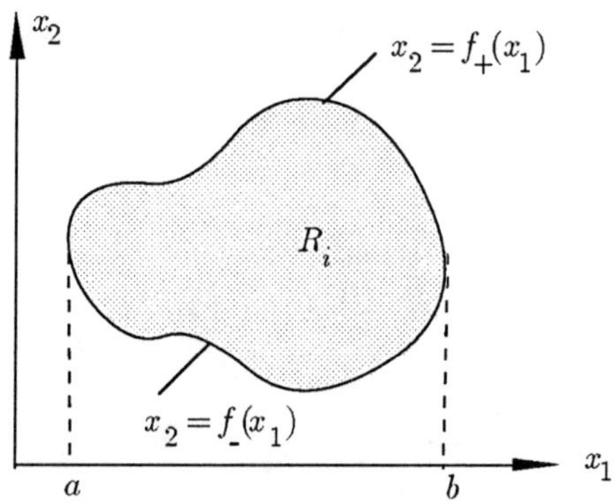

Figure 7.3

The resultant force on the end of the cylinder is zero and so the resultant couple has the same value about any point. We choose to the determine the couple at the point which is the intersection of the x_1- and x_2-axes on the right hand end of the cylinder. The resultant couple is (see Section 5.3 on beam bending):

$$M_i = \iint_D \eta_{ijk}\overline{x}_j \sigma_{k\alpha} m_\alpha dx_1 dx_2 \qquad \overline{x}_j = (x_1, x_2, 0)$$

The only non-zero component is M_3:

$$M \equiv M_3 = \iint_D (x_1\sigma_{23} - x_2\sigma_{13})dx_1 dx_2$$
$$= -\mu\tau \iint_D \left[x_1 \frac{\partial \Psi}{\partial x_1} + x_2 \frac{\partial \Psi}{\partial x_2} \right] dx_1 dx_2$$

Thus, as expected, the resultant couple is a torsional couple parallel to the axis of the cylinder of magnitude τT_R where

$$T_R = \left| \mu \iint_D \left[x_1 \frac{\partial \Psi}{\partial x_1} + x_2 \frac{\partial \Psi}{\partial x_2} \right] dx_1 dx_2 \right| \qquad \textbf{(7.28)}$$

(taking the modulus to ensure T_R is always positive)

T_R is called the **torsional rigidity** of the cylinder as it measures the couple required to produce unit twist. This confirms that the expression for the displacement field proposed in (7.2) is indeed the solution to the torsion problem of a uniform cylinder with cross-section D twisted by torsional couples applied at either end.

It will prove convenient to rewrite (7.28). Clearly

$$T_R = \left| \mu \iint_D \left[\frac{\partial}{\partial x_1}(x_1 \Psi) + \frac{\partial}{\partial x_2}(x_2 \Psi) - 2\Psi \right] dx_1 dx_1 \right|$$
$$= \left| -2\mu \iint_D \Psi dx_1 dx_2 + \mu \sum_{i=0}^{k} \oint_{C_i} \Psi(-x_2 dx_1 + x_1 dx_2) \right|$$

in which we have again employed Stokes' theorem and also taken account of the specified directions on C_i. But, from (7.14) $\Psi = k_i$ on C_i and so

$$T_R = \left| -2\mu \iint_D \Psi dx_1 dx_2 + \mu \sum_{i=0}^{k} k_i \oint_{C_i} (-x_2 dx_1 + x_1 dx_2) \right|$$
$$= \left| -2\mu \iint_D \Psi dx_1 dx_2 - 2\mu \sum_{i=1}^{k} k_i A_i \right| \qquad \textbf{(7.29)}$$

where we have used the choice $k_0 = 0$.

Although the detailed solution for the stress, strain and displacement components are of importance in the torsion problem to a stress engineer it is a knowledge of the torsional rigidity, for a given cross-section, that is of greatest interest. It will prove useful to realise that, for simply connected cross-sections, the torsional rigidity is directly proportional to the volume under the Ψ surface, a conclusion which follows immediately from the standard interpretation of a double integral.

7.3 Shearing Stress in the Torsion Problem

At any point on a plane x_3=constant the stress vector $T_i^{(m)}$ takes the form (from (7.25)):

$$T_i^{(m)} = (\sigma_{13}, \sigma_{23}, 0) = \mu\tau(\frac{\partial\Psi}{\partial x_2}, -\frac{\partial\Psi}{\partial x_1}, 0)$$

This vector is inclined to the positive x_1-axis at an angle θ:

$$\theta = \tan^{-1}\left[-\frac{\partial\Psi/\partial x_1}{\partial\Psi/\partial x_2}\right] \tag{7.30}$$

However, from (7.9) we also have

$$n_1\frac{\partial\Psi}{\partial x_2} - n_2\frac{\partial\Psi}{\partial x_1} = 0 \quad \text{on } C_i \;\; i = 0, 1, 2, ...k$$

which indicates that the stress vector is perpendicular to the outward pointing normal $n_i = (n_1, n_2, 0)$. Hence the shearing stress is tangential to each curve forming the boundary. (This could have been anticipated since the boundary is stress free, implying that the normal component of the stress vector is zero).

The magnitude of the shearing stress is, from (7.4) or alternatively from (7.25):

$$\mu\tau\sqrt{\left[\frac{\partial\phi}{\partial x_1} - x_2\right]^2 + \left[\frac{\partial\phi}{\partial x_2} + x_1\right]^2} = \mu\tau\sqrt{\left(\frac{\partial\Psi}{\partial x_2}\right)^2 + \left(\frac{\partial\Psi}{\partial x_1}\right)^2} \tag{7.31}$$

An important result in the torsion problem is that the shearing stress magnitude assumes its maximum value somewhere on the boundary. This result is based on the following property of harmonic functions:

An harmonic function ϕ in a region D cannot have an extremum (maximum or minimum) within D.

The surprisingly easy proof is achieved by assuming the contrary, namely that the function has such an extremum within D. Suppose $\phi(x_1, x_2)$ has a maximum at a point P within D. Then it is a maximum along any line passing through P, in particular along lines parallel to the x_1- and x_2-axes. For a maximum this requires that

$$\frac{\partial^2\phi}{\partial x_1^2} < 0 \quad \text{and} \quad \frac{\partial^2\phi}{\partial x_2^2} < 0 \quad \text{at } P$$

Hence

$$\frac{\partial^2\phi}{\partial x_1^2} + \frac{\partial^2\phi}{\partial x_2^2} < 0 \quad \text{at } P$$

which contradicts the hypothesis that ϕ is an harmonic function. A similar approach shows that ϕ cannot have a minimum anywhere in D. If D is bounded by a curve C then the maximum or minimum must occur somewhere on C.

We can apply this result to the torsion problem. To show that the shearing stress magnitude cannot assume a maximum value within D we assume the contrary. That is, there exists a point P in the interior of the cylinder at which the resulting shearing stress is a maximum.

We first orientate the axes so that the x_1'-axis is parallel to the direction of the resultant shearing stress at P. (That is, rotate axes through angle θ of (7.30)). Then, with respect to this new system of coordinates, the only non-zero stress component (see (4.16)) is σ_{13}' with magnitude $\mu\tau|\dfrac{\partial\phi}{\partial x_1'} - x_2'|$. Now

$$\frac{\partial^2}{\partial(x_1')^2}\left[\frac{\partial\phi}{\partial x_1'} - x_2'\right] + \frac{\partial^2}{\partial(x_2')^2}\left[\frac{\partial\phi}{\partial x_1'} - x_2'\right] = \frac{\partial}{\partial x_1'}\left[\frac{\partial^2\phi}{\partial(x_1')^2} + \frac{\partial^2\phi}{\partial(x_2')^2}\right]$$

showing that if ϕ is harmonic then so is $\left[\dfrac{\partial\phi}{\partial x_1'} - x_2'\right]$. (If ϕ is harmonic in an unprimed system then since $\phi,_{ii} = 0$, is a tensor equation then ϕ is harmonic in all coordinate systems obtained from the unprimed system by rotation). It follows that $\left[\dfrac{\partial\phi}{\partial x_1'} - x_2'\right]$ is neither a maximum nor a minimum at P and so there exists a point Q, in the close neighbourhood of P at which

$$\left|\frac{\partial\phi}{\partial x_1'} - x_2'\right|_Q \geq \left|\frac{\partial\phi}{\partial x_1'} - x_2'\right|_P$$

The resultant shearing stress magnitude at Q, denoted by R_Q say, is

$$\begin{aligned} R_Q &= \mu\tau\sqrt{\left[\frac{\partial\phi}{\partial x_1'} - x_2'\right]_Q^2 + \left[\frac{\partial\phi}{\partial x_2'} + x_1'\right]_Q^2} \\ &\geq \mu\tau\left|\frac{\partial\phi}{\partial x_1'} - x_2'\right|_Q \\ &\geq \mu\tau\left|\frac{\partial\phi}{\partial x_1'} - x_2'\right|_P = R_P \end{aligned}$$

Hence $R_Q \geq R_P$ which contradicts the assumption that the resultant shearing stress is a maximum at P. It follows that if P is a point at which the resultant shearing stress is a maximum it cannot be in the interior – it must be somewhere on the boundary.

The result that the shearing stress is a maximum on the boundary has a practical application. Plasticity may occur at any point where the resultant shearing stress surpasses a certain critical value known as the yield stress. If the stress engineer does not wish plasticity to occur but wishes to apply the greatest possible torsional couple within this limit, s/he must determine the maximum resultant shearing stress magnitude. If the maximum stress is $\alpha\tau$ where τ is the twist then the applied couple is $T_R\tau$ where T_R is the torsional rigidity of the cross-section. If the yield stress is Y, then $\alpha\tau$ must not exceed Y. Hence the torsional couple $(T_R\tau = \dfrac{T_R}{\alpha}\alpha\tau)$ must be less than $T_R\dfrac{Y}{\alpha}$ for the cylinder to remain elastic.

7.4 Simple Exact Solutions in the Torsion Problem

- **Cylinder of elliptic cross section**

If the x_3-axis is chosen along the axis of the cylinder and the x_1- and x_2-axes oriented to coincide with the major and minor axes of the elliptic cross section then the equation of the boundary of the cylinder is (see Figure 7.4):

$$\frac{x_1^2}{a^2} + \frac{x_2^2}{b^2} = 1 \qquad b < a$$

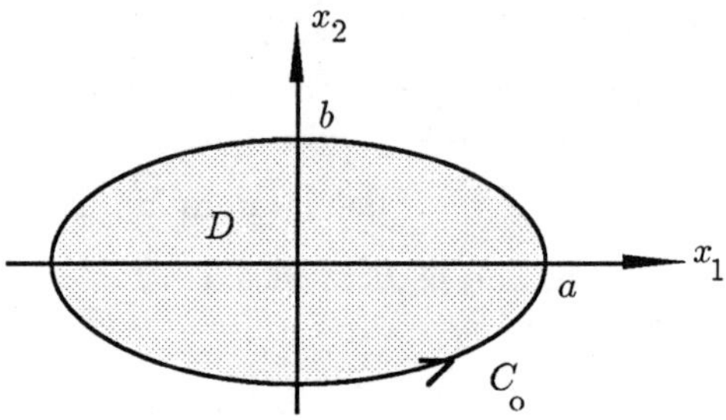

Figure 7.4

Since the cross-section is simply connected (no holes) the torsion problem ((7.21), (7.22) and (7.23)) reduces to (the compatibility conditions (7.23) are redundant)

$$\frac{\partial^2 \Psi}{\partial x_1^2} + \frac{\partial^2 \Psi}{\partial x_2^2} + 2 = 0 \quad \text{within } D \tag{7.32}$$

and

$$\Psi = 0 \quad \text{on the boundary } C_0 \tag{7.33}$$

As a solution to this problem we shall propose to take

$$\Psi = K\left[\frac{x_1^2}{a^2} + \frac{x_2^2}{b^2} - 1\right] \tag{7.34}$$

where K is a constant. This obviously satisfies the boundary condition (7.33). Now, with this choice,

$$\frac{\partial^2 \Psi}{\partial x_1^2} + \frac{\partial^2 \Psi}{\partial x_2^2} = K\left[\frac{2}{a^2} + \frac{2}{b^2}\right]$$

and so (7.32) implies $K = -\dfrac{a^2b^2}{a^2+b^2}$. Hence the solution to the problem is

$$\Psi = -\frac{a^2b^2}{a^2+b^2}\left[\frac{x_1^2}{a^2} + \frac{x_2^2}{b^2} - 1\right] \tag{7.35}$$

therefore

$$\begin{aligned}\psi &= \Psi + \frac{1}{2}(x_1^2 + x_2^2) \\ &= \frac{1}{2}\left[\frac{a^2 - b^2}{a^2 + b^2}\right](x_1^2 - x_2^2) + \frac{a^2 b^2}{a^2 + b^2}\end{aligned}$$

Now the warping function ϕ can be determined from (7.8):

$$\frac{\partial \phi}{\partial x_1} = -\left[\frac{a^2 - b^2}{a^2 + b^2}\right] x_2 \qquad \frac{\partial \phi}{\partial x_2} = -\left[\frac{a^2 - b^2}{a^2 + b^2}\right] x_1 \tag{7.36}$$

We find, easily,

$$\phi = -\left[\frac{a^2 - b^2}{a^2 + b^2}\right] x_1 x_2 \tag{7.37}$$

The (possible) additive constant in ϕ is taken as zero to eliminate rigid body translations. This solution, (7.37), shows that particles on the coordinate axes suffer no warping as a result of the application of the torsional couples and (obviously) the further away a particle is from the coordinate axes the greater will be the displacement. Particles in the first and third quadrants ($x_1 x_2 > 0$) are displaced in the negative x_3-direction whilst particles in the second and fourth quadrants ($x_1 x_2 < 0$) are displaced in the positive x_3-direction. See Figure 7.5.

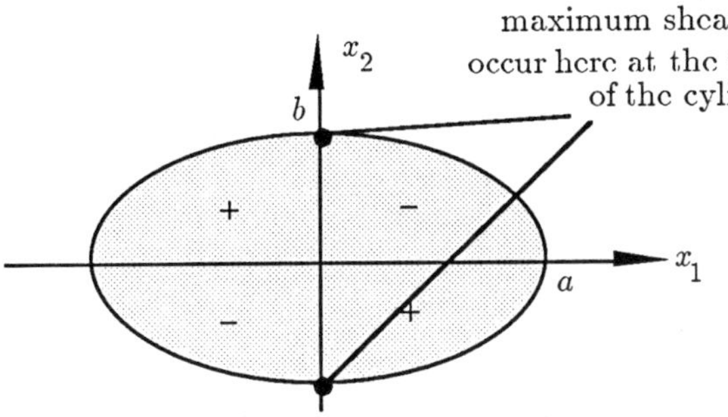

Figure 7.5

From the previous section we know that the maximum shearing stress is somewhere on the boundary. This position is easily located. The shearing stress magnitude is:

$$\begin{aligned}\mu\tau\sqrt{\left(\frac{\partial \Psi}{\partial x_1}\right)^2 + \left(\frac{\partial \Psi}{\partial x_2}\right)^2} &= \frac{2a^2 b^2}{a^2 + b^2}\mu\tau\sqrt{\frac{x_1^2}{a^4} + \frac{x_2^2}{b^4}} \\ &= \frac{2a^2 b^2}{a^2 + b^2}\mu\tau\sqrt{\frac{1}{a^2} + x_2^2\left[\frac{-b^2 + a^2}{a^2 b^4}\right]} \quad \text{on the boundary}\end{aligned}$$

But, since $b < a$ this expression takes its greatest value when x_2 takes its greatest boundary value; that is, when $x_2 = \pm b$ (see Figure 7.5). The greatest shearing stress has magnitude $\dfrac{2a^2b}{a^2+b^2}\mu\tau$.

The reader should confirm that the results obtained here for the elliptic cross-section are entirely consistent with those obtained for the circular cross-section $(a = b)$.

Finally we calculate the torsional rigidity of the elliptic cross-section. We use (7.29) with $k_i = 0 \quad i = 1, 2, \ldots k$.

$$T_R = \left| 2\mu \iint_D \Psi dx_1 dx_2 \right|$$

$$= \left| \frac{2\mu a^2 b^2}{a^2+b^2} \iint_D \left[\frac{x_1^2}{a^2} + \frac{x_2^2}{b^2} - 1 \right] dx_1 dx_2 \right|$$

To evaluate this double integral it is simplest to transform to elliptical polar coordinates:

$$x_1 = ar\cos\theta \qquad x_2 = br\sin\theta$$

The Jacobian of the transformation is easily found to be abr and so

$$T_R = \left| \frac{2\mu a^2 b^2}{a^2+b^2} \iint_{D_{r\theta}} (r^2-1) abr\,dr\,d\theta \right|$$

where $D_{r\theta}$ is the region D expressed in terms of elliptical polars. A straightforward evaluation of this double integral now gives:

$$T_R = \mu \frac{a^3b^3}{a^2+b^2}\pi \quad (= \frac{\mu a^4 \pi}{2} \quad \text{when } a = b)$$

It is seen that the torsional rigidity of an elliptical cylinder is always less than that of a circular cylinder with radius equal to the half major length. In fact we can make a stronger statement than this: The circular shape is stronger, in torsion, than the elliptical shape. Consider an ellipse with cross-sectional area A then $A = \pi ab$ with torsional rigidity

$$T_R^{(e)} = \mu \frac{a^3b^3\pi}{a^2+b^2} = \mu \frac{A^3b^2}{\pi^2b^4 + A^2}$$

where we have eliminated the parameter a. For a circle with the *same* cross-sectional area $A = \pi a^2$ then its rigidity is $T_R^{(c)} = \mu A^2/2\pi$. It is a straightforward exercise to show that $T_R^{(c)} \geq T_R^{(e)}$ equality only occurring if $a = b$.

- **Hollow Elliptical Cylinder**

The geometrical configuration is described in Figure 7.6. We shall assume that the boundary of D is comprised of two confocal ellipses:

$$C_0 : \quad \frac{x_1^2}{a^2} + \frac{x_2^2}{b^2} = 1 \qquad \text{and} \qquad C_1 : \quad \frac{x_1^2}{a^2} + \frac{x_2^2}{b^2} = m^2$$

where $b < a$ and $m < 1$.

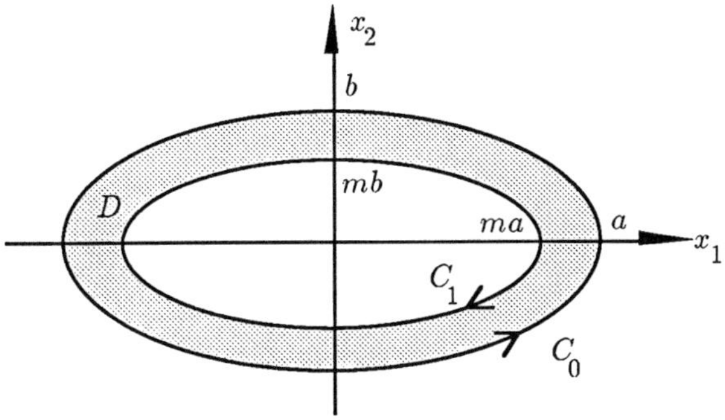

Figure 7.6

The cross-section is not simply connected, having one hole, and so the torsion problem (7.21), (7.22) and (7.23) reduces to

$$\frac{\partial^2 \Psi}{\partial x_1^2} + \frac{\partial^2 \Psi}{\partial x_2^2} + 2 = 0 \quad \text{within } D \tag{7.38}$$

$$\Psi = 0 \quad \text{on } C_0 \qquad \Psi = k_1 \quad \text{on } C_1 \tag{7.39}$$

with k_1 chosen to satisfy

$$\oint_{C_1} \left[\frac{\partial \Psi}{\partial x_2} dx_1 - \frac{\partial \Psi}{\partial x_1} dx_2 \right] + 2A_1 = 0 \tag{7.40}$$

We can attempt a solution similar to that proposed for the previous problem:

$$\Psi = \frac{k_1}{m^2 - 1} \left[\frac{x_1^2}{a^2} + \frac{x_2^2}{b^2} - 1 \right] \tag{7.41}$$

This certainly satisfies the boundary conditions (7.39). Substituting into (7.38) we find

$$k_1 = -(m^2 - 1) \frac{a^2 b^2}{a^2 + b^2} \tag{7.42}$$

We need to check that (7.40) is satisfied. Now, from (7.41)

$$\oint_{C_1} \left[\frac{\partial \Psi}{\partial x_2} dx_1 - \frac{\partial \Psi}{\partial x_1} dx_2 \right] = \frac{k_1}{m^2 - 1} \oint_{C_1} \left[\frac{2x_2}{b^2} dx_1 - \frac{2x_1}{a^2} dx_2 \right]$$

Applying Stokes' theorem and realising that C_1 is traversed in the negative (clockwise) direction we have:

$$\begin{aligned}
\oint_{C_1} \left[\frac{\partial \Psi}{\partial x_2} dx_1 - \frac{\partial \Psi}{\partial x_1} dx_2 \right] &= \frac{k_1}{m^2 - 1} \left[\frac{2}{b^2} + \frac{2}{a^2} \right] \iint_{R_1} dx_1 dx_2 \\
&= \frac{k_1}{m^2 - 1} \left[\frac{2}{b^2} + \frac{2}{a^2} \right] A_1 \\
&= -2A_1
\end{aligned}$$

Hence (somewhat fortituously) (7.40) is satisfied.

The solution now progresses as in the first example. We find, easily,

$$\phi = -\left[\frac{a^2 - b^2}{a^2 + b^2}\right] x_1 x_2$$

so that the solution for ϕ and Ψ are identical to that for a solid ellipse. The shearing stress:

$$\frac{2\mu\tau a^2 b^2}{a^2 + b^2}\sqrt{\frac{x_1^2}{a^4} + \frac{x_2^2}{b^4}}$$

has a value

$$\frac{2\mu\tau a^2 b^2}{a^2 + b^2}\sqrt{\frac{1}{a^2} + x_2^2\left[\frac{-b^2 + a^2}{a^2 b^4}\right]}$$

on the outer boundary C_0 and a value

$$\frac{2\mu\tau a^2 b^2}{a^2 + b^2}\sqrt{\frac{m^2}{a^2} + x_2^2\left[\frac{-b^2 + a^2}{a^2 b^4}\right]}$$

on the inner boundary C_1. Since $m < 1$ the maximum value of the shearing stress *always* occurs on the outer boundary at $x_2 = \pm b$ (where the cylinder is thinnest) and the minimum value of the shearing stress always occurs on the inner boundary at $x_2 = 0$ (where the cylinder is thickest).

The torsional rigidity T_R for the hollow elliptic cylinder is (see previous calculation and (7.29) with $k = 1$),

$$\begin{aligned} T_R &= \left|\frac{2\mu\tau a^2 b^2}{a^2 + b^2}\int_m^1\int_0^{2\pi}(r^2 - 1)abr\,d\theta dr - 2\mu k_1 A_1\right| \\ &= \frac{\mu\pi a^3 b^3}{a^2 + b^2}(1 - m^4) \end{aligned}$$

Hence for an elliptic cylinder of *fixed dimensions* the presence of the hole always *weakens* the cylinder the torsion.

• Solid Rectangular Cylinder

We consider a cylinder with dimensions defined in Figure 7.7. The torsion problem is the boundary value problem listed in (7.32) and (7.33).

Unlike the last two examples it is not possible to simply 'guess' what form the solution is likely to take. A more methodical approach is required. We will adopt a Fourier-series technique and assume that the torsion function Ψ may be expanded in a Fourier-cosine series:

$$\Psi = \sum_r \overline{X}_r(x_1)\cos r x_2 \qquad \textbf{(7.43)}$$

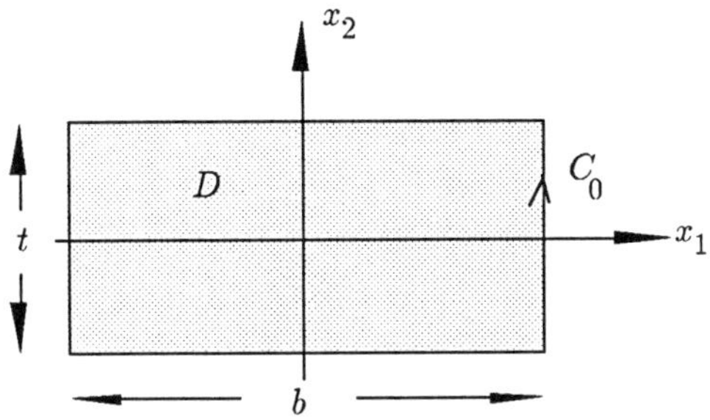

Figure 7.7

A cosine series is used since, due to the symmetry in the problem we expect Ψ to be an even function of x_2. Also Ψ is symmetric in the x_1-variable and so we demand that $\overline{X}_r(x_1)$ is an even function of x_1.

If Ψ is to vanish at $x_2 = \pm\frac{t}{2}$ for all x_1 such that $-\frac{b}{2} < x_1 < \frac{b}{2}$ (as required in (7.33)) then

$$\cos\frac{rt}{2} = 0$$

implying that $r = \frac{2}{t}(2m+1)\frac{\pi}{2}$ $\quad m = 0, 1, 2, ...$ and so (7.43) can be written:

$$\Psi = \sum_{m=0}^{\infty} \overline{X}_m(x_1)\cos(2m+1)\frac{\pi x_2}{t}$$

or, more conveniently,

$$\Psi = \sum_{\substack{n=1 \\ n \text{ odd}}}^{\infty} X_n(x_1)\cos\frac{n\pi x_2}{t} \tag{7.44}$$

where $\overline{X}_m(x_1) = X_{2m+1}(x_1)$.

Substituting (7.44) into the torsion equation (7.32) implies

$$\sum_{\substack{n=1 \\ n \text{ odd}}}^{\infty} \left[-(\frac{n\pi}{t})^2 X_n(x_1) + X_n''(x_1) \right] \cos\frac{n\pi x_2}{t} + 2 = 0 \tag{7.45}$$

We can readily extract the coefficient of the cosine terms in (7.45) by multiplying through by $\cos\frac{p\pi x_2}{t}$ and integrating between $\pm t/2$ to give:

$$\sum_{\substack{n=1 \\ n \text{ odd}}}^{\infty} \left[-(\frac{n\pi}{t})^2 X_n(x_1) + X_n''(x_1) \right] \int_{-t/2}^{t/2} \cos\frac{n\pi x_2}{t} \cos\frac{p\pi x_2}{t} dx_2$$
$$+ 2\int_{-t/2}^{t/2} \cos\frac{p\pi x_2}{t} dx_2 = 0 \tag{7.46}$$

But, using the identity (easily proven) when n and p are positive integers:

$$\int_{-t/2}^{t/2} \cos\frac{n\pi x_2}{t}\cos\frac{p\pi x_2}{t}dx_2 = \begin{cases} 0 & \text{if } n \neq p \\ t/2 & \text{if } n = p \end{cases}$$

gives immediately from (7.46):

$$\frac{t}{2}\left[-(\frac{p\pi}{t})^2 X_p(x_1) + X_p''(x_1)\right] + \frac{4t}{p\pi}(-1)^{(1-p)/2} = 0$$

so that $X_p(x_1)$ satisfies the ordinary differential equation:

$$X_p''(x_1) - (\frac{p\pi}{t})^2 X_p(x_1) + \frac{8}{p\pi}(-1)^{(1-p)/2} = 0 \qquad \textbf{(7.47)}$$

with solution

$$X_p(x_1) = A_p \cosh\frac{p\pi}{t}x_1 + (\frac{t}{p\pi})^2(-1)^{(1-p)/2}(\frac{8}{p\pi})$$

where A_p is a constant. (We have ignored the 'sinh' term since X_p is even). Now imposing the remaining boundary condition that $X_p(b/2) = 0$ we deduce

$$A_p = -\frac{8t^2}{(p\pi)^3}\frac{(-1)^{(1-p)/2}}{\cosh\frac{p\pi b}{2t}}$$

so that finally,

$$\Psi = \sum_{\substack{n=1 \\ n \text{ odd}}}^{\infty} \frac{8t^2(-1)^{(1-n)/2}}{(n\pi)^3}\left[1 - \frac{\cosh\frac{n\pi x_1}{t}}{\cosh\frac{n\pi b}{2t}}\right]\cos\frac{n\pi x_2}{t} \qquad \textbf{(7.48)}$$

This solution can be expressed in an alternative form which we shall find useful in the next section. This is obtained by noting that the Fourier-cosine series expansion of $(-x_2^2 + \frac{t^2}{4})$ over the range $(-t, t)$ is:

$$-x_2^2 + \frac{t^2}{4} = \sum_{\substack{n=1 \\ n \text{ odd}}}^{\infty} \frac{8t^2}{(n\pi)^3}(-1)^{(1-n)/2}\cos\frac{n\pi x_2}{t} \qquad \textbf{(7.49)}$$

so that (7.48) can be written:

$$\Psi = -x_2^2 + \frac{t^2}{4} - \sum_{\substack{n=1 \\ n \text{ odd}}}^{\infty} \frac{8t^2}{(n\pi)^3}(-1)^{(1-n)/2}\frac{\cosh\frac{n\pi x_1}{t}}{\cosh\frac{n\pi b}{2t}}\cos\frac{n\pi x_2}{t} \qquad \textbf{(7.50)}$$

We leave it as an exercise for the reader to verify that the maximum shear stress magnitude occurs at $x_1 = 0,\ x_2 = \pm t/2$ with value

$$\sigma_{\max} = \mu\tau\left[t - \sum_{\substack{n=1 \\ n \text{ odd}}}^{\infty} \frac{8t}{(n\pi)^2}\text{sech}\frac{n\pi b}{2t}\right] \qquad \textbf{(7.50a)}$$

The torsional rigidity can be determined in the usual way using (7.50) (from (7.29) with $k_i = 0 \;\; i = 1, 2, \ldots k$). We find

$$T_R = \frac{\mu t^3 b}{3} - \frac{64\mu}{\pi^5} \sum_{\substack{n=1 \\ n \text{ odd}}}^{\infty} \frac{t^4}{n^5} \tanh \frac{n\pi b}{2t} \qquad \textbf{(7.51)}$$

The infinite series in (7.51) converges quickly due to the fifth power of n in the denominator and so only the first few terms of the series need be taken when calculating T_R.

- **Hollow Rectangular Cylinder**

The geometrical configuration is described in Figure 7.8

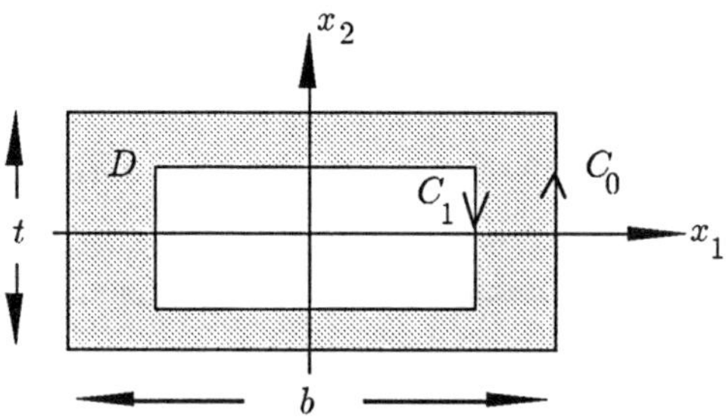

Figure 7.8

In this case the torsion problem is the determination of the torsion function satisfying (7.38), (7.39) and (7.40). Unfortunately the solution cannot be written down in 'closed form'. As shown by B L Abramyam in 1950 (Prikladnayn Mathematika i Mekhanika, vol 14, no.3) the solution depends upon being able to solve an *infinite* system of linear equations. This difficulty can only be surmounted by treating the problem numerically.

The analysis of Abramyam serves to remind us all that the equations of infinitessimal elasticity, though linear, can only be solved exactly in special cases in which the underlying geometry of the problem has a high degree of symmetry.

In the next section we will consider a more practical approach to the analysis of the torsion problem in which the exact analysis is relegated in favour of a numerical treatment.

EXERCISES

7.1 (a) A compound shaft is formed from two cylinders of differing materials welded together, with lengths and radii ℓ_1, r_1 and ℓ_2, r_2 respectively (see Figure 7.9). The shaft is held fixed at either end and a torsional couple T applied at the weld

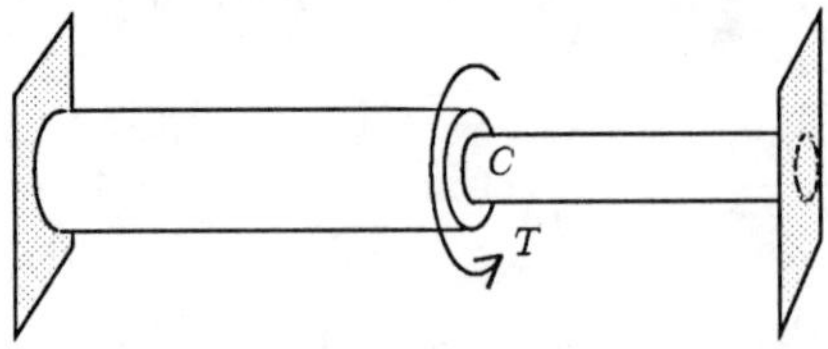

Figure 7.9

Show that the angle of twist at the point C is $\dfrac{2}{\pi}\dfrac{\ell_1\ell_2 T}{(\mu_1\ell_2 r_1^4+\mu_2\ell_1 r_2^4)}$.
(b) if $\mu_1 = 7.7\times 10^{10}$ N m^{-2} (steel) with a maximum permissible shear stress magnitude of 140 MN m^{-2} and if $\mu_2 = 2.6\times 10^{10}$ N m^{-2} (aluminium) with a maximum permissible shear stress magnitude of 90 MN m^{-2} determine the maximum value of T that is allowed consistent with these restrictions. (Use $\ell_1=\ell_2=1$m, $r_1=0.04$m and $r_2=0.05$m). How would your conclusions change if there was a hole of diameter 0.06m running through the shaft?

7.2 (a) Determine the position of the maximum shear stress magnitude on a rectangular beam of breadth b and thickness t if $b>t$ (see Figure 7.7).
(b) A cylinder with generators parallel to the x_3-axis and cross-section bounded by $x_1^2+x_2^2-6ax_1=0$ has a fillet cut out so as to exclude from the cross-section the area contained within the circle $x_1^2+x_2^2=a^2$. Verify that the torsional stress system in the cylinder may be expressed in terms of the function:

$$\Psi = \frac{1}{2}(1-\frac{a^2}{x_1^2+x_2^2})(6ax_1 - x_1^1 - x_2^2)$$

(hint: $f\equiv \dfrac{\cos\theta}{r}$ satisfies $\nabla^2 f = 0$).

7.3 The torsion function Ψ satisfies $\nabla^2\Psi+2=0$ and $\Psi=0$ on the boundary of a simply connected region. ∇^2 is the two dimensional Laplacian operator, which in plane polars takes the form

$$\nabla^2 \equiv \frac{\partial^2}{\partial r^2}+\frac{1}{r}\frac{\partial}{\partial r}+\frac{1}{r^2}\frac{\partial^2}{\partial\theta^2}$$

If $\Psi = Ar\cos\theta + \dfrac{B\cos\theta}{r} - \dfrac{1}{2}r^2 + C$ in which A, B and C are constants, show that $\nabla^2\Psi+2=0$.
If a beam under torsion has cross-section $b\le r\le 2a\cos\theta,\quad 2a>b>0$ sketch the cross-section and show with a suitable choice for A, B and C the given form of Ψ is the torsion function for this problem. Determine the position of the maximum shear stress magnitude.

7.4 Show that the potential energy functional of a uniform beam deformed by pure torsional couples is

$$U = \mu\tau^2 L\iint_D \left[2\Psi - \frac{\partial}{\partial x_1}(x_1\Psi) - \frac{\partial}{\partial x_2}(x_2\Psi)\right]dx_1dx_2$$

where L is the length of the beam and D is the cross-section.

7.5 Approximate Formulae in the Torsion Problem

The approximate formulae to be discussed in the first part of this section are, to a large extent, based on the exact solution to the torsion problem, obtained in the previous section, for a cylinder with rectangular cross-section (see (7.50) and (7.51)).

Indeed, a straightforward analysis of (7.50) shows that, for sections with a breadth to thickness ratio greater than about 1.5, for the region away from the ends ($|x_1| < \frac{b}{2} - \frac{t}{2}$) the solution for Ψ is essentially a parabolic function of x_2. Expressing this dependence explicitly we find:

$$\Psi = -x_2^2 + \frac{t^2}{4}(1 - s(x_1, x_2)) \tag{7.52}$$

in which

$$s(x_1, x_2) \equiv \frac{32}{\pi^3} \sum_{\substack{n=1 \\ n \text{ odd}}}^{\infty} \frac{(-1)^{(1-n)/2}}{n^3} \frac{\cosh \frac{n\pi x_1}{t}}{\cosh \frac{n\pi b}{2t}} \cos \frac{n\pi x_2}{t} \tag{7.53}$$

The modulus of $s(x_1, x_2)$ is indicated for various ratios of breadth to thickness in table 7.1

$b : t$	$\max\|s(x_1, x_2)\|$	$\max\|s(0, x_2)\|$
1.0	0.103	0.103
1.5	0.064	0.048
2.0	0.056	0.022
2.5	0.054	0.010

Table 7.1

The information to be gleaned from this analysis is that, for a rectangular cylinder for which $b >> t$ the solution for Ψ (away from the ends) is, to a very good approximation given by taking $s = 0$ in (7.52):

$$\Psi \approx -x_2^2 + \frac{t^2}{4} \tag{7.54}$$

Using (7.54) one can calculate the torsional rigidity of a thin-section rectangular cylinder and we find, by direct integration:

$$T_R = 2\mu \iint_D \Psi dx_1 dx_2 \approx \frac{\mu b t^3}{3} \tag{7.55}$$

which can be deduced directly from the exact result (7.51) when $b >> t$ (essentially all the terms in the summation are negligible in comparison to the first).

- **Torsion of Thin-walled Open Sections**

A thin-walled section is said to be *open* when the locus of points defining the centre line of the walls is not a closed curve. Channels, angles, I-beams, T-beams are examples of common structural shapes which are fabricated by combining thin-walled rectangular elements in which the thickness to length ratio of each component element is small (see Figure 7.10).

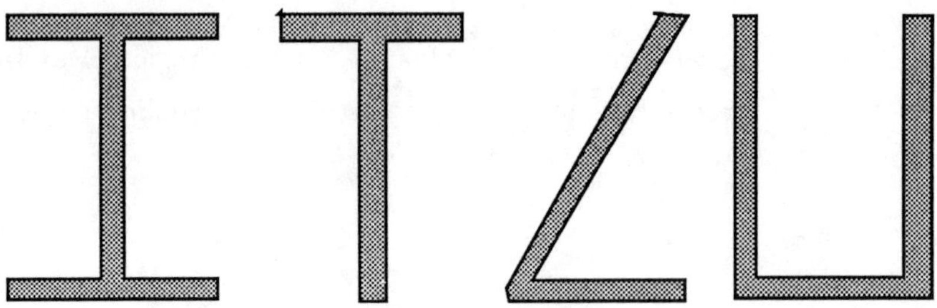

Figure 7.10

Since, for each of these examples, the torsion function Ψ must vanish on the boundary, it is a reasonable assumption that within each section the behaviour in Ψ is parabolic *across* the section as in (7.54). This assumption ignores any interaction effects which occur at corners and at element intersections.

A reasonable approximation to the torsional rigidity of any of these sections (which as we have noted earlier is, to a stress engineer, the most important torsional characteristic) is obtained by adding together the torsional rigidity of each element. For example for the thin-section beam with cross section shown in Figure 7.11 the torsional rigidity would be approximated by

$$T_R \approx \frac{1}{3}\mu(b_1 t_1^3 + b_2 t_2^3 + b_3 t_3^3 + b_4 t_4^3) \qquad \textbf{(7.56)}$$

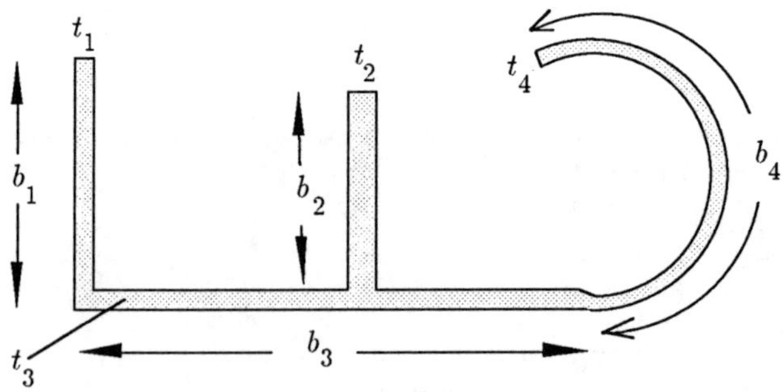

Figure 7.11

The maximum shearing stress magnitude for a general thin-walled open section with k elements may also be determined. We imagine, as above, that the total torsional rigidity T_R is distributed through each element so that if $(T_R)_i$ is the torsional rigidity arising from the ith element then

$$T_R = \sum_{i=1}^{k} (T_R)_i$$

The twist τ_i developed on each element is still related to the torsional couple M_i sustained across this element and its torsional rigidity as before:

$$\tau_i = \frac{M_i}{(T_R)_i}$$

However, as the twist (being directly related to the rotation of the section as a whole) on each element must be the same and be equal to the overall twist τ of the cross-section:

$$\tau \equiv \frac{M}{T_R} = \frac{M_i}{(T_R)_i} \quad \text{each } i$$

in which M is the overall applied torsional couple.

On the ith element the maximum shearing stress magnitude is, from (7.50a) (here we ignore *all* the terms in the summation since $b >> t$)

$$\begin{aligned}(\sigma_{\max})_i = \mu_i \tau_i t_i &= \mu_i \frac{M_i}{(T_R)_i} t_i \\ &= \mu_i \frac{M}{T_R} t_i \quad i \text{ not summed}\end{aligned}$$

Overall, the element with the largest shearing stress magnitude has the largest value of $\mu_i t_i$ (i not summed). For an open section made from the same material throughout the largest shear stress magnitude occurs on the *thickest* element. This should be contrasted with the corresponding result, obtained below, for thin-walled closed sections.

In some cases it is not appropriate to use the thin-section formula (7.55), for torsional rigidity. This applies particularily to cross-sections which have many 'intersections' or 'ends' or 'corners'. A typical selection is shown in Figure 7.12. For cross-sections such as these a more accurate estimation of torsional rigidity of each element may be desirable. It will be shown how greater accuracy may be achieved with little increase in computational cost or complication over the simple thin-section formulae of (7.55).

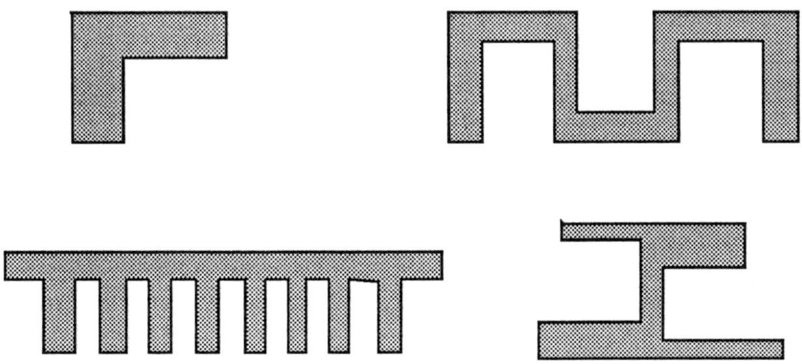

Figure 7.12

We note that each of the complicated sections of Figure 7.12 (and, of course, of many others) can be constructed from just three simple cut-sections shown in Figure 7.13.

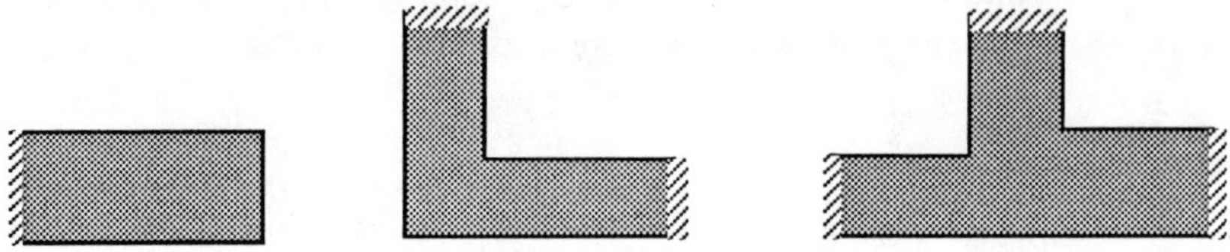

Figure 7.13

If we can develop simple formulae for the torsional rigidity of cut-sections such as those of Figure 7.13 we can, by 'glueing' together sections of this type construct the more complex sections of Figure 7.12. Clearly the torsional rigidity of a complete section is the sum of the torsional rigidities of the cut-sections which go to form it.

Along the cuts the boundary condition is taken to be $\Psi = -z^2 + \frac{t^2}{4}$ in which z is the variable across the cut and t is the cut width. Whilst this assumption has been shown to be valid to a good approximation, using the exact solution, for section (a) of Figure 7.13 it has been confirmed following many numerical experiments, to be also approximately true for sections (b) and (c) also. The only constraint is that the cuts should not be too close to the corner in (b) or to the intersection in (c) of Figure 7.13.

These new approximate formulae for torsional rigidity are derived following a detailed analysis of the cut-section rectangle (CR) of Figure 7.13(a). Since there exists an exact solution for a complete rectangular section (see (7.50)) we can, by direct integration, determine the torsional rigidity arising from any region of the rectangle. In particular, for the cut rectangle shown shaded in Figure 7.14 we find

$$
\begin{aligned}
T_{CR} &= 2\mu \int_{b/2-a}^{b/2} \int_{-t/2}^{t/2} \left[-x_2^2 + \frac{t^2}{4} \right] dx_1 dx_2 \\
&\quad - 2\mu \int_{b/2-a}^{b/2} \int_{-t/2}^{t/2} \left[\sum_{\substack{n=1 \\ n \text{ odd}}}^{\infty} \frac{8t^2}{(n\pi)^3} (-1)^{(1-n)/2} \frac{\cosh \frac{n\pi x_1}{t}}{\cosh \frac{n\pi b}{2t}} \cos \frac{n\pi x_2}{t} \right] dx_1 dx_2 \\
&= 2\mu \left[\frac{at^3}{6} - \frac{16t^4}{\pi^5} \sum_{\substack{n=1 \\ n \text{ odd}}}^{\infty} \frac{1}{n^5} \frac{1}{\cosh \frac{n\pi b}{2t}} \left(\sinh \frac{n\pi b}{2t} - \sinh \frac{n\pi}{t} (\frac{b}{2} - a) \right) \right]
\end{aligned}
$$

which, if only the first term of the series is considered, implies

$$
T_{CR} \approx \mu \left[\frac{at^3}{3} - 32 \frac{t^4}{\pi^5} \right] \qquad \text{if } a = \frac{b}{2} \text{ and } b > 2t \tag{7.57}
$$

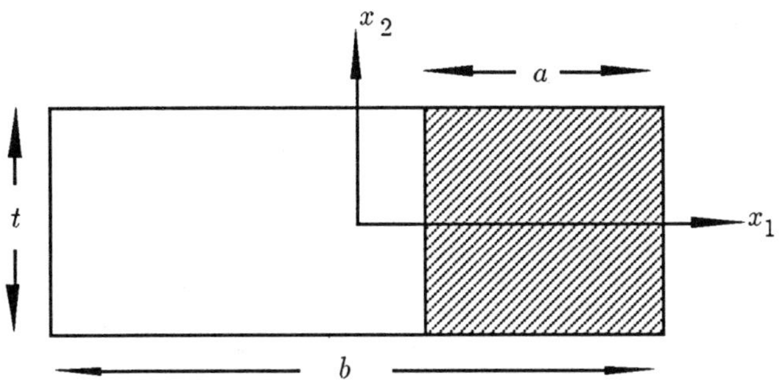

Figure 7.14

The largest relative error between this and the exact result is less than 0.10% for $0 < t/a < 0.5$.

To construct an approximate formula for the torsional rigidity of the cut L-section (CL) of Figure 7.13(b) we imagine this section to be fabricated from two CR sections as shown in Figure 7.15. Of course the shaded region is then counted twice so it is reasonable to subtract a quantity related to the torsional rigidity due to a rectangle of sides t_1 and t_2. This line of thinking relies heavily on the interpretation of rigidity as being proportional to the volume under the Ψ surface.

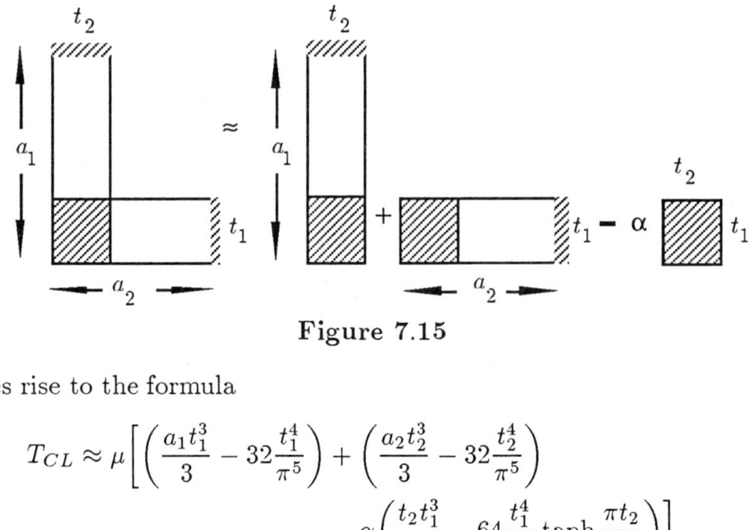

Figure 7.15

This gives rise to the formula

$$T_{CL} \approx \mu \left[\left(\frac{a_1 t_1^3}{3} - 32 \frac{t_1^4}{\pi^5} \right) + \left(\frac{a_2 t_2^3}{3} - 32 \frac{t_2^4}{\pi^5} \right) - \alpha \left(\frac{t_2 t_1^3}{3} - 64 \frac{t_1^4}{\pi^5} \tanh \frac{\pi t_2}{2 t_1} \right) \right] \tag{7.58}$$

In (7.58) the last term in parentheses assumes $t_2 > t_1$ and is a good approximation for the torsional rigidity of a rectangle with sides t_1 and t_2 deduced from the exact result of (7.51). The constant α is chosen to minimise the relative error between T_{CL} and the 'exact' result (obtained using an accurate finite element analysis) for a cross-section of this kind. I have concluded that a value of $\alpha = 0.41$ gives the best approximations over a large range of the ratios $a_1 : t_1$ and $a_2 : t_2$. Indeed over the range $0 < \dfrac{t_1}{a_1} < 0.5$ and $0 < \dfrac{t_2}{a_2} < 0.5$ the largest relative error was found to be 0.22% with an average value of 0.05%.

A similar approach is used to develop an approximate formula for the torsional rigidity of the cut T-section (CT) of Figure 7.13(c). Here we suppose the section to be formed as described in Figure 7.16.

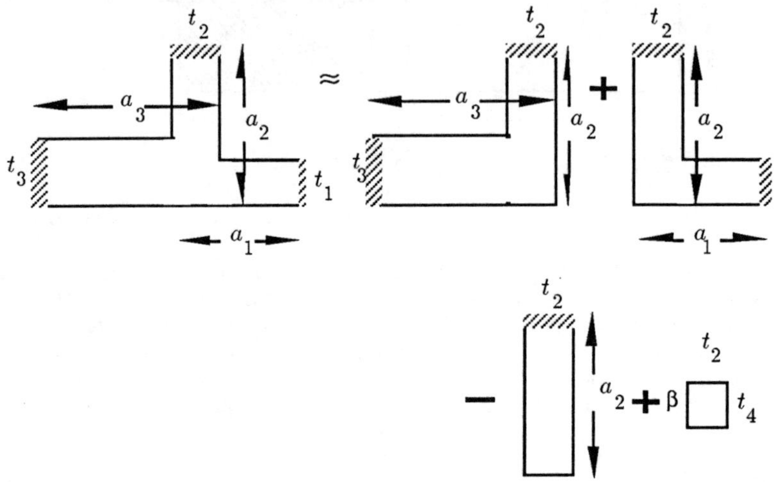

Figure 7.16

With respect to this figure t_4 is the larger of t_3 and t_1 and β is a factor chosen to minimise the relative errors over a large range of $t_i : a_i \quad i = 1, 2, 3$. We find $\beta = 0.07$ to give the largest relative error of 0.67% (average 0.33%) over the range $0 < \dfrac{t_i}{a_i} < 0.5 \quad i = 1, 2, 3$. The approximate formula for the cut T-section is:

$$\begin{aligned} T_{CR} \approx& \mu\left[\frac{1}{3}\left(a_1 t_1^3 + a_2 t_2^3 + a_3 t_3^3\right) - \frac{32}{\pi^5}\left(t_1^4 + t_2^4 + t_4^4\right)\right. \\ &- 0.41\left\{\frac{1}{3}\left(t_2 t_1^3 + t_2 t_1^3\right) - \frac{64}{\pi^5} t_1^4 \tanh\frac{\pi t_2}{2t_1} - \frac{64}{\pi^5} t_3^4 \tanh\frac{\pi t_2}{2t_3}\right\} \\ &\left. + 0.07\left\{\frac{1}{3} t_2 t_4^3 - \frac{64}{\pi^5} t_4^4 \tanh\frac{\pi t_2}{2t_4}\right\}\right] \end{aligned} \qquad \textbf{(7.59)}$$

In this formula it is assumed in the terms in curly brackets $t_2 > t_1$, $t_2 > t_3$ and $t_2 > t_4$. If any of these inequalities are reversed then the relevant parameters must be intechanged in the formula.

As an application of the formulae developed here we apply the approximate formulae of (7.57) and (7.59) to determine the torsional rigidity of a typical I-beam detailed in Figure 7.17(a) (not to scale).

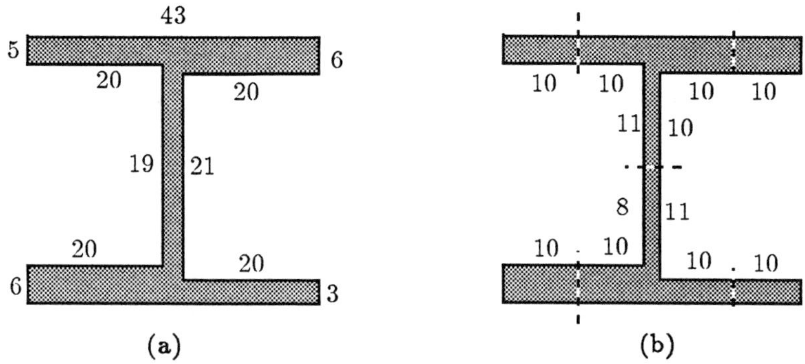

Figure 7.17

The cross-section is clearly formed from four CR-sections and two CT-sections with dimensions as in Figure 7.17(b). A simple calculation using the formulae developed above gives an approximate torsional rigidity of 3999μ. The 'exact' value obtained using a careful finite element analysis is 4060μ giving a 1.5% relative error between approximate and 'exact' values

- **Torsion of Thin-walled Closed Sections**

Here we assume the locus of points defining the centre line of the walls is a closed curve. See Figure 7.18

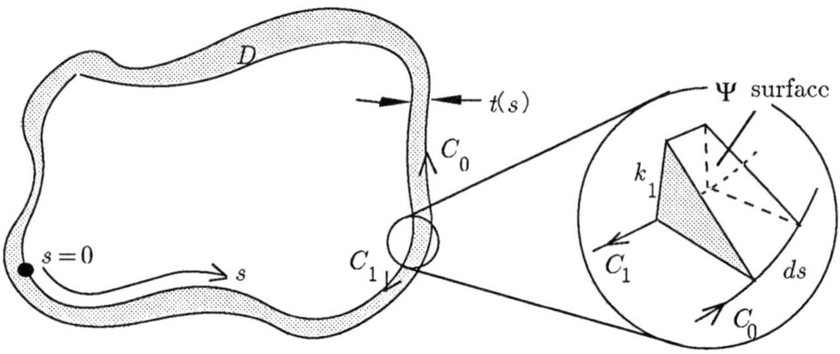

Figure 7.18

If the tube is thin we can assume that the torsion function Ψ varies linearly across the thickness t. Although t is small we shall not necessarily assume that it is a constant. We shall consider it as a function of s the arc-length parameter whose origin is taken at some convenient point.

Unlike the analysis of open sections we cannot assume Ψ vanishes on both C_0 and C_1. We take:

$$\Psi = 0 \quad \text{on } C_0 \qquad \text{and} \qquad \Psi = k_1 \quad \text{on } C_1 \tag{7.60}$$

The constant k_1 is chosen to satisfy (7.23a) (with $k = 1$):

$$\oint_{C_1} \frac{d\Psi}{dn} ds + 2A_1 = 0 \qquad \textbf{(7.61)}$$

in which, we remember, C_1 is described in the negative (clockwise) sense and A_1 is the area enclosed by C_1.

Now since Ψ varies linearly through the cross-section its derivative *across* the thickness is:

$$\frac{d\Psi}{dn} = \frac{\Psi(C_1) - \Psi(C_0)}{t} = \frac{k_1}{t}$$

and so (7.61) gives

$$k_1 = 2A_1 \Big/ \oint_{C_1^+} \frac{ds}{t(s)} \qquad \textbf{(7.62)}$$

where C_1^+ is the same contour as C_1 but described in the positive (anti-clockwise) sense. Also $\iint_D \Psi dx_1 dx_2$ represents the volume above the cross-section D and under the Ψ-surface and so, to a good approximation, (see inset Figure 7.18)

$$\iint_D \Psi dx_1 dx_2 = \frac{1}{2} k_1 \oint_{C_1^+} t(s) ds = \frac{1}{2} k_1 \overline{A}$$

in which $\overline{A}$ is the area of D (the shaded region in Figure 7.18). Thus the torsional rigidity of this closed section is, by (7.29) with $k = 1$:

$$\begin{aligned} T_R &= 2\mu \iint_D \Psi dx_1 dx_2 + 2\mu k_1 A_1 \\ &= 2\mu \left[\frac{1}{2} k_1 \overline{A} + k_1 A_1 \right] \approx 2\mu k_1 A_1 \end{aligned}$$

since, as the tube is thin $\overline{A} << A_1$. Finally:

$$T_R \approx 4\mu A_1^2 \Big/ \oint_{C_1^+} \frac{ds}{t(s)} \qquad \textbf{(7.63)}$$

or, in the case of a cylinder of constant thickness t, $T_R \approx 4\mu A_1^2 t/L$ where L is the length of the centre line.

We can better appreciate the accuracy (or otherwise) of the approximate formulae developed here by comparing with exact results obtained earlier for the hollow circular cylinder. In the exact solution for an elliptic cylinder choose $a = b$ and $m = 0.9$ to give:

$$(T_R)_{\text{exact}} = \mu\pi \frac{a^2}{2}(1 - 0.9^4) \approx 0.54\mu a^4 \qquad (T_R)_{\text{approx}} \approx 0.434\mu a^4$$

with a 20% relative error whilst for a thinner tube $m = 0.99$ (for which we expect closer correspondence)

$$(T_R)_{\text{exact}} = \mu\pi \frac{a^2}{2}(1 - 0.99^4) \approx 0.0619\mu a^4 \qquad (T_R)_{\text{approx}} \approx 0.0606\mu a^4$$

with a 2% relative error.

Also of interest to a stress engineer is the value of the maximum shearing stress magnitude over the cylinder. If n_i is normal to the curves Ψ=constant (that is, in the direction of $\underline{\nabla}\Psi$ and *across* the thickness of the cylinder) then $\dfrac{d\Psi}{dn} = \hat{n}.\underline{\nabla}\Psi$ and

$$\left|\frac{d\Psi}{dn}\right| = |\underline{\nabla}\Psi| = \sqrt{(\frac{\partial\Psi}{\partial x_1})^2 + (\frac{\partial\Psi}{\partial x_2})^2}$$

But by (7.31) this is proportional to the shearing stress at any point. Hence if σ_s denotes the shearing stress magnitude then

$$\sigma_s = \mu\tau\left|\frac{d\Psi}{dn}\right| = \mu\tau\frac{k_1}{t(s)} \tag{7.64}$$

which shows that, over the complete section σ_s is a maximum at the minimum value of $t(s)$ at those points at which the cylinder is thinnest. (This deduction is supported by the exact result, seen above, for a hollow elliptical cylinder).

For hybrid sections comprised of a closed cell plus 'open' fin elements we would estimate the torsional rigidity of the complete cross-section by adding together the torsional rigidities of the separate elements. For example for the cylinder shown in Figure 7.19 with four fins then

$$T_R \approx 4\mu A_1^2 \Big/ \oint_{C_1^+} \frac{ds}{t(s)} + \frac{1}{3}\mu\sum_{i=1}^{4} b_i t_i^3$$

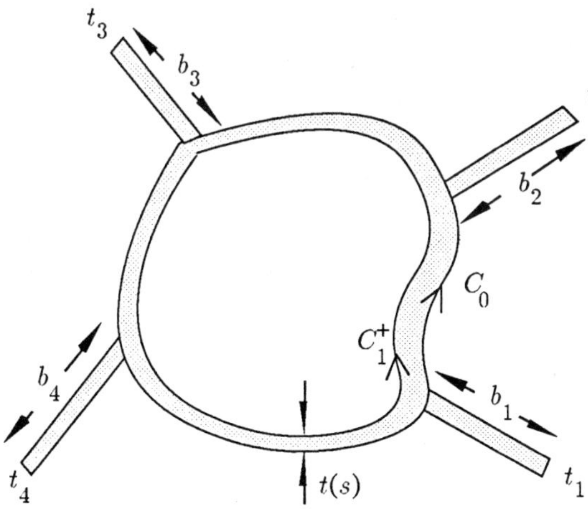

Figure 7.19

EXERCISES

7.5 An 8cm wide flange steel beam (Figure 7.20) is to be subjected to a torsional couple M, and the maximum shearing stress is to be limited to 140 $\mathrm{MN/m^2}$. Determine the maximum value M can have and the twist which will be produced by that couple. For steel $\mu = 7.7 \times 10^{10}\mathrm{N/m^2}$

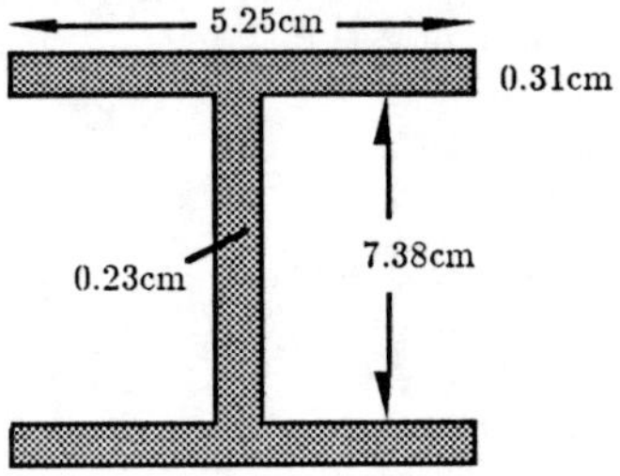

Figure 7.20

7.6 Consider a bar of thin-walled closed section having the geometry illustrated in Figure 7.21. Compare the twist and the maximum shear stress magnitude of this bar with those of the same bar if it was split open at any point in the cross-section along the axial direction and subjected to the same torsional couple M.

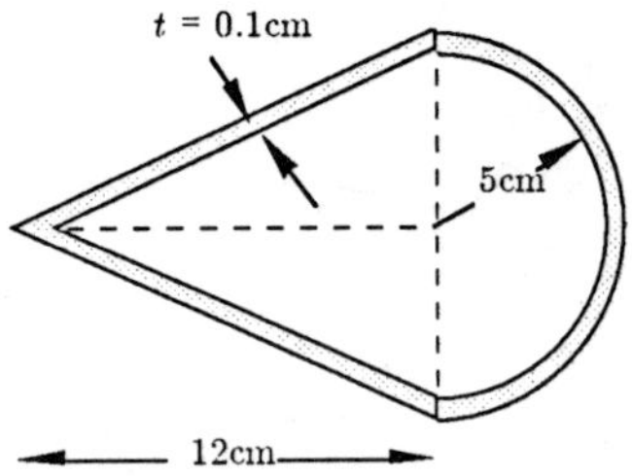

Figure 7.21

Can you generalise your analysis to apply to *any* cross-section?

7.7 Show that $\psi = -\dfrac{1}{6a}(x_1^3 - 3x_1x_2^2) + \dfrac{2}{3}a^2$ is an harmonic function. Hence by finding those curves which make $\psi - \dfrac{1}{2}(x_1^2 + x_2^2)$ vanish determine the solution of the torsion problem with respect to the equaliteral triangle with sides $x_1 = a$, $x_1 + \sqrt{3}x_2 = -2a$ and $x_1 - \sqrt{3}x_2 = -2a$. Show that the resultant shearing stress is a maximum at the centres of each side and has magnitude $\dfrac{3}{2}\mu\tau a$. Show that the resultant couple has magnitude

$$\frac{9\sqrt{3}}{5}\mu\tau a^4.$$

7.8 A hollow cylinder, with generators parallel to the x_3-axis is bounded by the surfaces $x_1^2 + 4x_2^2 = 4$ and $x_1^2 + 4x_2^2 = 1$. Find a suitable torsion function Ψ for this cylinder and hence determine its torsional rigidity.

Chapter 8

THE MATRIX ANALYSIS OF STRUCTURES

8.1 Introduction

As we have seen in previous chapters the exact determination of a deformation of an elastic body resulting from an applied loading is only a realistic possibility if the body and/or the loading takes an especially simple form. However, there are a number of practical situations in which the precise detail on stress and strain levels and on displacement values available from an exact solution are not generally required. This suggests that, for these situations, we approximate, whenever possible, the physical system being modelled with simpler systems of equations than those of the exact theory of linear infinitessimal elasticity.

This approach has already been used in the modelling of a slender beam by the Euler-Bernoulli equation (6.39). In the Euler-Bernoulli approximation we *accept* that the transverse deformation (due to the action of *transverse* forces) of a slender beam is adequately characterised by the deformation of the centre line alone. In this we are simply *ignoring* all the detailed information on the deformation of particles not on the centre line. What we gain in this process is the reduction of the full system of linear elasticity equations ((5.9) to (5.14)) to a single fourth order linear boundary value problem which can be (relatively speaking) easily solved.

If, instead of the transverse deformation of a slender beam, we model the longitudinal deformation due to longitudinally directed forces alone, using an Euler-Bernoulli approach, an even simpler boundary value problem is obtained.

Consider a uniform beam of length L and cross-sectional area A subjected to longitudinal forces P_0, P_L applied at either end. See Figure 8.1. In modelling the deformation a reasonable assumption in this case is to imagine each plane x_3=constant is displaced by a certain amount $u_3(x_3)$. That is, we approximate the exact displacement (which, from the exact treatment of the simple tension experiment described in Section 5.3, must also include a 'thinning') by the displacement field:

$$u_i = (0, 0, u_3(x_3))$$

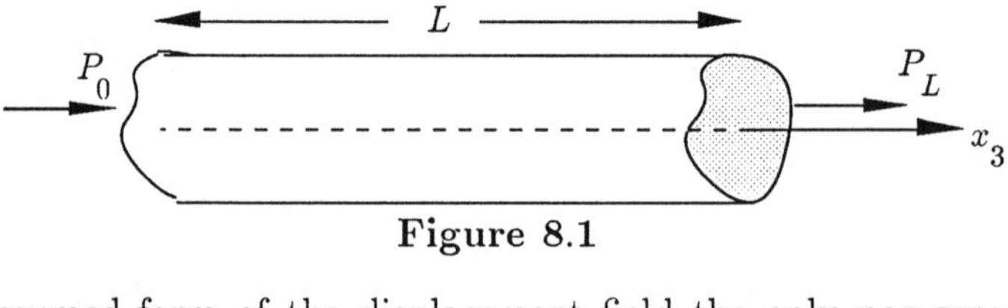

Figure 8.1

From this assumed form of the displacement field the only non-zero component of the strain tensor is $\varepsilon_{33} = \dfrac{du_3}{dx_3}$. Also, making the same approximation here as was made following equation (6.34) in the Euler-Bernoulli theory, (that the longitudinal stress σ_{33} is the dominant stress so that σ_{11} and σ_{22} can be ignored) then the longitudinal stress is related to longitudinal strain as in the simple tension experiment: $\sigma_{33} = E\varepsilon_{33}$. Using this we find the strain energy to be:

$$W = \frac{1}{2}\iiint_V \sigma_{ij}\varepsilon_{ij}dV = \frac{1}{2}EA\int_0^L \left(\frac{du_3}{dx_3}\right)^2 dx_3$$

The work done by the deforming forces P_0 and P_L is $P_0u_3(0) + P_Lu_3(L)$. Hence the total potential energy (see (6.24)) is

$$U = \frac{1}{2}EA\int_0^L \left(\frac{du_3}{dx_3}\right)^2 dx_3 - P_0u_3(0) - P_Lu_3(L) \tag{8.1}$$

Now, by the theorem of minimum potential energy, U is minimised by the true value of u_3 (that is, 'true' to the level of approximation being considered). The problem of finding the form that $u_3(x_3)$ must take in order that U takes its minimum value is a standard problem in the calculus of variations. It is easily shown that $u_3(x_3)$ satisfies the boundary value problem (see Appendix 1):

$$\frac{d}{dx_3}\left[EA\frac{du_3}{dx_3}\right] = 0 \qquad 0 < x_3 < L \tag{8.2}$$

with boundary conditions

$$-EAu_3'(0) = P_0 \qquad \text{and} \qquad EAu_3'(L) = P_L \tag{8.3}$$

In the approximation considered here u_3 is the solution to a *second-order* boundary value problem. (The boundary conditions may change depending on the initial information on $u_3(x_3)$ at the end-points $x_3 = 0$ and $x_3 = L$. If, instead of forces being specified at $x_3 = 0, L$ as in Figure 8.1, displacements were specified then the corresponding boundary condition in (8.3) would be redundant. For example if the left-hand end of the beam is held fixed so that $u_3(0) = 0$ and a stretching force P_L applied at the other end then the boundary conditions (8.3) would be replaced by:

$$u_3(0) = 0 \qquad \text{and} \qquad EAu_3'(L) = P_L \tag{8.3a}$$

Although we are making aproximations from the full theory of linear elasticity we are nonetheless still modelling the beam in a *continuous* manner (hence the appearance of differential equations). We shall find it convenient to model the longitudinal deformation of a beam in a *discrete* manner, in terms of its measured displacements at each end.

Assuming A is a constant along the beam (8.2) can be integrated to obtain

$$u_3 = c_1x_3 + c_2 \tag{8.4}$$

where c_1, c_2 are constants. These constants can be expressed in terms of the end-displacements $u_3(0)$ and $u_3(L)$. Since, from (8.4), $u_3(0) = c_2$ and $u_3(L) = c_1L + c_2$ we have, after solving these two relations for c_1 and c_2:

$$
\begin{aligned}
u_3(x_3) &= \left[\frac{u_3(L) - u_3(0)}{L}\right] x_3 + u_3(0) \\
&= u_3(0)[1 - \frac{x_3}{L}] + u_3(L)[\frac{x_3}{L}]
\end{aligned}
\tag{8.5}
$$

This relation shows that if the end-displacements are known then the displacement throughout the beam is known (to the approximation that we think appropriate for such a problem).
Note that, from (8.3)

$$
P_L = EAu_3'(L) = \frac{EA}{L}[u_3(L) - u_3(0)]
$$

and

$$
P_0 = -EAu_3'(0) = -\frac{EA}{L}[u_3(L) - u_3(0)]
$$

These two equations can be written in matrix form as

$$
\begin{bmatrix} P_0 \\ P_L \end{bmatrix} = \frac{EA}{L} \begin{bmatrix} 1 & -1 \\ -1 & 1 \end{bmatrix} \begin{bmatrix} u_3(0) \\ u_3(L) \end{bmatrix}
\tag{8.6}
$$

The relation (8.6) is the *discrete* model of the longitudinal deformation of a slender beam. We shall find the discrete model to be of far greater use in the analysis of structures, comprised of many beam elements, than the continuous variable model seen earlier. The reason for this will become plain as we proceed through this chapter though, on a superficial level, it is likely to be easier to deal with large combinations of matrices than to deal with large combinations of boundary value problems.

An alternative method of deriving the relation (8.6) utilises the Castigliano relations. Using (8.5):

$$
\begin{aligned}
W = \frac{1}{2}EA \int_0^L \left[\frac{du_3}{dx_3}\right]^2 dx_3 &= \frac{1}{2}EA \int_0^L \left[\frac{u_3(L) - u_3(0)}{L}\right]^2 dx_3 \\
&= \frac{1}{2}\frac{EA}{L}\left[u_3(L) - u_3(0)\right]^2
\end{aligned}
\tag{8.7}
$$

Hence, using (6.23)

$$
P_0 = \frac{\partial W}{\partial u_3(0)} = -\frac{EA}{L}[u_3(L) - u_3(0)]
$$

and

$$
P_L = \frac{\partial W}{\partial u_3(L)} = \frac{EA}{L}[u_3(L) - u_3(0)]
$$

which are identical to the equations contained in (8.6)

8.2 Pin-jointed Elements

A pin-jointed element is a slender beam whose major deformation is longitudinal and one which can only sustain longitudinal forces. The end-points (see Figure 8.2) are universal joints not capable of transmitting either torsional couples, bending couples or shear forces.

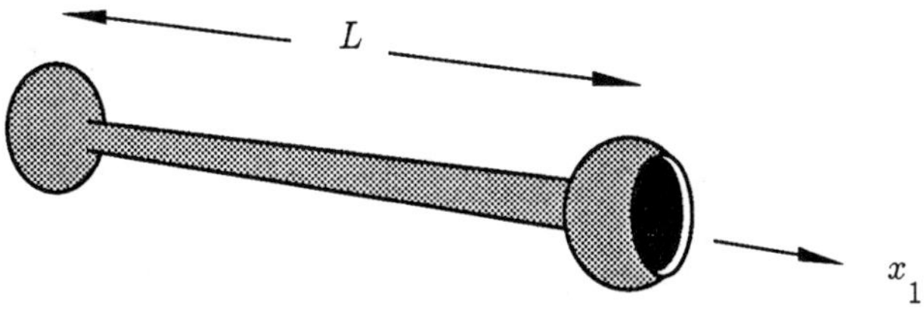

Figure 8.2

A pin-joint element can only be stretched or compressed. As we have seen in the previous section it is fully characterised by the displacement measured at its end points (called nodes) and by the forces applied there. This information is displayed in Figure 8.3.

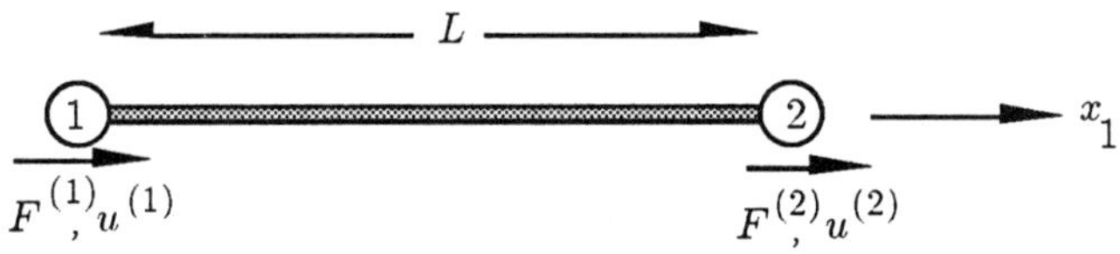

Figure 8.3

We find it convenient to change notation. The axis of the element is along the x_1-axis. The coordinates $x_i \quad i = 1, 2, 3$ are called the **local** coordinate axes of the element. The left-hand node is labelled 1 and the right-hand node labelled 2. Also, the longitudinal forces applied are labelled $F^{(1)}, F^{(2)}$ and the longitudinal displacements labelled $u^{(1)}, u^{(2)}$ at nodes 1 and 2 respectively.

Incorporating this change of notation, (8.6) becomes

$$\begin{bmatrix} F^{(1)} \\ F^{(2)} \end{bmatrix} = \frac{EA}{L} \begin{bmatrix} 1 & -1 \\ -1 & 1 \end{bmatrix} \begin{bmatrix} u^{(1)} \\ u^{(2)} \end{bmatrix} \tag{8.8}$$

or

$$\mathbf{F} = \mathbf{k}\mathbf{u} \tag{8.9}$$

in which

$$\mathbf{F} = \begin{bmatrix} F^{(1)} \\ F^{(2)} \end{bmatrix} \qquad \mathbf{u} = \begin{bmatrix} u^{(1)} \\ u^{(2)} \end{bmatrix} \quad \text{and} \quad \mathbf{k} = \frac{EA}{L} \begin{bmatrix} 1 & -1 \\ -1 & 1 \end{bmatrix}$$

The matrix $\mathbf{k}$ relating the applied forces to the measured displacements is called the **local stiffness matrix** of the pin-jointed element referred to local coordinates (x_1, x_2, x_3). The stiffness matrix can be regarded as the basic building block of structural analysis. The factor $c = EA/L$ is known as the stiffness of the element.

Note that the stiffness matrix is *singular* (since det**k**=0), is symmetric and has non-negative terms on the leading diagonal. These are properties enjoyed by all stiffness matrices.

• Assemblages of Pin-jointed Elements

We now combine pin-jointed elements together into an assemblage. The simplest grouping of just two pin-jointed elements of stiffnesses k_a and k_b connected at a node is shown in Figure 8.4. The elements can be made from different materials, may have different cross-sectional areas and different lengths. From these values, the element stiffnesses can be determined: $k_a = E_a A_a / L_a$ and $k_b = E_b A_b / L_b$.

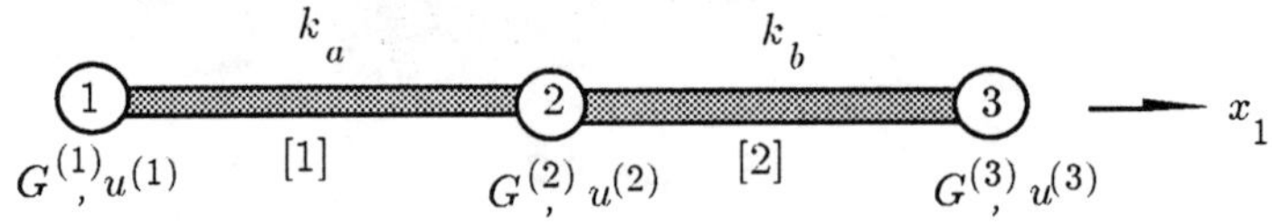

Figure 8.4

We now have two elements, labelled [1] and [2] as shown. At each node a force may be applied and a corresponding displacement measured (each is directed along the positive x_1-axis).

The forces $G^{(1)}, G^{(2)}$ and $G^{(3)}$ are the **global forces** acting at the nodes. These are to be distinguished from the **internal forces** developed within each element which are (for a pin-jointed element) related to the displacements measured at each node by (8.8). The global forces give rise to the internal forces. At each node the internal forces within those elements adjoining that node combine (vectorially) to equal the global force at that node. This is to be contrasted to the displacement at a node, which is the *same* for every element having the node in common.

The extension of element [1] is $(u^{(2)} - u^{(1)})$ whilst that of element [2] is $(u^{(3)} - u^{(2)})$. Thus the strain energy of the assemblage is:

$$W = \frac{1}{2}k_a(u^{(2)} - u^{(1)})^2 + \frac{1}{2}k_b(u^{(3)} - u^{(2)})^2 \tag{8.10}$$

The Castigliano relation (6.23) may be used to determine the relationship between nodal forces and nodal displacements

$$G^{(1)} = \frac{\partial W}{\partial u^{(1)}} = -k_a(u^{(2)} - u^{(1)}) \tag{8.11a}$$

$$G^{(2)} = \frac{\partial W}{\partial u^{(2)}} = k_a(u^{(2)} - u^{(1)}) - k_b(u^{(3)} - u^{(2)}) \tag{8.11b}$$

$$G^{(3)} = \frac{\partial W}{\partial u^{(3)}} = k_b(u^{(3)} - u^{(2)}) \tag{8.11c}$$

which, in matrix form, is

$$\begin{bmatrix} G^{(1)} \\ G^{(2)} \\ G^{(3)} \end{bmatrix} = \begin{bmatrix} k_a & -k_a & 0 \\ -k_a & k_a + k_b & -k_b \\ 0 & -k_b & k_b \end{bmatrix} \begin{bmatrix} u^{(1)} \\ u^{(2)} \\ u^{(3)} \end{bmatrix} \tag{8.12}$$

or

$$\mathbf{G} = \mathbf{K}\mathbf{U} \tag{8.13}$$

where

$$\mathbf{K} = \begin{bmatrix} k_a & -k_a & 0 \\ -k_a & k_a + k_b & -k_b \\ 0 & -k_b & k_b \end{bmatrix} \begin{matrix} 1 \\ 2 \\ 3 \end{matrix} \tag{8.14}$$

The matrix $\mathbf{K}$ connecting the forces to the displacements is called the **overall stiffness matrix** of the structure. We must emphasize that the stiffness matrix is meaningful only when written with respect to the force/displacement equation (8.12). If we want to refer to the stiffness matrix alone we should always indicate (unless it is clear) which nodes each stiffness matrix component refers to. This is the reason for attaching to each row of $\mathbf{K}$ in (8.14) the corresponding node number.

In setting up this problem there is a certain amount of freedom in the choice of node numbers and element numbers. For example, instead of the choices shown in Figure 8.4, we could choose the node and element numbers shown in Figure 8.5.

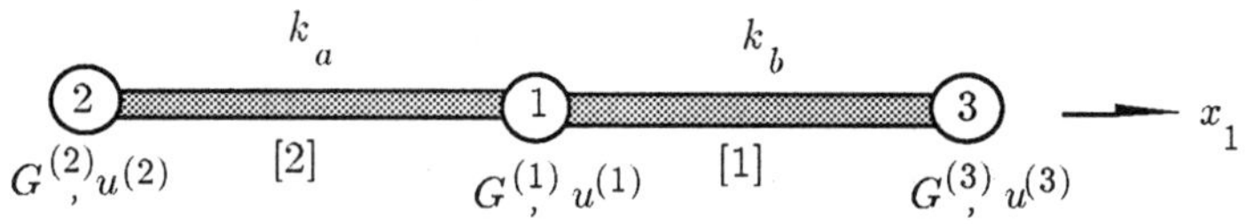

Figure 8.5

For this configuration we find:

$$W = \frac{1}{2}k_a(u^{(1)} - u^{(2)})^2 + \frac{1}{2}k_b(u^{(3)} - u^{(1)})^2 \tag{8.15}$$

and, repeating the analysis above,

$$\begin{bmatrix} G^{(1)} \\ G^{(2)} \\ G^{(3)} \end{bmatrix} = \begin{bmatrix} k_a + k_b & -k_a & -k_b \\ -k_a & k_a & 0 \\ -k_b & 0 & k_b \end{bmatrix} \begin{bmatrix} u^{(1)} \\ u^{(2)} \\ u^{(3)} \end{bmatrix} \tag{8.16a}$$

or

$$\mathbf{G} = \mathbf{K}\mathbf{U} \tag{8.16b}$$

where, this time

$$\mathbf{K} = \begin{bmatrix} k_a + k_b & -k_a & -k_b \\ -k_a & k_a & 0 \\ -k_b & 0 & k_b \end{bmatrix} \begin{matrix} 1 \\ 2 \\ 3 \end{matrix} \tag{8.17}$$

The information contained in (8.16b) is exactly the same as that contained in (8.13): only the form has changed.

We note that the overall stiffness matrix (either (8.14) or (8.17)), displays the properties outlined earlier: each is singular, each is symmetric and each have non-negative terms on the leading diagonal.

One of the major problems faced when considering larger assemblages is the construction of the overall stiffness matrix of the structure. It is a rather laborious procedure to write down the strain energy each time and then use the Castigliano relation to deduce the overall stiffness matrix. An alternative procedure for constructing the overall stiffness matrix may be deduced by observing that this matrix is formed by simply 'adding in' local stiffness matrices for each pin-jointed element forming the asssemblage. We simply take each pin-joint member in turn and superimpose the elements of the local stiffness matrix into appropriate positions in the overall matrix. An example will help to explain the procedure to be followed.

Consider, as an example, a three-element pin-jointed assemblage with individual element stiffnesses k_a, k_b and k_c. See Figure 8.6.

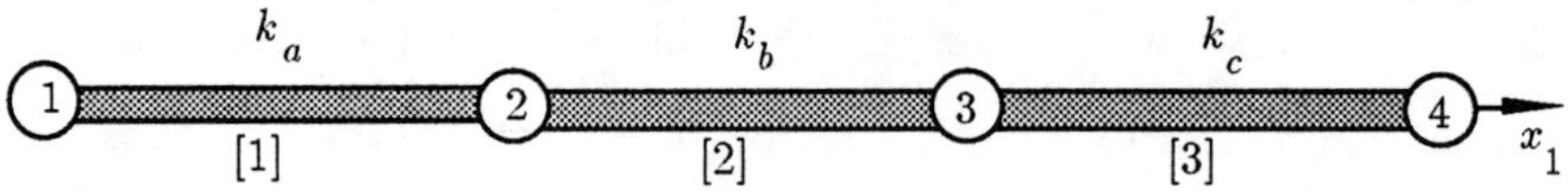

Figure 8.6

The local element stiffness matrices are

$$k_{[1]} = \begin{bmatrix} k_a & -k_a \\ -k_a & k_a \end{bmatrix} \begin{matrix} 1 \\ 2 \end{matrix} \qquad k_{[2]} = \begin{bmatrix} k_b & -k_b \\ -k_b & k_b \end{bmatrix} \begin{matrix} 2 \\ 3 \end{matrix} \qquad k_{[3]} = \begin{bmatrix} k_c & -k_c \\ -k_c & k_c \end{bmatrix} \begin{matrix} 3 \\ 4 \end{matrix}$$

We expect the overall stiffness matrix to be a 4×4 square matrix. We begin with a 4×4 zero matrix and superimpose $k_{[1]}, k_{[2]}$ and $k_{[3]}$ into this matrix as shown in Figure 8.7:

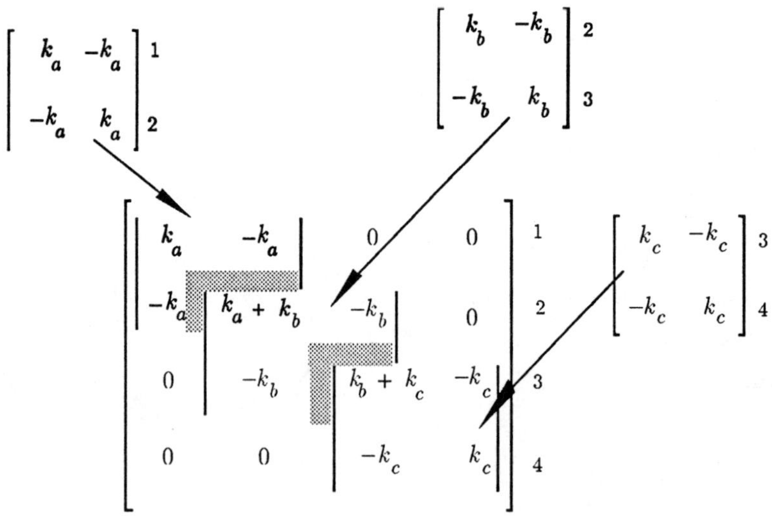

Figure 8.7

The superposition procedure outlined here may be applied generally to any structure. The overall stiffness matrix in a problem may always be obtained by superimposing the element stiffness matrices (in appropriate form and position) together. This is one of the main reasons for the popularity of the stiffness matrix method for analysing structural problems.

As mentioned earlier the internal forces (noted in (8.8)) also combine together to produce the global forces. This is, in effect, a superposition of the element force vectors. For the simple assemblage described in Figure 8.6 the force vectors in elements [1], [2] and [3] are respectively, (together with their displacement 'labels')

$$\begin{bmatrix} F^{(1)} \\ F^{(2)} \end{bmatrix} \begin{matrix} 1 \\ 2 \end{matrix} \qquad \begin{bmatrix} H^{(2)} \\ H^{(3)} \end{bmatrix} \begin{matrix} 2 \\ 3 \end{matrix} \qquad \begin{bmatrix} J^{(3)} \\ J^{(4)} \end{bmatrix} \begin{matrix} 3 \\ 4 \end{matrix}$$

These matrices 'superimpose' in an obvious way to give a global force matrix:

$$\begin{bmatrix} F^{(1)} \\ F^{(2)} + H^{(2)} \\ H^{(3)} + J^{(3)} \\ J^{(4)} \end{bmatrix} \begin{matrix} 1 \\ 2 \\ 3 \\ 4 \end{matrix}$$

In our analysis these components are *relabelled* $G^{(1)}, G^{(2)}, G^{(3)}$ and $G^{(4)}$. In effect we need not concern ourselves with the superposition of the 'forces' since, in structural analysis, they are either given as initial information or are unknowns which need to be determined.

• Problem Solving

It has already been noted that the stiffness matrix is singular. This is not surprising because if the stiffness matrix $\mathbf{K}$ in (8.14) were not singular we would be able to determine a unique set of displacements $\mathbf{U}$ given a set of applied forces $\mathbf{G}$ since, in this case, $\mathbf{U} = (\mathbf{K})^{-1}\mathbf{G}$. But, for a pin-jointed element not fixed in some way this would be physically unreasonable as the solution for $\mathbf{U}$ is only unique up to a rigid body motion. (To see this consider a 'spring' which is an example of a very flexible pin-jointed element. Imagine deforming the spring across your chest. You can move your feet – thereby moving the spring as a rigid body – changing the nodal displacements of the spring without affecting the deformation you have induced in it). For one dimensional problems this means that to each set of values for $\mathbf{U} = [u^{(1)}, u^{(2)}, ..., u^{(n)}]^T$ we can always add an arbitrary constant α to each component of $\mathbf{U}$: $\mathbf{U} = [u^{(1)}+\alpha, u^{(2)}+\alpha, ..., u^{(n)}+\alpha]^T$ without affecting the physical situation in any way. It is only the difference between nodal displacements that matter physically not their absolute values. This two-element problem (and indeed any other structural problem may be solved for the unknowns in terms of the known quantities as long as sufficient physical conditions (boundary conditions) are imposed to prevent rigid body motions.

Forces and displacements are, in a certain sense, complimentary. If, at a node, a given force is applied then the corresponding displacement is an unknown to be determined from the elasticity equations. On the other hand if, at a node, a displacement is specified then the corresponding force is an unknown. This 'duality' becomes clearer if one considers the workings of a spring-type weighing scale. If one applies a given weight to the scale it is not known, at the outset, (before the scale is calibrated) what the corresponding spring displacement will be. Similarly, if one decides to depress the weighing scale by a given amount one does not know, at the outset, what weight is required to acomplish this. These considerations imply, as is generally exhibited in structural problems, that in systems of equations of the type (8.13) the unknowns in the problem may appear on either side of the equation. However, in a given problem, the number of unknowns is fixed: for a pin-jointed element (in one dimension) there are two unknowns whether they be two unknown reaction forces, or two unknown displacements or one unknown force and one unknown displacement.

Example 8.1
Consider the pin-jointed system shown in Figure 8.4 with node 3 held fixed and with given forces $P^{(1)}, P^{(2)}$ applied at nodes 1 and 2 respectively (see Figure 8.8). Determine all the unknowns.

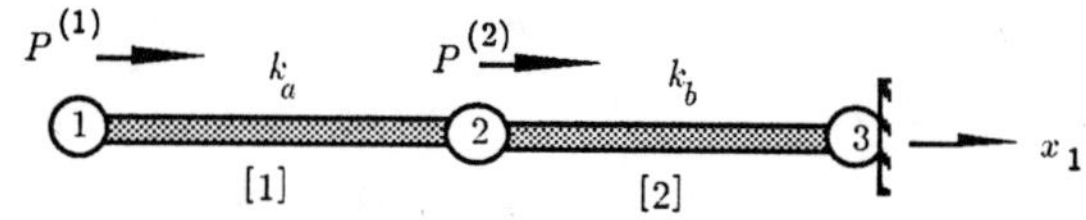

Figure 8.8

Solution

For this problem the unknowns are the displacements of nodes 1 and 2, $u^{(1)}, u^{(2)}$ and $G^{(3)}$ the reaction force at node 3. The known quantities are $u^{(3)} = 0$, $G^{(1)} = P^{(1)}$, $G^{(2)} = P^{(2)}$.
We have from (8.13)

$$\begin{bmatrix} P^{(1)} \\ P^{(2)} \\ G^{(3)} \end{bmatrix} = \begin{bmatrix} k_a & -k_a & 0 \\ -k_a & k_a + k_b & -k_b \\ 0 & -k_b & k_b \end{bmatrix} \begin{bmatrix} u^{(1)} \\ u^{(2)} \\ 0 \end{bmatrix} \tag{8.18}$$

The three equations contained in (8.18) may be written as

$$\begin{bmatrix} P^{(1)} \\ P^{(2)} \end{bmatrix} = \begin{bmatrix} k_a & -k_a \\ -k_a & k_a + k_b \end{bmatrix} \begin{bmatrix} u^{(1)} \\ u^{(2)} \end{bmatrix} \tag{8.19}$$

and

$$G^{(3)} = -k_b u^{(2)} \tag{8.20}$$

Equation (8.19) may be solved by simply inverting the coefficient matrix (which is not singular) to obtain

$$\begin{bmatrix} u^{(1)} \\ u^{(2)} \end{bmatrix} = \frac{1}{k_a k_b} \begin{bmatrix} k_a + k_b & k_a \\ k_a & k_a \end{bmatrix} \begin{bmatrix} P^{(1)} \\ P^{(2)} \end{bmatrix} \tag{8.21}$$

that is,

$$u^{(1)} = \frac{1}{k_a k_b}[P^{(1)}(k_a + k_b) + P^{(2)} k_a] \tag{8.22}$$

and

$$u^{(2)} = \frac{1}{k_b}(P^{(1)} + P^{(2)}) \tag{8.23}$$

Now, having solved for the unknown displacements, we can substitute (8.23) into (8.20) to find the unknown reaction force

$$G^{(3)} = -P^{(1)} - P^{(2)} \tag{8.24}$$

confirming that, globally, all three nodal forces are in equilibrium.

Example 8.2
Solve the two-element problem, shown in Figure 8.9, in which node 1 is held fixed and node 3 is displaced a distance u. Node 2 is free with no external force applied at this point.

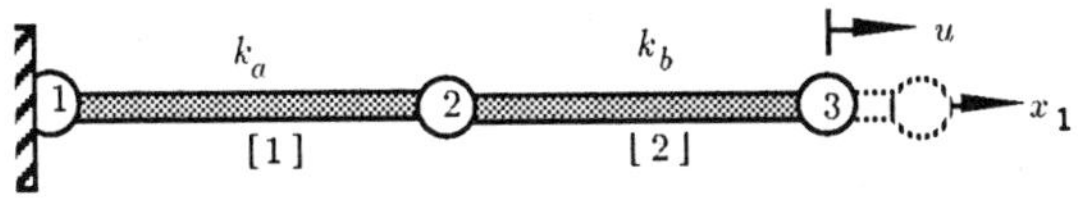

Figure 8.9

Solution

The problem we have to solve is

$$\begin{bmatrix} G^{(1)} \\ 0 \\ G^{(3)} \end{bmatrix} = \begin{bmatrix} k_a & -k_a & 0 \\ -k_a & k_a + k_b & -k_b \\ 0 & -k_b & k_b \end{bmatrix} \begin{bmatrix} 0 \\ u^{(2)} \\ u \end{bmatrix} \tag{8.25}$$

in which the unknowns are $G^{(1)}$, $u^{(2)}$ and $G^{(3)}$. We find, from (8.25)

$$G^{(1)} = -k_a u^{(2)} \tag{8.26}$$

$$G^{(3)} = -k_b u^{(2)} + k_b u \tag{8.27}$$

and

$$0 = (k_a + k_b)u^{(2)} - k_b u \tag{8.28}$$

which gives

$$u^{(2)} = \frac{k_b u}{(k_a + k_b)} \quad G^{(1)} = -\frac{k_a k_b u}{(k_a + k_b)} \quad G^{(3)} = \frac{k_a k_b u}{(k_a + k_b)}$$

That is, a force of magnitude $\dfrac{k_a k_b u}{(k_a + k_b)}$ has to be applied to displace node 3 in this two-element system a distance u.

The examples given above involve a small number of coupled pin-jointed elements and are quite straightforward to analyse. However, realistic assemblages of structural elements may involve large numbers of unknowns and the above procedures, for rearranging and hence solving the equations, become unworkable. There are however, a number of alternative numerical procedures for implementing the displacement boundary conditions and then solving problems of this type. We outline the most popular method.

For illustration purposes consider a 3×3 system of equations of the form $\mathbf{G} = \mathbf{KU}$ in which one of the displacements, $u^{(2)}$ say, is given a fixed value h and the other two are unknowns.

The system of equations might be:

$$\begin{bmatrix} G^{(1)} \\ G^{(2)} \\ G^{(3)} \end{bmatrix} = \begin{bmatrix} a & b & c \\ b & d & e \\ c & e & f \end{bmatrix} \begin{bmatrix} u^{(1)} \\ h \\ u^{(3)} \end{bmatrix} \tag{8.29}$$

with $G^{(1)}$, $G^{(3)}$ known and $G^{(2)}$ unknown. Expanding (8.29) directly, the system is clearly equivalent to:

$$\begin{bmatrix} G^{(1)} - bh \\ G^{(3)} - eh \end{bmatrix} = \begin{bmatrix} a & c \\ c & f \end{bmatrix} \begin{bmatrix} u^{(1)} \\ u^{(3)} \end{bmatrix} \tag{8.30}$$

and

$$G^{(2)} = bu^{(1)} + dh + eu^{(3)} \tag{8.31}$$

The numerical procedure involves (temporarily) treating $u^{(2)}$ as an unknown and replacing the diagonal component of the stiffness matrix, corresponding to $u^{(2)}$, namely d, by a very large number say 10^{30}. The corresponding component of the $\mathbf{G}$ matrix is similarly replaced by $10^{30}h$ so that the amended system of equations is:

$$\begin{bmatrix} G^{(1)} \\ 10^{30}h \\ G^{(3)} \end{bmatrix} = \begin{bmatrix} a & b & c \\ b & 10^{30} & e \\ c & e & f \end{bmatrix} \begin{bmatrix} u^{(1)} \\ u^{(2)} \\ u^{(3)} \end{bmatrix} \tag{8.32}$$

The system of equations (8.32) is now in the usual form with all the unknowns on the right hand side and all the known quantities on the other side. Once amended in this way we can use any standard numerical method to solve for the unknowns. It is easy to see that (numerically) the solution to the amended system of equations (8.32) is the same as the solution to the original system (8.30),(8.31). Written out in full, the three equations in (8.32) are:

$$G^{(1)} = au^{(1)} + bu^{(2)} + cu^{(3)} \tag{8.33a}$$

$$10^{30}h = bu^{(1)} + 10^{30}u^{(2)} + eu^{(3)} \tag{8.33b}$$

$$G^{(3)} = cu^{(1)} + eu^{(2)} + fu^{(3)} \tag{8.33c}$$

Since 10^{30} is likely, in any practical problem, to be very much bigger than the terms $bu^{(1)}$, $eu^{(3)}$ equation (8.33b) is, to a very good numerical approximation, equivalent to

$$10^{30}h = 10^{30}u^{(2)}$$

that is

$$u^{(2)} = h$$

as required.

The solution of (8.32) gives (to a good numerical approximation) the correct solution for $u^{(2)}$. It follows that the solutions for $u^{(1)}$, $u^{(3)}$ from (8.33a,c) will also be correct to a good approximation.

This numerical procedure of replacing appropriate diagonal terms in the stiffness matrix is carried through for each node which is held fixed. Using this procedure in Example 1.2 would imply replacing equation (8.25) with

$$\begin{bmatrix} 0 \\ 0 \\ u10^{30} \end{bmatrix} = \begin{bmatrix} 10^{30} & -k_a & 0 \\ -k_a & k_a + k_b & -k_b \\ 0 & -k_b & 10^{30} \end{bmatrix} \begin{bmatrix} u^{(1)} \\ u^{(2)} \\ u^{(3)} \end{bmatrix}$$

giving (numerically)

$$\begin{aligned} 0 &= 10^{30}u^{(1)} \\ 0 &= -k_a u^{(1)} + (k_a + k_b)u^{(2)} - k_b u^{(3)} \\ u10^{30} &= 10^{30}u^{(3)} \end{aligned}$$

or

$$u^{(1)} = 0, \quad u^{(3)} = u \quad \text{and then} \quad u^{(2)} = \frac{k_b u}{(k_a + k_b)} \quad \text{as before.}$$

Clearly this procedure enables us to calculate the unknown displacements. Once these are determined the unknown forces are found from the *original* force/displacement equation $\mathbf{G} = \mathbf{KU}$ with the *original* overall stiffness matrix of the problem

EXERCISES

8.1 (a) Obtain the stiffness matrix for the assemblage in Figure 8.10 by (i) using the Castigliano relation and (ii) by superposition
(b) If nodes 1 and 4 are held fixed calculate the unknown displacements, reactions and member forces, assuming applied loads act at nodes 2 and 3 (in the x_1-direction).
(c) If node 2 of the assemblage is subjected to a displacement $u^{(2)} = 1$, and assuming nodes 1 and 4 are held fixed, determine $u^{(3)}$, $G^{(2)}$ and the unknown reactions.

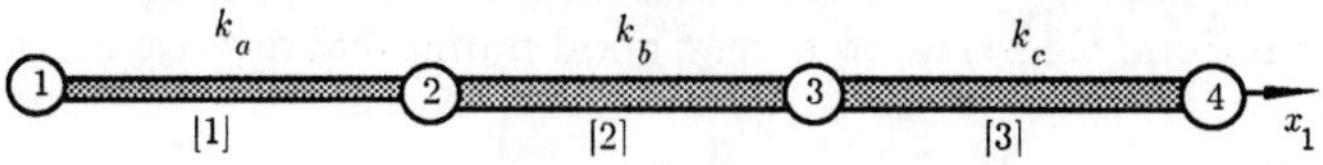

Figure 8.10

8.2 Determine the stiffness combination laws for pin-jointed elements connected together in series and in parallel. (Hint: For series connections consider, as in Figure 8.11, a given force F applied to the end node of two systems and in both cases, using the stiffness method, determine the relation between F and the end-node displacements).

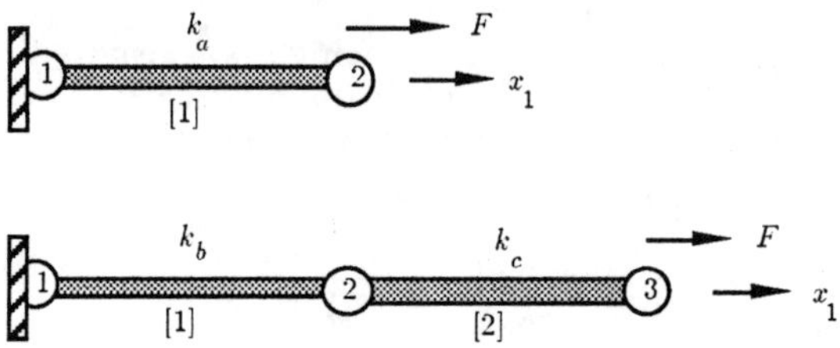

Figure 8.11

8.3 For the assemblage of pin-jointed elements and stiffnesses, applied forces and boundary conditions as shown in Figure 8.12 verify that the displacements at nodes 2 and 3 are 1.67 and 6.67 respectively.

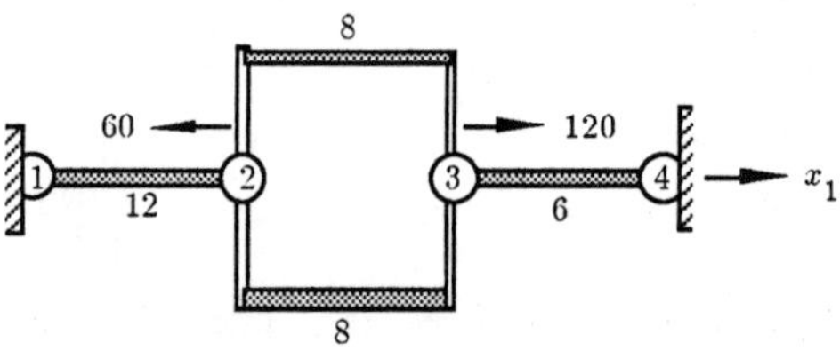

Figure 8.12

8.3 Two and Three Dimensional Pin-jointed Structures

A number of pin-jointed elements connected together to form a two- or three-dimensional structure is called a **truss**. In Figure 8.13 two such assemblages are shown.

Although the transmission tower in Figure 8.13(b) is not normally comprised of pin-jointed elements it can nonetheless be modelled quite accurately as if it were. This will also be the case with other structures (bridge supports, roof supports etc), as long as bending, torsional and shearing stresses are small in comparison to longitudinal stresses.

It is an implicit assumption in the analysis of trusses that the loading on the structure is applied at the joints only.

In real structures, pin-joints (particularly in three dimensions) are rarely used and the connections that are used *do* transmit moments from one element to another. Because of this we can regard the work of this section as applying to *ideal* trusses. One would regard the modelling of real structures by a truss with members connected together with frictionless hinges as being a first approximation. Often, this approach leads to sufficiently accurate results as to not warrant further study. Should greater accuracy (or realism) be required then the structure could be modelled with beam elements which may be rigidly connected together. The (more complicated) analysis of these elements is considered in the next section.

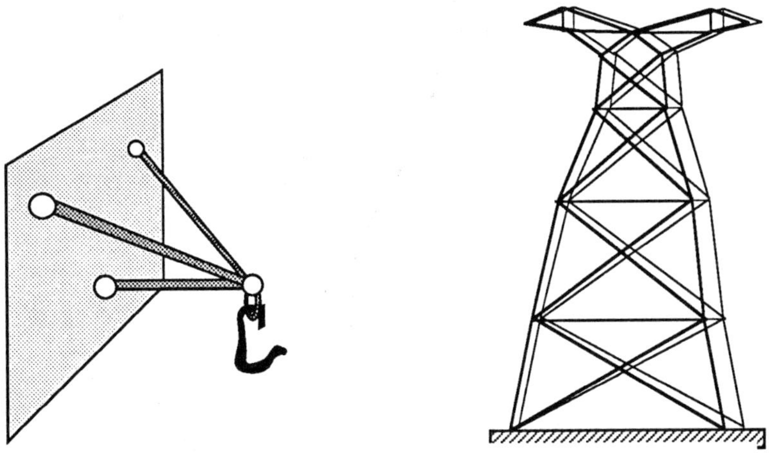

(a) (b)

Figure 8.13

If we are to develop an expression for the overall stiffness matrix of two and three dimensional structures we need first to construct the stiffness matrix of a pin-jointed element in a more general coordinate system. Thus far we have, quite reasonably, chosen a special coordinate system x_i $i = 1, 2, 3$ with the x_1-axis lying along the axis of the element in which to derive the element stiffness matrix. Obviously in a general structure individual elements need not be aligned in the same direction. We need to deal with two distinct coordinate systems: a *global coordinate system* x'_i $i = 1, 2, 3$ chosen for convenience as far as the complete structure is concerned and, for each individual element, a *local coordinate system* x_i $i = 1, 2, 3$ discussed above.

The situation we have in mind is described in Figure 8.14.

We consider an element $[e]$. The element has nodes labelled α and β. Each node may also be attached to other elements ($[a], [b]$ at node α and $[c], [d], [f]$ at node β in the figure). In the general case nodes α and β may be subjected to global forces $G_i^{(\alpha)}$ and $G_i^{(\beta)}$ respectively and at each node displacements $u_i^{(\alpha)}$, $u_i^{(\beta)}$ may be measured. (See Figure 8.15. To aid clarity in this diagram, we have omitted the other elements adjoining the nodes). Each of the vectors may be referred to either the x_i or the x'_i system of coordinates.

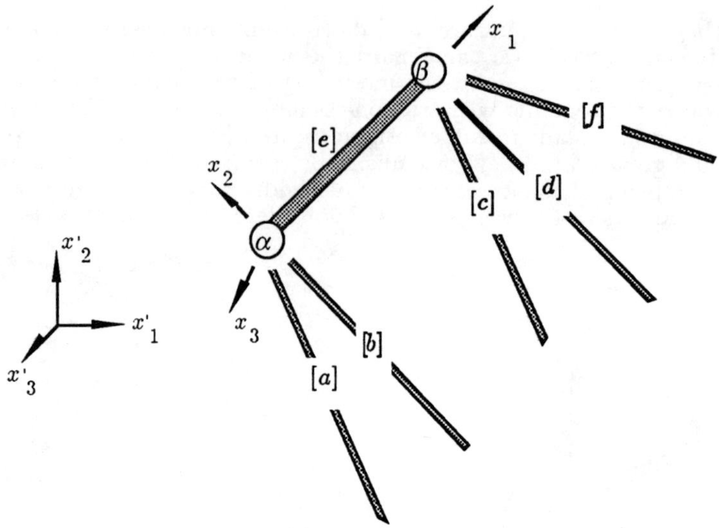

Figure 8.14

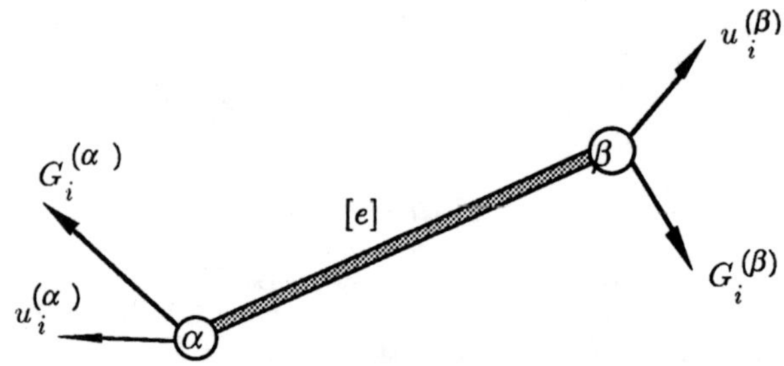

Figure 8.15

At each node the global forces will be distributed between the various elements connected to that node. The vector sum of the forces developed in each element (the member forces) is equal to the global force applied at that node. We shall use the same notation, introduced earlier, and denote the internal force developed within the element at node β by $F_i^{(\beta)} = (F_1^{(\beta)}, F_2^{(\beta)}, F_3^{(\beta)})$ and the internal force developed within this element at node α by $F_i^{(\alpha)} = (F_1^{(\alpha)}, F_2^{(\alpha)}, F_3^{(\alpha)})$. The components of these forces with respect to the global coordinate system are $F'^{(\beta)}_i$ and $F'^{(\alpha)}_i$ respectively. It ought to be emphasised that, for a pin-jointed element, the components of the member forces satisfy the relations, in the unprimed system:

$$F_1^{(\alpha)} = -F_1^{(\beta)} \qquad F_2^{(\alpha)} = F_3^{(\alpha)} = 0 \qquad F_2^{(\beta)} = F_3^{(\beta)} = 0$$

These characteristics of the pin-jointed element are contained in the amended form of (8.8) to apply to members located in three dimensional space:

$$\begin{bmatrix} F_1^{(\alpha)} \\ F_2^{(\alpha)} \\ F_3^{(\alpha)} \\ F_1^{(\beta)} \\ F_2^{(\beta)} \\ F_3^{(\beta)} \end{bmatrix} = \frac{EA}{L} \begin{bmatrix} 1 & 0 & 0 & -1 & 0 & 0 \\ 0 & 0 & 0 & 0 & 0 & 0 \\ 0 & 0 & 0 & 0 & 0 & 0 \\ -1 & 0 & 0 & 1 & 0 & 0 \\ 0 & 0 & 0 & 0 & 0 & 0 \\ 0 & 0 & 0 & 0 & 0 & 0 \end{bmatrix} \begin{bmatrix} u_1^{(\alpha)} \\ u_2^{(\alpha)} \\ u_3^{(\alpha)} \\ u_1^{(\beta)} \\ u_2^{(\beta)} \\ u_3^{(\beta)} \end{bmatrix} \tag{8.34a}$$

or

$$\mathbf{F} = \mathbf{k}_{[e]}\mathbf{u} \tag{8.34b}$$

Since force is a 1st order tensor the components with respect to the primed and unprimed coordinate systems are related through (see (2.7)):

$$F_i^{(\alpha)} = \alpha_{ji}F'^{(\alpha)}_j \qquad \text{and} \qquad F_i^{(\beta)} = \alpha_{ji}F'^{(\beta)}_j \tag{8.35}$$

These two relations may be expressed in a single matrix equation:

$$\begin{bmatrix} F_p^{(\alpha)} \\ F_p^{(\beta)} \end{bmatrix} = \begin{bmatrix} \alpha_{qp} & 0 \\ 0 & \alpha_{qp} \end{bmatrix} \begin{bmatrix} F'^{(\alpha)}_q \\ F'^{(\beta)}_q \end{bmatrix} \tag{8.36}$$

with an exactly similar relation connecting the displacement components at each node. Although we shall not refer to it each time in what follows the 'zero' terms in the square matrices refer to *zero matrices* of whatever order is required to make sense of the matrix equation in which they appear. In this particular case they represent 3×3 zero matrices. Now, from (8.34a)

$$\begin{bmatrix} F_p^{(\alpha)} \\ F_p^{(\beta)} \end{bmatrix} = \frac{EA}{L} \begin{bmatrix} k_{pq} & -k_{pq} \\ -k_{pq} & k_{pq} \end{bmatrix} \begin{bmatrix} u_q^{(\alpha)} \\ u_q^{(\beta)} \end{bmatrix} \tag{8.37}$$

in which

$$k_{pq} = \begin{bmatrix} 1 & 0 & 0 \\ 0 & 0 & 0 \\ 0 & 0 & 0 \end{bmatrix} = \delta_{p1}\delta_{q1} \tag{8.38}$$

But, using (8.36) in (8.37) we have, (changing dummy indices where necessary)

$$\begin{bmatrix} \alpha_{ip} & 0 \\ 0 & \alpha_{ip} \end{bmatrix} \begin{bmatrix} F'^{(\alpha)}_i \\ F'^{(\beta)}_i \end{bmatrix} = \frac{EA}{L} \begin{bmatrix} k_{pq} & -k_{pq} \\ -k_{pq} & k_{pq} \end{bmatrix} \begin{bmatrix} \alpha_{jq} & 0 \\ 0 & \alpha_{jq} \end{bmatrix} \begin{bmatrix} u'^{(\alpha)}_j \\ u'^{(\beta)}_j \end{bmatrix}$$

We can isolate the force matrix on the left-hand side if we multiply this relation by the matrix:

$$\begin{bmatrix} \alpha_{mp} & 0 \\ 0 & \alpha_{mp} \end{bmatrix}$$

and use the identity $\alpha_{mp}\alpha_{ip} = \delta_{mi}$ (see (2.8)b). This implies:

$$\begin{bmatrix} F'^{(\alpha)}_m \\ F'^{(\beta)}_m \end{bmatrix} = \frac{EA}{L} \begin{bmatrix} \alpha_{mp} & 0 \\ 0 & \alpha_{mp} \end{bmatrix} \begin{bmatrix} k_{pq} & -k_{pq} \\ -k_{pq} & k_{pq} \end{bmatrix} \begin{bmatrix} \alpha_{jq} & 0 \\ 0 & \alpha_{jq} \end{bmatrix} \begin{bmatrix} u'^{(\alpha)}_j \\ u'^{(\beta)}_j \end{bmatrix} \tag{8.39}$$

The matrix

$$\mathbf{k}'_{[e]} = \frac{EA}{L} \begin{bmatrix} \alpha_{mp} & 0 \\ 0 & \alpha_{mp} \end{bmatrix} \begin{bmatrix} k_{pq} & -k_{pq} \\ -k_{pq} & k_{pq} \end{bmatrix} \begin{bmatrix} \alpha_{jq} & 0 \\ 0 & \alpha_{jq} \end{bmatrix} \tag{8.40}$$

is the stiffness matrix of the element $[e]$ expressed with respect to global coordinates $x'_i \; i = 1, 2, 3$. We can write this in the more elegant form:

$$\mathbf{k}'_{[e]} = \mathbf{T}^T \mathbf{k}_{[e]} \mathbf{T} \tag{8.40a}$$

where $\mathbf{T}$ is the 6×6 *transformation matrix*

$$\mathbf{T} = \begin{bmatrix} t_{ij} & 0 \\ 0 & t_{ij} \end{bmatrix} \qquad t_{ij} = \alpha_{ji}$$

By direct expansion of (8.40) we find (using (8.38)) that

$$\mathbf{k}'_{[e]} = \frac{EA}{L} \begin{bmatrix} \alpha_{j1}\alpha_{m1} & -\alpha_{j1}\alpha_{m1} \\ -\alpha_{j1}\alpha_{m1} & \alpha_{j1}\alpha_{m1} \end{bmatrix} \tag{8.41}$$

in which

$$\alpha_{j1}\alpha_{m1} = \begin{bmatrix} \alpha_{11}^2 & \alpha_{11}\alpha_{21} & \alpha_{11}\alpha_{31} \\ \alpha_{21}\alpha_{11} & \alpha_{21}^2 & \alpha_{21}\alpha_{31} \\ \alpha_{31}\alpha_{11} & \alpha_{31}\alpha_{21} & \alpha_{31}^2 \end{bmatrix}$$

The rotation cosines are easily determined in terms of the nodal coordinates; $x'^{(\beta)}_i$ and $x'^{(\alpha)}_i$ through the relations:

$$\alpha_{11} = \frac{x'^{(\beta)}_1 - x'^{(\alpha)}_1}{L} \qquad \alpha_{21} = \frac{x'^{(\beta)}_2 - x'^{(\alpha)}_2}{L} \qquad \alpha_{31} = \frac{x'^{(\beta)}_3 - x'^{(\alpha)}_3}{L} \tag{8.42}$$

Equation (8.39) contains the local information for each element. If we develop an expression for the strain energy of the complete structure and again appeal to the Castigliano relations we shall find that the global forces, held in the matrix $\mathbf{G}$, are related to the nodal displacements, held in the matrix $\mathbf{U}$, through the relation:

$$\mathbf{G} = \mathbf{K}\mathbf{U} \tag{8.43}$$

where $\mathbf{K}$ is the overall stiffness matrix. An examination of the formation of (8.43), by this process, shows that, as previously, $\mathbf{K}$ may be formed by directly superimposing the separate stiffness matrices, $\mathbf{k}$, (expressed in global coordinates) for each element and that $\mathbf{G}$ may be constructed by similarly superimposing the force matrices $\mathbf{F}$ of each element. In reality, of course, in structures we can ignore the force superposition since either the elements of $\mathbf{G}$ are given as initial information (in which case we can write them into $\mathbf{G}$ directly) or else the elements of $\mathbf{G}$ are unknowns of the problem, in which case (8.43) needs to be solved to determine them.

Example 8.3

Determine all the unknowns in the three element pin-jointed structure shown in Figure 8.16. A force P is applied to node 1 in the direction indicated. Assume each element has the same Young's modulus E and cross-sectional area A. Nodes 2,3 and 4 are held fixed in position against linear displacements.

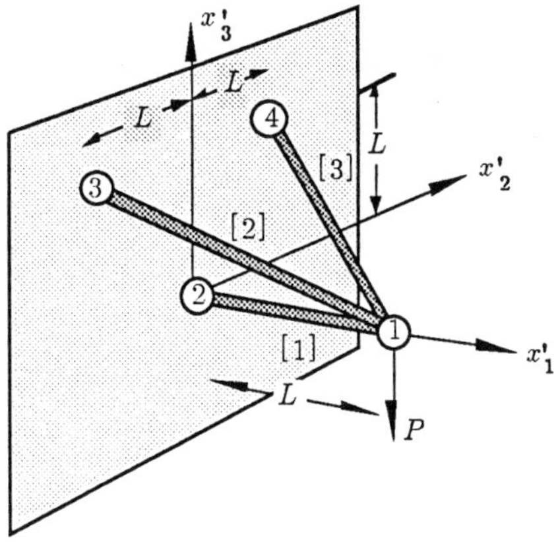

Figure 8.16

Solution

In this problem the nodal coordinates are:

$$x'^{(1)}_i = (L, 0, 0) \quad x'^{(2)}_i = (0, 0, 0) \quad x'^{(3)}_i = (0, -L, L) \quad x'^{(4)}_i = (0, L, L)$$

We now proceed to determine the element stiffness matrices in global coordinates.

- Element [1] nodes $\alpha = 1, \beta = 2$ length $= L$

From (8.42) $\alpha_{11} = -1 \quad \alpha_{21} = 0 \quad \alpha_{31} = 0$. Hence

$$\mathbf{k}'_{[1]} = \frac{EA}{L}\begin{bmatrix} k'_{pq} & -k'_{pq} \\ -k'_{pq} & k'_{pq} \end{bmatrix}\begin{matrix} 1 \\ 2 \end{matrix} \qquad \text{where} \quad k'_{pq} = \begin{bmatrix} 1 & 0 & 0 \\ 0 & 0 & 0 \\ 0 & 0 & 0 \end{bmatrix}$$

- Element [2] nodes $\alpha = 1, \beta = 3$ length $= \sqrt{3}L$

From (8.42) $\alpha_{11} = -\frac{1}{\sqrt{3}} \quad \alpha_{21} = -\frac{1}{\sqrt{3}} \quad \alpha_{31} = \frac{1}{\sqrt{3}}$. Hence

$$\mathbf{k}'_{[2]} = \frac{EA}{L}\begin{bmatrix} k'_{pq} & -k'_{pq} \\ -k'_{pq} & k'_{pq} \end{bmatrix}\begin{matrix} 1 \\ 3 \end{matrix} \qquad \text{where} \quad k'_{pq} = \frac{1}{3\sqrt{3}}\begin{bmatrix} 1 & 1 & -1 \\ 1 & 1 & -1 \\ -1 & -1 & 1 \end{bmatrix}$$

- Element [3] nodes $\alpha = 1, \beta = 4$ length $= \sqrt{3}L$

From (8.42) $\alpha_{11} = -\frac{1}{\sqrt{3}} \quad \alpha_{21} = \frac{1}{\sqrt{3}} \quad \alpha_{31} = \frac{1}{\sqrt{3}}$. Hence

$$\mathbf{k}'_{[3]} = \frac{EA}{L}\begin{bmatrix} k'_{pq} & -k'_{pq} \\ -k'_{pq} & k'_{pq} \end{bmatrix} \begin{matrix} 1 \\ 4 \end{matrix} \qquad \text{where} \quad k'_{pq} = \frac{1}{3\sqrt{3}}\begin{bmatrix} 1 & -1 & -1 \\ -1 & 1 & 1 \\ -1 & 1 & 1 \end{bmatrix}$$

Superimposing these three matrices would give an overall stiffness $\mathbf{K}$ matrix of size 12×12 and the force/displacement equations:

$$\mathbf{G} = \mathbf{K}\mathbf{U} \tag{8.44}$$

can now be analysed. (The reader should construct this matrix. S/he will then begin to appreciate that this technique, though extremely useful, requires a powerful computer when analysing realistic structural problems which may have hundreds of pin-jointed elements).
The first three equations of the set (8.44) are

$$\begin{bmatrix} 0 \\ 0 \\ -P \end{bmatrix} = \frac{EA}{L}\begin{bmatrix} 1+\frac{2\sqrt{3}}{9} & 0 & -\frac{2\sqrt{3}}{9} \\ 0 & \frac{2\sqrt{3}}{9} & 0 \\ -\frac{2\sqrt{3}}{9} & 0 & \frac{2\sqrt{3}}{9} \end{bmatrix}\begin{bmatrix} u'^{(1)}_1 \\ u'^{(1)}_2 \\ u'^{(1)}_3 \end{bmatrix} \tag{8.45}$$

where we have utilised the boundary conditions on displacement:

$$u'^{(\alpha)}_i = 0 \quad \alpha = 2, 3, 4$$

The second equation of (8.45) immediately implies that $u'^{(1)}_2 = 0$ (as we expect from considerations of symmetry). The other two equations of (8.45) are easily solved to give

$$u'^{(1)}_i = (-\frac{PL}{EA}, 0, -[1 + \frac{9}{2\sqrt{3}}]\frac{PL}{EA})$$

In realistic problems the displacement will be extremely small. For example if we use steel rods, with a Young's modulus $2 \times 10^{11}\text{Nm}^{-2}$ of cross-sectional area $A = 0.005\text{m}^2$ (radius $\approx$ 4cm) and $L = 1$m then for a load of magnitude 10^4N (equivalent to a weight of about 1000kg):

$$u'^{(1)}_i = (-10^{-5}\text{m}, 0, -3.6 \times 10^{-5}\text{m})$$

The reaction forces at the supports can now be obtained. For example, the second three equations of the set (8.44) are (again taking account of the displacement boundary conditions)

$$\begin{bmatrix} G^{(2)}_1 \\ G^{(2)}_2 \\ G^{(2)}_3 \end{bmatrix} = \frac{EA}{L}\begin{bmatrix} -1 & 0 & 0 \\ 0 & 0 & 0 \\ 0 & 0 & 0 \end{bmatrix}\begin{bmatrix} u'^{(1)}_1 \\ u'^{(1)}_2 \\ u'^{(1)}_3 \end{bmatrix} = P\begin{bmatrix} -1 & 0 & 0 \\ 0 & 0 & 0 \\ 0 & 0 & 0 \end{bmatrix}\begin{bmatrix} -1 \\ 0 \\ -[1 + \frac{9}{2\sqrt{3}}] \end{bmatrix}$$

implying

$$G^{(2)}_i = (P, 0, 0)$$

In an exactly similar way, using the third set of three equations from (8.44) we have

$$\begin{bmatrix} G_1^{(3)} \\ G_2^{(3)} \\ G_3^{(3)} \end{bmatrix} = \frac{EA}{3\sqrt{3}L} \begin{bmatrix} -1 & -1 & 1 \\ -1 & -1 & 1 \\ 1 & 1 & -1 \end{bmatrix} \begin{bmatrix} u'^{(1)}_1 \\ u'^{(1)}_2 \\ u'^{(1)}_3 \end{bmatrix} = -\frac{P}{2} \begin{bmatrix} 1 \\ 1 \\ -1 \end{bmatrix}$$

and, from the final set of three equations in (8.44):

$$\begin{bmatrix} G_1^{(4)} \\ G_2^{(4)} \\ G_3^{(4)} \end{bmatrix} = \frac{EA}{3\sqrt{3}L} \begin{bmatrix} -1 & 1 & 1 \\ 1 & -1 & -1 \\ 1 & -1 & -1 \end{bmatrix} \begin{bmatrix} u'^{(1)}_1 \\ u'^{(1)}_2 \\ u'^{(1)}_3 \end{bmatrix} = -\frac{P}{2} \begin{bmatrix} 1 \\ -1 \\ -1 \end{bmatrix}$$

As a check on the working the global forces should be in equilibrium : $\sum_{\alpha=1}^{4} G_i^{(\alpha)} = 0$.

This is easily confirmed from the results obtained above.

We now turn our attention to the internal forces developed within each member as a result of the loading. These are obtained from the local force/displacement relations in (8.37). For example, for element [2] say, with nodes $\alpha = 1$, $\beta = 3$ and length $\sqrt{3}L$

$$\begin{bmatrix} F_p^{(1)} \\ F_p^{(3)} \end{bmatrix} = \frac{EA}{\sqrt{3}L} \begin{bmatrix} k_{pq} & -k_{pq} \\ -k_{pq} & k_{pq} \end{bmatrix} \begin{bmatrix} u_q^{(1)} \\ u_q^{(3)} \end{bmatrix} \qquad \textbf{(8.46)}$$

with k_{pq} defined in (8.38). But using the transformation relations (which are exactly those determined for the force matrix in (8.36)):

$$\begin{bmatrix} u_q^{(1)} \\ u_q^{(3)} \end{bmatrix} = \begin{bmatrix} \alpha_{jq} & 0 \\ 0 & \alpha_{jq} \end{bmatrix} \begin{bmatrix} u'^{(1)}_j \\ u'^{(3)}_j \end{bmatrix} = \begin{bmatrix} \alpha_{jq} & 0 \\ 0 & \alpha_{jq} \end{bmatrix} \begin{bmatrix} u'^{(1)}_j \\ 0 \end{bmatrix}$$

$$= \begin{bmatrix} \alpha_{jq} u'^{(1)}_j \\ 0 \end{bmatrix}$$

in which we have again utilised the boundary conditions. Hence for this element (8.46) becomes (using (8.38))

$$\begin{bmatrix} F_p^{(1)} \\ F_p^{(3)} \end{bmatrix} = \frac{EA}{\sqrt{3}L} \begin{bmatrix} k_{pq}\alpha_{jq} u'^{(1)}_j \\ -k_{pq}\alpha_{jq} u'^{(1)}_j \end{bmatrix} = \frac{EA}{\sqrt{3}L} \begin{bmatrix} \delta_{p1}\alpha_{j1} u'^{(1)}_j \\ -\delta_{p1}\alpha_{j1} u'^{(1)}_j \end{bmatrix}$$

$$= \frac{EA}{\sqrt{3}L} \begin{bmatrix} \delta_{p1}(\alpha_{11} u'^{(1)}_1 + \alpha_{31} u'^{(1)}_3) \\ -\delta_{p1}(\alpha_{11} u'^{(1)}_1 + \alpha_{31} u'^{(1)}_3) \end{bmatrix}$$

and, using the solution for the displacements we find:

$$F_p^{(1)} = (-\frac{3}{2\sqrt{3}}P, 0, 0) \qquad \text{and} \qquad F_p^{(3)} = -F_p^{(1)}$$

showing (see Figure 8.17) that element [2] is in tension.

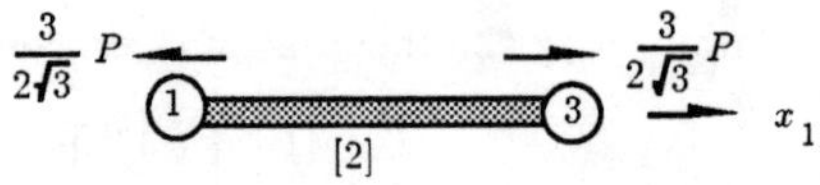

Figure 8.17

The internal state of the other two elements can be analysed in a similar way.

EXERCISES

8.4 Repeat Exercise 6.1 using the stiffness matrix approach.

8.5 The three element pin-jointed structure in Figure 8.18 is loaded at node 1 as shown. All elements have the same cross-sectional area and Young's modulus. Using the stiffness matrix method show that the displacement at node 1 is

$$u_1^{(1)} = \frac{L}{AE}\frac{1}{3(\sqrt{3}+1)}[(\sqrt{3}+3)P_1^{(1)} + (\sqrt{3}-3)P_2^{(1)}]$$

$$u_2^{(1)} = \frac{L}{AE}\frac{1}{3(\sqrt{3}+1)}[(\sqrt{3}-3)P_1^{(1)} + (3\sqrt{3}+9)P_2^{(1)}]$$

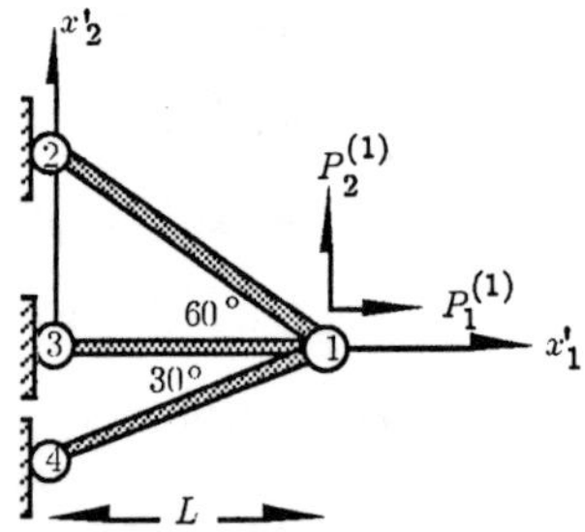

Figure 8.18

8.6 A seven-element truss (see Figure 8.19) is loaded at nodes 2 and 3 as shown. All elements have the same cross-sectional area and Young's modulus. Calculate all the nodal displacements. (Hint: utilise the geometrical symmetry in the structure to reduce the number of unknowns to manageable proportions). Also determine the internal forces developed within each member.

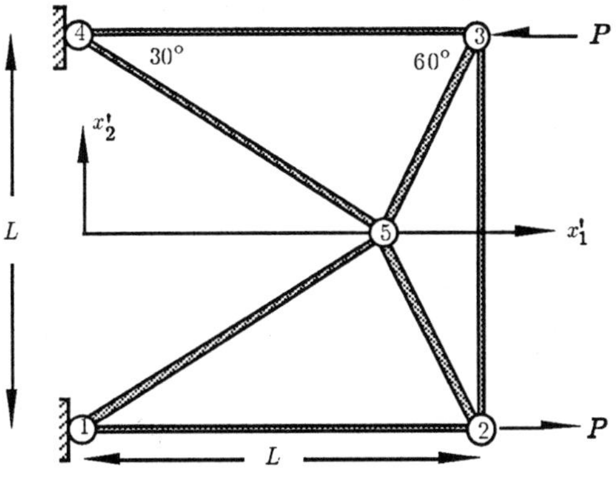

Figure 8.19

8.7 Determine the maximum load the three-element truss (see Figure 8.20) can carry if neither member is to exceed the yield stress Y.

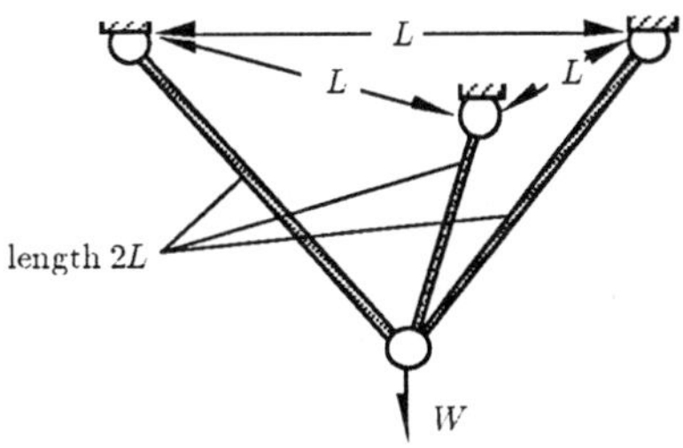

Figure 8.20

8.4 Beam Elements

A beam element, unlike a pin-jointed element, may be *rigidly* connected to other elements and so can transmit torsional couples, bending moments and shearing forces from one element to another. It is a more realistic building element than the pin-jointed element but is, correspondingly, a great deal more complicated.

Consider the beam described in Figure 8.21. As with the pin-joined element the x_1-axis is chosen to lie along the line of centroids of the cross-section. The x_2- and the x_3-axes are chosen to lie along the principal axes of the second moment of area of the cross-section. (See the beam bending analysis of Section 5.3 where, essentially, the same choices for directions of the coordinate axes were made).

In Figure 8.21, to aid clarity, we have only shown the moment and rotation vectors (indicated by a double arrow) $M_i^{(1)}$, $\theta_i^{(1)}$ at node 1 and the force and displacement vectors $F_i^{(2)}$, $u_i^{(2)}$ at node 2. We emphasize that at each node we can measure six generalised displacement components (u_i, θ_i) $i = 1, 2, 3$ and we can apply six generalised force components (F_i, M_i) $i = 1, 2, 3$.

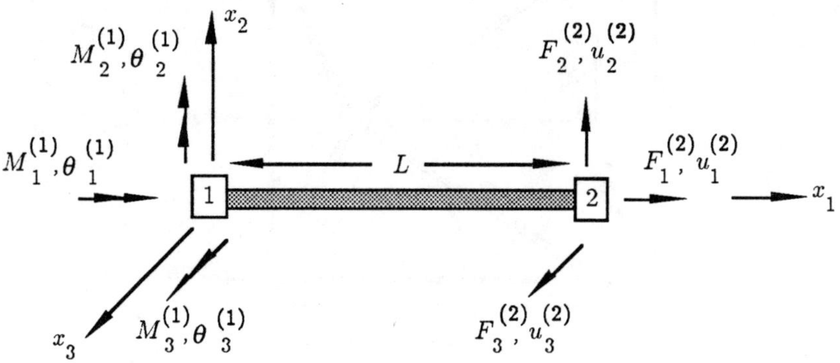

Figure 8.21

Our aim is to determine the stiffness matrix with respect to local coordinates x_i $i = 1, 2, 3$. At a later stage we will need to express the stiffness matrix with respect to a set of global coordinates x'_i $i = 1, 2, 3$ so that, for a three dimensional structure comprised of many beam elements, an overall stiffness matrix of the structure can be obtained.

As we have already noted the beam may be stretched or compressed along its axis by longitudinal forces, it may bend in any one of the two principal planes by bending couples and/or shearing forces and it may also be twisted about its axis by torsional couples. We shall examine the force/displacement relation for each separate deformation in turn.

- **Axial Deformation**

The forces and displacements pertinant to this deformation are $F_1^{(1)}, F_1^{(2)}$ and $u_1^{(1)}, u_1^{(2)}$ respectively. The force/displacement relation has already been deduced in the analysis of the pin-joint element:

$$\begin{bmatrix} F_1^{(1)} \\ F_1^{(2)} \end{bmatrix} = \frac{EA}{L} \begin{bmatrix} 1 & -1 \\ -1 & 1 \end{bmatrix} \begin{bmatrix} u_1^{(1)} \\ u_1^{(2)} \end{bmatrix} \tag{8.47}$$

- **Torsional Deformation**

The forces and displacements of concern in this deformation are $M_1^{(1)}, M_1^{(2)}$ and $\theta_1^{(1)}, \theta_1^{(2)}$ respectively. The strain energy in torsion is (see (7.24), (7.25))

$$\begin{aligned} W_T = \frac{1}{2} \iiint_V \varepsilon_{ij}\sigma_{ij} dV &= \iiint_V (\varepsilon_{21}\sigma_{21} + \varepsilon_{31}\sigma_{31}) dV \\ &= \frac{1}{2}\mu\tau^2 \iiint_V \left[(\frac{\partial \Psi}{\partial x_2})^2 + \frac{\partial \Psi}{\partial x_3})^2 \right] dV \end{aligned} \tag{8.48}$$

However, Stokes' Theorem (in fact a special case of Green's Theorem in the x_2x_3

plane: see (1.65)) with vector field $\underline{A} = (0, -\Psi\frac{\partial\Psi}{\partial x_3}, \Psi\frac{\partial\Psi}{\partial x_2})$ implies

$$\oint_C \left[-\Psi\frac{\partial\Psi}{\partial x_3}dx_2 + \Psi\frac{\partial\Psi}{\partial x_2}dx_3\right] = \iint_A \left[(\frac{\partial\Psi}{\partial x_2})^2 + (\frac{\partial\Psi}{\partial x_3})^2 + \Psi\left(\frac{\partial^2\Psi}{\partial x_2^2} + \frac{\partial^2\Psi}{\partial x_3^2}\right)\right] dx_2 dx_3$$

Since, for simply connected cross-sections (henceforth we will assume all beam elements are of this kind) $\Psi = 0$ on the boundary C of the cross-section A we have,

$$\iint_A \left[(\frac{\partial\Psi}{\partial x_2})^2 + (\frac{\partial\Psi}{\partial x_3})^2\right] dx_2 dx_3 = -\iint_A \Psi\left(\frac{\partial^2\Psi}{\partial x_2^2} + \frac{\partial^2\Psi}{\partial x_3^2}\right) dx_2 dx_3$$

Now for the volume integral on the right hand side of (8.48):

$$\begin{aligned}\frac{1}{2}\mu\tau^2 \iiint_V \left[(\frac{\partial\Psi}{\partial x_2})^2 + \frac{\partial\Psi}{\partial x_3})^2\right] dV &= \frac{1}{2}\mu\tau^2 L \iint_A \left[(\frac{\partial\Psi}{\partial x_2})^2 + (\frac{\partial\Psi}{\partial x_3})^2\right] dx_2 dx_3 \\ &= \frac{1}{2}\mu\tau^2 L \iint_A 2\Psi dx_2 dx_3 \qquad \text{from (7.21)}\end{aligned}$$

and so

$$W_T = \frac{1}{2}\tau^2 L T_R \tag{8.49}$$

where T_R is the torsional rigidity (see (7.29) with $k_i = 0 \quad i = 1, 2, ...k$) of the cross-section.

The twist τ is related to the angle through which each plane $x_1 =$ constant is rotated with respect to one end:

$$\theta_1 = \theta_1^{(1)} + \tau x_1$$

and when $x_1 = L$ then $\theta_1 \equiv \theta_1^{(2)}$ to give

$$\tau = \frac{\theta_1^{(2)} - \theta_1^{(1)}}{L}$$

and finally:

$$W_T = \frac{1}{2}(\theta_1^{(2)} - \theta_1^{(1)})^2 \frac{T_R}{L} \tag{8.50}$$

Now using the generalised Castigliano relations (see Exercise 6.5) we deduce

$$M_1^{(1)} = \frac{\partial W_T}{\partial \theta_1^{(1)}} = -(\theta_1^{(2)} - \theta_1^{(1)})\frac{T_R}{L} \tag{8.51a}$$

and

$$M_1^{(2)} = \frac{\partial W_T}{\partial \theta_1^{(2)}} = (\theta_1^{(2)} - \theta_1^{(1)})\frac{T_R}{L} \tag{8.51b}$$

(These could have been obtained from the expression for the torsional couple immediately preceeding (7.28): $M = \tau T_R$)

The relations in (8.51) can obviously be written in matrix form:

$$\begin{bmatrix} M_1^{(1)} \\ M_1^{(2)} \end{bmatrix} = \frac{T_R}{L} \begin{bmatrix} 1 & -1 \\ -1 & 1 \end{bmatrix} \begin{bmatrix} \theta_1^{(1)} \\ \theta_1^{(2)} \end{bmatrix} \tag{8.52}$$

- **Bending Deformation in the x_1x_2 Plane**

Bending in the x_1x_2 plane can be produced by forces $F_2^{(1)}, F_2^{(2)}, M_3^{(1)}, M_3^{(2)}$. The corresponding displacements are $u_2^{(1)}, u_2^{(2)}$ and $\theta_3^{(1)}, \theta_3^{(2)}$ respectively.

The strain energy in bending in the x_1x_2 plane, W_{B3} is (see (6.36)):

$$W_{B3} = \frac{1}{2}EI_3 \int_0^L (\frac{d^2u_2}{dx_1^2})^2 dx_1 \tag{8.53}$$

where I_3 is the principal second moment of area of the cross-section about the x_3-axis.

If linear transverse forces are absent then, according to the Euler-Bernoulli theory (see (6.39))

$$\frac{d^4u_2}{dx_1^4} = 0$$

and so

$$u_2 = b_0 + b_1x_1 + b_2x_1^2 + b_3x_1^3 \tag{8.54}$$

in which b_0, b_1, b_2 and b_3 are constants. This implies, by direct integration of the integral in (8.53) that

$$W_{B3} = \frac{1}{2}EI_3(4b_2^2L + 12b_2b_3L^2 + 12b_3^2L^3) \tag{8.55}$$

We can express u_2 (and hence W_{B3}) in terms of the displacements and rotations measured at each end of the beam. First, however, we must relate the rotations to the derivative of u_2 with respect to x_1. Consider Figure 8.22 which indicates the typical deformation induced in the beam as a result of the application of positive bending moments at each node.

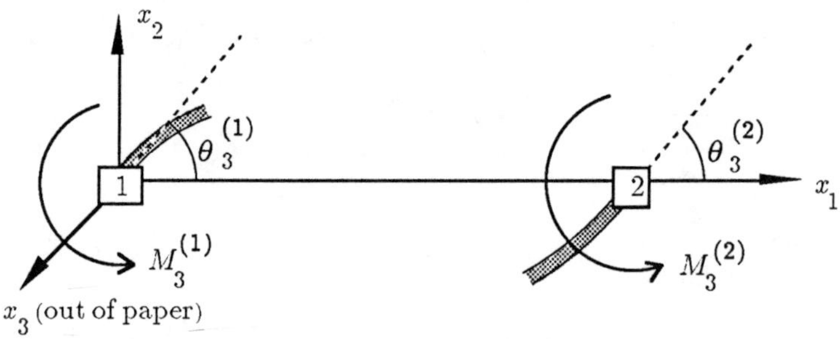

Figure 8.22

The angle $\theta_3^{(2)}$ (in radians) is always measured from the positive x_1-axis in the same direction as the applied couple $M_3^{(2)}$. Realising that the angle of rotation will always, in any practical situation, be extremely small we have, from the calculus:

$$\theta_3^{(2)} \approx \tan\theta_3^{(2)} = \frac{du_2}{dx_1}(x_1 = L)$$

and

$$\theta_3^{(1)} \approx \tan\theta_3^{(1)} = \frac{du_2}{dx_1}(x_1 = 0)$$

Also (obviously) $u_2(0) = u_2^{(1)}$ and $u_2(L) = u_2^{(2)}$. Hence, using (8.54) we find:

$$\begin{aligned}
u_2^{(1)} &= b_0 \\
\theta_3^{(1)} &= b_1 \\
u_2^{(2)} &= b_0 + b_1 L + b_2 L^2 + b_3 L^3 \\
\theta_3^{(2)} &= b_1 + 2b_2 L + 3b_3 L^2
\end{aligned}$$

The solution of these four equations implies, in particular:

$$\begin{aligned}
b_2 &= \frac{3}{L^2}(u_2^{(2)} - u_2^{(1)}) - \frac{1}{L}(2\theta_3^{(1)} + \theta_3^{(2)}) \\
b_3 &= \frac{2}{L^3}(u_2^{(1)} - u_2^{(2)}) + \frac{1}{L^2}(\theta_3^{(1)} + \theta_3^{(2)})
\end{aligned} \tag{8.56}$$

These values can now be used in (8.55) to express W_{B3} in terms of the nodal displacements and rotations. Now using the generalised Castigliano relations we find

$$\begin{aligned}
F_2^{(1)} = \frac{\partial W_{B3}}{\partial u_2^{(1)}} &= \frac{1}{2}EI_3[8Lb_2(-\frac{3}{L^2}) + 12b_3L^2(-\frac{3}{L^2}) + 12b_2L^2(\frac{2}{L^3}) + 24b_3L^3(\frac{2}{L^3})] \\
&= 6EI_3b_3
\end{aligned}$$

Similarly, for the other nodal forces and moments:

$$\begin{aligned}
M_3^{(1)} &= \frac{\partial W_{B3}}{\partial \theta_3^{(1)}} = -\frac{2EI_3}{L}b_2 \\
F_2^{(2)} &= \frac{\partial W_{B3}}{\partial u_2^{(2)}} = -6EI_3b_3 \\
M_3^{(2)} &= \frac{\partial W_{B3}}{\partial \theta_3^{(2)}} = EI_3(2b_2 + 6b_3L)
\end{aligned}$$

Hence, using (8.56), we obtain the force displacement relation as:

$$\begin{bmatrix} F_2^{(1)} \\ M_3^{(1)} \\ F_2^{(2)} \\ M_3^{(2)} \end{bmatrix} = \frac{EI_3}{L^3} \begin{bmatrix} 12 & 6L & -12 & 6L \\ 6L & 4L^2 & -6L & 2L^2 \\ -12 & -6L & 12 & -6L \\ 6L & 2L^2 & -6L & 4L^2 \end{bmatrix} \begin{bmatrix} u_2^{(1)} \\ \theta_3^{(1)} \\ u_2^{(2)} \\ \theta_3^{(2)} \end{bmatrix} \tag{8.57}$$

- **Bending Deformation in the x_1x_3 Plane**

Bending in the x_1x_3 plane can be produced by forces $F_3^{(1)}, F_3^{(2)}, M_2^{(1)}, M_2^{(2)}$. The corresponding displacements are $u_3^{(1)}, u_3^{(2)}$ and $\theta_2^{(1)}, \theta_2^{(2)}$ respectively.

The strain energy in bending in the x_1x_3 plane, W_{B2} is (see (6.36)):

$$W_{B2} = \frac{1}{2}EI_2 \int_0^L (\frac{d^2u_3}{dx_1^2})^2 dx_1 \tag{8.58}$$

where I_2 is the principal second moment of area of the cross-section about the x_2-axis.

Proceeding as in the previous analysis:

$$u_2 = a_0 + a_1x_1 + a_2x_1^2 + a_3x_1^3 \tag{8.59}$$

in which a_0, a_1, a_2 and a_3 are constants, implying (from (8.58)) that

$$W_{B2} = \frac{1}{2}EI_2(4a_2^2L + 12a_2a_3L^2 + 12a_3^2L^3) \tag{8.60}$$

We again express the constants a_2, a_3 in terms of the displacements and rotations measured at each node. In this respect we must examine the relation between angle of rotation measured at each end of the beam and the derivative of u_3 with respect to x_1. Consider Figure 8.23 which sketches the deformation induced in the beam as a result of the application of *positive* bending moments at each node.

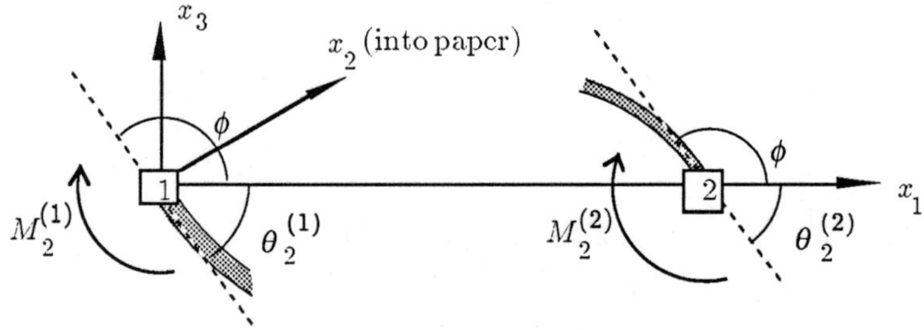

Figure 8.23

At each node the angle θ_2 (which is extremely small) is measured from the positive x_1-axis in the same direction as the positive couple M_2 applied at that point.

At node [1]

$$\frac{du_3}{dx_1}(x_1 = 0) = \tan\phi = \tan(180 - \theta_2^{(1)}) = -\tan(\theta_2^{(1)}) \approx -\theta_2^{(1)}$$

whilst at node [2], using the same argument

$$\frac{du_3}{dx_1}(x_1 = L) = -\theta_2^{(2)}$$

Hence

$$\begin{aligned}
u_3^{(1)} &= a_0 \\
\theta_2^{(1)} &= -\frac{du_3}{dx_1}(x_1 = 0) = -a_1 \\
u_3^{(2)} &= a_0 + a_1 L + a_2 L^2 + a_3 L^3 \\
\theta_2^{(2)} &= -\frac{du_3}{dx_1}(x_1 = L) = -a_1 - 2a_2 L - 3a_3 L^2
\end{aligned}$$

Solving, we obtain:

$$\begin{aligned}
a_2 &= \frac{3}{L^2}(u_3^{2)} - u_3^{(1)}) + \frac{1}{L}(2\theta_2^{(1)} + \theta_2^{(2)}) \\
a_3 &= \frac{2}{L^3}(u_3^{1)} - u_3^{(2)}) - \frac{1}{L^2}(\theta_2^{(1)} + \theta_2^{(2)})
\end{aligned}$$

So again using the Castigliano relations we find:

$$\begin{bmatrix} F_3^{(1)} \\ M_2^{(1)} \\ F_3^{(2)} \\ M_2^{(2)} \end{bmatrix} = \frac{EI_2}{L^3} \begin{bmatrix} 12 & -6L & -12 & -6L \\ -6L & 4L^2 & 6L & 2L^2 \\ -12 & 6L & 12 & 6L \\ -6L & 2L^2 & 6L & 4L^2 \end{bmatrix} \begin{bmatrix} u_3^{(1)} \\ \theta_2^{(1)} \\ u_3^{(2)} \\ \theta_2^{(2)} \end{bmatrix} \tag{8.61}$$

The full relationship connecting generalised forces,

$$\overline{F}_i^{(\alpha)} = (F_1^{(\alpha)}, F_2^{(\alpha)}, F_3^{(\alpha)}, M_1^{(\alpha)}, M_2^{(\alpha)}, M_3^{(\alpha)})$$

to generalised displacements

$$\overline{u}_i^{(\alpha)} = (u_1^{(\alpha)}, u_2^{(\alpha)}, u_3^{(\alpha)}, \theta_1^{(\alpha)}, \theta_2^{(\alpha)}, \theta_3^{(\alpha)})$$

at each node ($\alpha = 1, 2$) for a beam element in local coordinates $x_i \quad i = 1, 2, 3$ can now be written in matrix form:

$$\begin{bmatrix} \overline{F}_p^{(\alpha)} \\ \overline{F}_p^{(\beta)} \end{bmatrix} = \begin{bmatrix} K_{pq}^{(\alpha\alpha)} & K_{pq}^{(\alpha\beta)} \\ K_{pq}^{(\beta\alpha)} & K_{pq}^{(\beta\beta)} \end{bmatrix} \begin{bmatrix} \overline{u}_q^{(\alpha)} \\ \overline{u}_q^{(\beta)} \end{bmatrix} \tag{8.62}$$

in which (using the relations in (8.47), (8.52), (8.57) and (8.61))

$$K_{pq}^{(\alpha\alpha)} = \frac{E}{L^3}\begin{bmatrix} AL^2 & 0 & 0 & 0 & 0 & 0 \\ 0 & 12I_3 & 0 & 0 & 0 & 6I_3L \\ 0 & 0 & 12I_2 & 0 & -6I_2L & 0 \\ 0 & 0 & 0 & \dfrac{T_RL^2}{E} & 0 & 0 \\ 0 & 0 & -6I_2L & 0 & 4I_2L^2 & 0 \\ 0 & 6I_3L & 0 & 0 & 0 & 4I_3L^2 \end{bmatrix} \tag{8.63a}$$

$$K_{pq}^{(\alpha\beta)} = \frac{E}{L^3}\begin{bmatrix} -AL^2 & 0 & 0 & 0 & 0 & 0 \\ 0 & -12I_3 & 0 & 0 & 0 & 6I_3L \\ 0 & 0 & -12I_2 & 0 & -6I_2L & 0 \\ 0 & 0 & 0 & -\dfrac{T_RL^2}{E} & 0 & 0 \\ 0 & 0 & 6I_2L & 0 & 2I_2L^2 & 0 \\ 0 & -6I_3L & 0 & 0 & 0 & 2I_3L^2 \end{bmatrix} \tag{8.63b}$$

$$K_{pq}^{(\beta\alpha)} = K_{pq}^{(\alpha\beta)}$$

$$K_{pq}^{(\beta\beta)} = \frac{E}{L^3}\begin{bmatrix} AL^2 & 0 & 0 & 0 & 0 & 0 \\ 0 & 12I_3 & 0 & 0 & 0 & -6I_3L \\ 0 & 0 & 12I_2 & 0 & 6I_2L & 0 \\ 0 & 0 & 0 & \dfrac{T_RL^2}{E} & 0 & 0 \\ 0 & 0 & 6I_2L & 0 & 4I_2L^2 & 0 \\ 0 & -6I_3L & 0 & 0 & 0 & 4I_3L^2 \end{bmatrix} \tag{8.63c}$$

The matrix

$$\mathbf{K} = \begin{bmatrix} K_{pq}^{(\alpha\alpha)} & K_{pq}^{(\alpha\beta)} \\ K_{pq}^{(\beta\alpha)} & K_{pq}^{(\beta\beta)} \end{bmatrix} \tag{8.64}$$

is the stiffness matrix of a beam element in local coordinates. We can obtain the stiffness matrix in global coordinates by proceeding as we did for pin-jointed elements. We find

$$\mathbf{K}' = \mathbf{T}^T\mathbf{K}\mathbf{T} \tag{8.65}$$

in which $\mathbf{T}$ is the 12×12 transformaton matrix:

$$\mathbf{T} = \begin{bmatrix} t_{mp} & 0 & 0 & 0 \\ 0 & t_{mp} & 0 & 0 \\ 0 & 0 & t_{mp} & 0 \\ 0 & 0 & 0 & t_{mp} \end{bmatrix} \qquad t_{mp} = \alpha_{pm} \tag{8.66}$$

Unless there is some special symmetry in the structure the algebraic manipulation implied in (8.65) is unpleasant. However, the rather complex matrix algebra seen here, is easily programmed. Indeed, partly because of the matrix approach structural problems can be fully automated. There exist sophisticated computer packages which can analyse complex structures with the absolute minimum of information – requiring only element material and geometrical constants and the global coordinates of each node.

However, in practice, we should exploit any inherent symmetry as far as possible since a computer analysis (particularly a three dimensional analysis) can be expensive. There are two cases in the analysis of beam structures in which the full 12 × 12 stiffness matrix need not be used. These structures are called **grids** and **planar rigid frames**.

• The Planar Rigid Frame

If a structure has all its elements lying in a single plane and if the forces to be applied are such as to induce deformations also in the plane then the structure is called a planar rigid frame. In such a structure the joints (nodes) are subject to rotation within the plane as well as to linear planar displacements but *not* to torsion.

Assuming the plane of the structure is the x_1x_2 plane the forces and moments, together with their corresponding displacements, appropriate to a structure of this kind are shown in Figure 8.24.

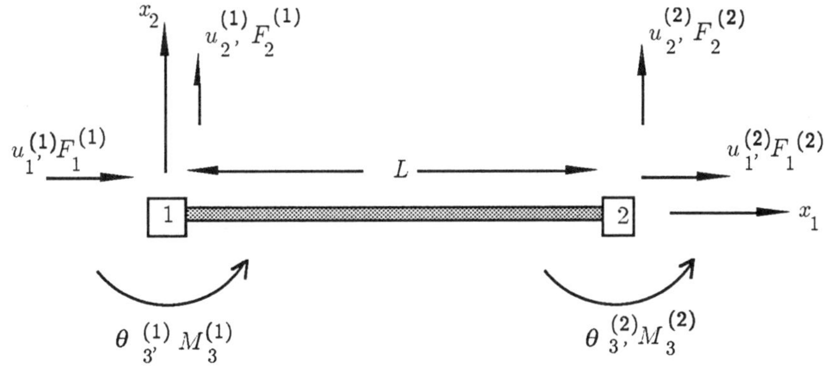

Figure 8.24

The force/displacement relation for this element is (extracted from the general 12 × 12 relation in (8.62)):

$$\begin{bmatrix} F_1^{(1)} \\ F_2^{(1)} \\ M_3^{(1)} \\ F_1^{(2)} \\ F_2^{(2)} \\ M_3^{(2)} \end{bmatrix} = \frac{E}{L^2}\begin{bmatrix} AL^2 & 0 & 0 & -AL^2 & 0 & 0 \\ 0 & 12I_3 & 6I_3L & 0 & -12I_3 & 6I_3L \\ 0 & 6I_3L & 4I_3L^2 & 0 & -6I_3L & 2I_3L^2 \\ -AL^2 & 0 & 0 & AL^2 & 0 & 0 \\ 0 & -12I_3 & -6I_3L & 0 & 12I_3L & -6I_3L \\ 0 & 6I_3L & 2I_3L^2 & 0 & -6I_3L & 4I_3L^2 \end{bmatrix}\begin{bmatrix} u_1^{(1)} \\ u_2^{(1)} \\ \theta_3^{(1)} \\ u_1^{(2)} \\ u_2^{(2)} \\ \theta_3^{(2)} \end{bmatrix} \quad \textbf{(8.67)}$$

which implicitly defines the stiffness matrix, $\mathbf{k}_p$, for a planar rigid element.

If we consider the plane of the structure to be in the $x'_1x'_2$ plane in global coordinates then the x'_i-axes and the x_i-axes are related by a simple planar rotation about the x_3-axis and so for an element with nodes α and β the transformation matrix is:

$$\mathbf{T}_p = \begin{bmatrix} t_{ij} & 0 \\ 0 & t_{ij} \end{bmatrix} \qquad \text{with } t_{ji} = \alpha_{ij} = \begin{bmatrix} \cos\theta & \sin\theta & 0 \\ -\sin\theta & \cos\theta & 0 \\ 0 & 0 & 1 \end{bmatrix} \tag{8.68}$$

in which

$$\cos\theta = \frac{(x'^{(\beta)}_1 - x'^{(\alpha)}_1)}{L} \qquad \sin\theta = \frac{(x'^{(\beta)}_2 - x'^{(\alpha)}_2)}{L} \tag{8.69}$$

(Evidently (F_1, F_2) transforms as a two-dimensional vector M_3 is unaffected by the rotation as it points along the x_3-axis). The stiffness matrix in global coordinates is:

$$\mathbf{k}'_p = \mathbf{T}^T_p \mathbf{k}_p \mathbf{T}_p \tag{8.70}$$

- **Grids**

A **grid** is a planar rigid frame in which the applied forces deform the frame in a direction *perpendicular* to the plane. Members of a grid may be subject to torsional forces, bending forces and also to shear forces. See Figure 8.25.

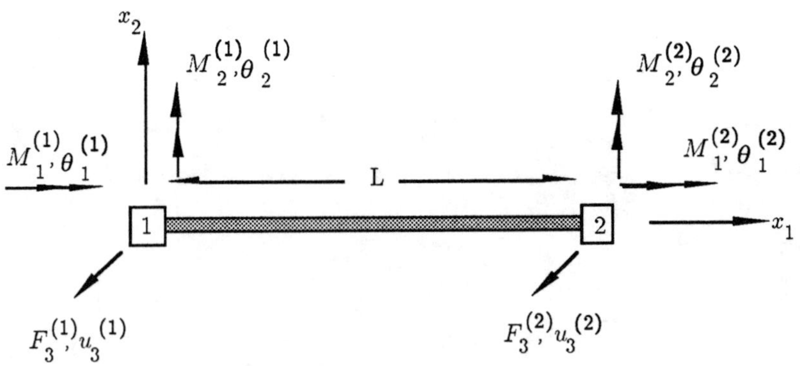

Figure 8.25

The force/displacement relation is again a 6×6 relation also extracted from the general relation (8.62):

$$\begin{bmatrix} F_3^{(1)} \\ M_1^{(1)} \\ M_2^{(1)} \\ F_3^{(2)} \\ M_1^{(2)} \\ M_2^{(2)} \end{bmatrix} = \frac{E}{L^3} \begin{bmatrix} 12I_2 & 0 & -6I_2 & -12I_2 & 0 & -6I_2L \\ 0 & \dfrac{T_RL^2}{E} & 0 & 0 & -\dfrac{T_RL^2}{E} & 0 \\ -6I_2 & 0 & 4I_2L^2 & 6I_2L & 0 & 2I_2L^2 \\ -12I_2 & 0 & 6I_2L & 12I_2 & 0 & 6I_2L \\ 0 & -\dfrac{T_RL^2}{E} & 0 & 0 & \dfrac{T_RL^2}{E} & 0 \\ -6I_2L & 0 & 2I_2L^2 & 6I_2L & 0 & 4I_2L^2 \end{bmatrix} \begin{bmatrix} u_3^{(1)} \\ \theta_1^{(1)} \\ \theta_2^{(1)} \\ u_3^{(2)} \\ \theta_1^{(2)} \\ \theta_2^{(2)} \end{bmatrix} \tag{8.71}$$

This relation defines the stiffness matrix, $\mathbf{k}_g$, for a grid element in local coordinates.

In this case, we again only consider coordinate transformations so that the x_3- and the x_3'-axes point in the same direction. Realising that F_3 points in the x_3'-direction and that (M_1, M_2) transform as a two-dimensional vector we obtain, for the transformation matrix for an element with nodes α, β

$$\mathbf{T}_g = \begin{bmatrix} t_{ij} & 0 \\ 0 & t_{ij} \end{bmatrix} \qquad \text{with } t_{ji} = \alpha_{ij} = \begin{bmatrix} 1 & 0 & 0 \\ 0 & \cos\theta & \sin\theta \\ 0 & -\sin\theta & \cos\theta \end{bmatrix} \tag{8.72}$$

in which

$$\cos\theta = \frac{(x_1'^{(\beta)} - x_1'^{(\alpha)})}{L} \qquad \sin\theta = \frac{(x_2'^{(\beta)} - x_2'^{(\alpha)})}{L} \tag{8.73}$$

The stiffness matrix with respect to global coordinates is obtained as in earlier cases:

$$\mathbf{k}_g' = \mathbf{T}_g^T \mathbf{k}_g \mathbf{T}_g \tag{8.74}$$

Example 8.4

Consider the simple grid of Figure 8.26 supporting a load P of magnitude 1000N $\approx$ weight of 10kg). Determine the generalised displacement components at node 2 and the reaction force components at nodes 1, 3. Both elements (solid circular rods of radius 0.02m) are formed from structural steel with $E = 10^{11}$ Nm^{-2} and $\mu = 7.7 \times 10^{10}$.

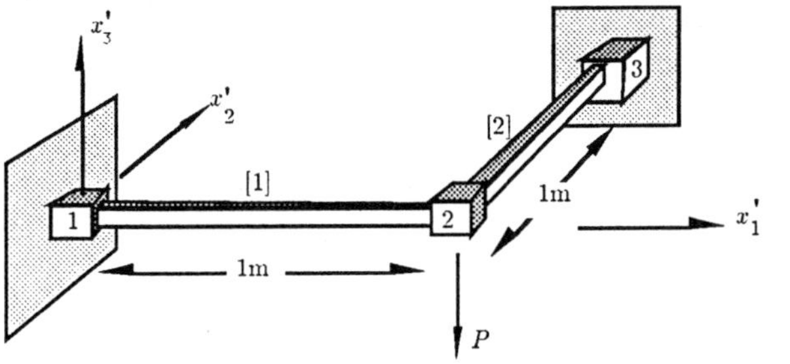

Figure 8.26

Solution

An easy calculation shows that

$$I_2 = \frac{1}{4}\pi a^4 = 1.25 \times 10^{-7} \qquad \text{and} \qquad T_R = \frac{1}{2}\mu\pi a^4 = 1.93 \times 10^4$$

Choosing global coordinates as in Figure 8.26 we construct the element stiffness matrices according to (8.71).

- **Element [1]**. $\alpha = 1,\ \beta = 2.$

For this element we can take the local coordinates in the same direction as the global coordinates so that $\mathbf{T}$ is the 6×6 unit matrix.

$$\mathbf{k}'_{g[1]} = EI_2 \begin{bmatrix} 12 & 0 & -6 & -12 & 0 & -6 \\ 0 & T_R/EI_2 & 0 & 0 & -T_R/EI_2 & 0 \\ -6 & 0 & 4 & 6 & 0 & 2 \\ -12 & 0 & 6 & 12 & 0 & 6 \\ 0 & -T_R/EI_2 & 0 & 0 & T_R/EI_2 & 0 \\ -6 & 0 & 2 & 6 & 0 & 4 \end{bmatrix} \begin{matrix} \\ \alpha = 1 \\ \\ \\ \beta = 2 \\ \\ \end{matrix}$$

- **Element [2]**. $\alpha = 2,\ \beta = 3 \quad x'^{(2)}_i = (1,0,0) \quad x'^{(3)}_i = (1,1,0).$

Hence using (8.72) and (8.73):

$$\mathbf{T} = \begin{bmatrix} 1 & 0 & 0 & 0 & 0 & 0 \\ 0 & 0 & 1 & 0 & 0 & 0 \\ 0 & -1 & 0 & 0 & 0 & 0 \\ 0 & 0 & 0 & 1 & 0 & 0 \\ 0 & 0 & 0 & 0 & 0 & 1 \\ 0 & 0 & 0 & 0 & -1 & 0 \end{bmatrix} \tag{8.75}$$

We find, from (8.74):

$$\mathbf{k}'_{g[2]} = \mathbf{T}^T\mathbf{k}_{g[2]}\mathbf{T} = EI_2 \begin{bmatrix} 12 & 6 & 0 & -12 & 6 & 0 \\ 6 & 4 & 0 & -6 & 2 & 0 \\ 0 & 0 & T_R/EI_2 & 0 & 0 & -T_R/EI_2 \\ -12 & -6 & 0 & 12 & -6 & 0 \\ 6 & 2 & 0 & -6 & 4 & 0 \\ 0 & 0 & -T_R/EI_2 & 0 & 0 & T_R/EI_2 \end{bmatrix} \begin{matrix} \\ \alpha = 2 \\ \\ \\ \beta = 3 \\ \\ \end{matrix}$$

The overall stiffness matrix will be a 9×9 formed by superimposing $\mathbf{k}'_{g[1]}$ and $\mathbf{k}'_{g[2]}$ (taking due account of node numbers). The force matrix is

$$\mathbf{G} = (G_3^{(1)}, N_1^{(1)}, N_2^{(1)}, -P, 0, 0, G_3^{(3)}, N_1^{(3)}, N_2^{(3)})^T$$

in which $(N_1^{(\alpha)}, N_2^{(\alpha)})$ represents the *global* moment at node α. Since nodes 1 and 3 are held fixed the displacement matrix has the form:

$$\mathbf{U} = (0,0,0,u'^{(2)}_3,\theta'^{(2)}_1,\theta'^{(2)}_2,0,0,0)^T$$

In the force/displacement relation

$$\mathbf{G} = \mathbf{K}\mathbf{U} \tag{8.76}$$

we first consider the second set of three equations:

$$\begin{bmatrix} -P \\ 0 \\ 0 \end{bmatrix} = EI_2 \begin{bmatrix} 24 & 6 & 6 \\ 6 & 4 + T_R/EI_2 & 0 \\ 6 & 0 & 4 + T_R/EI_2 \end{bmatrix} \begin{bmatrix} u'^{(2)}_3 \\ \theta'^{(2)}_1 \\ \theta'^{(2)}_2 \end{bmatrix} \tag{8.77}$$

Solving, we find

$$u'^{(2)}_3 = -\frac{P}{24EI_2}\left[\frac{4EI_2 + T_R}{EI_2 + T_R}\right] \qquad \theta'^{(2)}_2 = \theta'^{(2)}_1 = \frac{P}{4(EI_2 + T_R)} \tag{8.78}$$

For the particular set of material constants and applied forces in this question we calculate:

$$u'^{(2)}_3 = 1.2\text{cm} \qquad \theta'^{(2)}_2 = \theta'^{(2)}_1 = 0.007 \text{ radians } (\approx 0.04^0) \tag{8.79}$$

Using these values for the displacements the first and third sets of three equations of (8.76) provide solutions for the reaction forces (after some straightforward algebra):

$$\begin{bmatrix} G^{(1)}_3 \\ N^{(1)}_1 \\ N^{(1)}_2 \end{bmatrix} = \begin{bmatrix} \dfrac{P}{2} \\ -\dfrac{PT_R}{4(EI_2 + T_R)} \\ -\dfrac{P(2EI_2 + T_R)}{4(EI_2 + T_R)} \end{bmatrix} \quad \text{and} \quad \begin{bmatrix} G^{(3)}_3 \\ N^{(3)}_1 \\ N^{(3)}_2 \end{bmatrix} = \begin{bmatrix} G^{(1)}_3 \\ N^{(1)}_2 \\ N^{(1)}_1 \end{bmatrix} \tag{8.80}$$

Note that $G^{(1)}_1 + G^{(3)}_1 + (-P) = 0$ confirming that the global forces are in equilibrium. The components of the resultant couple can therefore be determined about any point. (It should be zero if our calculations are correct). We can evaluate about node 2. The x'_1-component is $N^{(1)}_1 + N^{(3)}_1 + (P/2) = 0$. Similarly the x'_2-component is $N^{(1)}_2 + N^{(3)}_2 + (P/2) = 0$ confirming that the system of forces and moments are in equilibrium.

The internal forces developed within each member are given by (8.71). For example for element [1] (for which the transformation matrix is the unit matrix) we have:

$$\begin{bmatrix} F_3^{(1)} \\ M_1^{(1)} \\ M_2^{(1)} \\ F_3^{(2)} \\ M_1^{(2)} \\ M_2^{(2)} \end{bmatrix} = EI_2 \begin{bmatrix} -12 & 0 & -6 \\ 0 & -T_R/EI_2 & 0 \\ 6 & 0 & 2 \\ 12 & 0 & 6 \\ 0 & T_R/EI_2 & 0 \\ 6 & 0 & 4 \end{bmatrix} \begin{bmatrix} u_3^{(2)} \\ \theta_1^{(2)} \\ \theta_2^{(2)} \end{bmatrix}$$

$$= \frac{1}{4(EI_2 + T_R)} \begin{bmatrix} 2P(EI_2 + T_R) \\ -PT_R \\ -P(2EI_2 + T_R) \\ -2P(EI_2 + T_R) \\ PT_R \\ -PT_R \end{bmatrix} \tag{8.81}$$

(The reader should verify, from (8.81), that the member forces and moments are in equilibrium).

The deformations at node 2 (highly exaggerated) resulting from these forces are displayed in Figure 8.27.

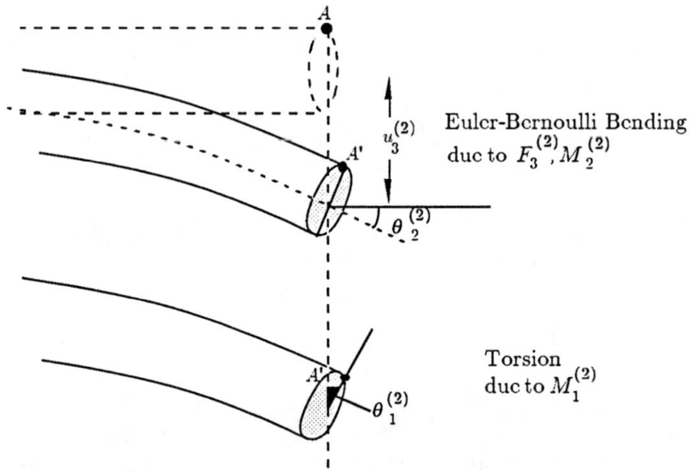

Figure 8.27

Each element can now be further analysed to determine its stress/strain levels. A longitudinal stress arises from F_3, M_3 (see Euler-Bernoulli bending, Section 6.5) and shearing stresses arise due to the torsional component M_1.

EXERCISES

8.8 The simple two-element structure of Figure 8.28 supports a load P. Element [1] is to be modelled as a pin-jointed element and element [2] as a beam element. The cross-sectional area of the elements are $A^{(1)}, A^{(2)}$ respectively. Both are made of the same material with Young's modulus E. Show that the displacement components at node 1 are:

$$u_1^{(1)} = \frac{P}{EA^{(2)}}\frac{\alpha}{(1+\alpha)} \qquad u_2^{(1)} = -\frac{P}{3(1+\alpha)EI} \qquad \theta_1^{(1)} = \frac{P}{2(1+\alpha)EI}$$

where

$$\alpha = \frac{A^{(1)}A^{(2)}}{I(6\sqrt{2}A^{(2)} + 3A^{(1)})}$$

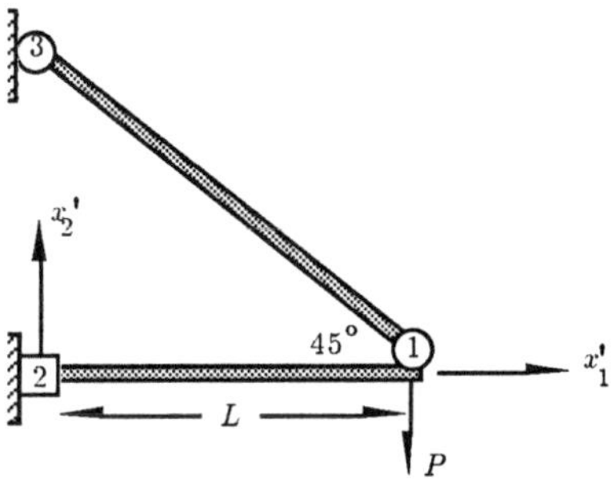

Figure 8.28

8.9 Repeat Exercise 7.1(a) using the stiffness matrix approach.

8.10 A grid, supporting a load P is shown in Figure 8.29. All members have the same length and material properties. Utilising symmetry wherever possible determine the unknown displacement components.

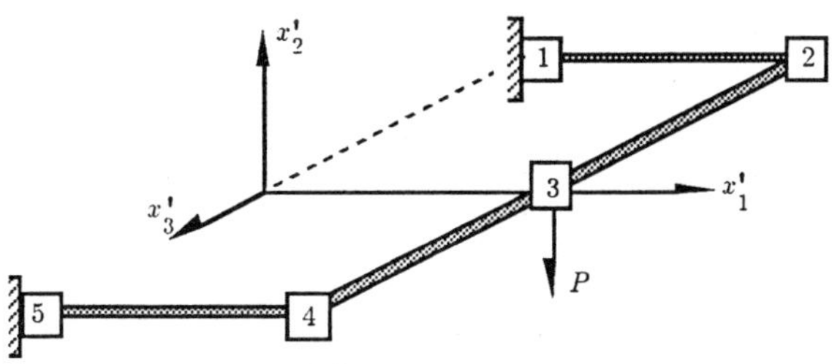

Figure 8.29

8.5 Equivalent Nodal Forces

In the previous examples and analyses we have always assumed that, in any structure, the applied forces are (a) concentrated and (b) always applied at the nodes. Any superficial examination of a real structure subjected to everyday forces indicates that these assumptions are rarely satisfied. Whilst it is true that, in buildings designed to carry furniture, people etc, the forces of most significance are due to the weight of the *fabric* of the building itself which is distributed through the joints (nodes) of the building, there are a number of circumstances in which different kinds of loading must be considered. For example, a tall slender building may be significantly affected by high winds which of course is a distributed force acting on each of the supporting beams and not just at the joints. In other examples individual beams may have to support large concentrated loads at non-nodal points. (A simple everyday example is a bookshelf. It must be designed so that the deformation due to the loading of several heavy books placed at its centre should not be detectable to the eye).

Fortunately, in cases like these, and in many others, the loadings, whether they be distributed or concentrated can be replaced by an *equivalent* system of concentrated nodal forces and, as such, can then be treated using the stiffness matrix approach. The equivalence is **not** in terms of the beam deflection.

For example, for a beam deformed by the action of a constant transverse pressure q we have, using the Euler-Bernoulli model (6.39):

$$\frac{d^2}{dx_1^2}\left(EI\frac{d^2w}{dx_1^2}\right) = q \tag{8.82}$$

which integrates to (E, I, q constants)

$$w = \frac{q}{24EI}x_1^4 + a_1x_1^3 + a_2x_1^2 + a_3x_1 + a_4 \tag{8.83}$$

where a_1, a_2, a_3 and a_4 are constants of integration. They may be determined depending on the boundary conditions. The important characteristic of this solution is that it is a quartic in x_1 (independent of the supports). If our intention is to replace the pressure load by equivalent joint forces then *it is impossible* to obtain the same deflection curve since (8.82) now becomes

$$\frac{d^2}{dx_1^2}\left(EI\frac{d^2w}{dx_1^2}\right) = 0 \tag{8.84}$$

the solution to which is a **cubic** in x_1 and so can never be identical to (8.84).

The equivalence, referred to above, is in terms of the work done by the original loading and by the replacement nodal loading.

To illustrate the procedure used to determine equivalent nodal loadings we consider a beam undergoing a constant transverse pressure load in the x_1x_3 plane. If the deformation is taken as that due to the equivalent set of nodal loads (shear forces $F_3^{(1)}, F_3^{(2)}$ and bending moments $M_2^{(1)}, M_2^{(2)}$) then according to Euler-Bernoulli theory the deformation is a cubic in x_1:

$$u_3(x_1) = c_0 + c_1x_1 + c_2x_1^2 + c_3x_1^3 \tag{8.85}$$

The work done by the equivalent set of forces is:

$$F_3^{(1)}u_3(0) + F_3^{(2)}u_3(L) - M_2^{(1)}u_3'(0) - M_2^{(2)}u_3'(L) \tag{8.86}$$

(The work done by the bending moment is $M_2\theta_2$ but, for bending in this plane, $\theta_2 = -u_3'$).

The work done by the pressure force, acting through the displacement in (8.85), is

$$\int_0^L q u_3(x_1)dx_1 \tag{8.87}$$

The constants c_0, c_1, c_2 and c_3 can be rewritten in terms of the nodal values $u_3(0)$, $u_3'(0)$, $u_3(L)$, $u_3'(L)$. After a lengthy calculation we find,

$$\begin{aligned} u_3(x_1) = u_3(0)\frac{(L-x_1)}{L^3}(L^2 + x_1L - 2x_1^2) + u_3'(0)\frac{x_1}{L^2}(L^2 - 2Lx_1 + x_1^2) \\ + u_3(L)\frac{x_1}{L^3}(3Lx_1 - 2x_1^2) + u_3'(L)\frac{x_1^2}{L^2}(x_1 - L) \end{aligned} \tag{8.88}$$

and, a direct evaluation of the integral in (8.87) provides an expression for the work done by the applied pressure loading:

$$u_3\frac{qL}{2} + u_3'(0)\frac{qL^2}{12} + u_3(L)\frac{qL}{2} - u_3'(L)\frac{qL^2}{12} \tag{8.89}$$

Comparing (8.89) with (8.86)

$$F_3^{(1)} = \frac{qL}{2} \quad F_3^{(2)} = \frac{qL}{2} \quad M_2^{(1)} = -\frac{qL^2}{12} \quad M_2^{(2)} = \frac{qL^2}{12} \tag{8.90}$$

as the **work-equivalent** nodal loadings for constant pressure load q. Other non-linear pressure loads can be considered in a similar way.

If instead of a transverse pressure load, a transverse point load of strength Q acting at a point with coordinate value $x_1 = x_0 \;\; 0 < x_0 < L$, is to be considered then we model this using a delta function: $Q\delta(x_0)$. The work done, by (8.87) is

$$\int_0^L Q\delta(x_0)u_3(x_1)dx_1 = Qu_3(x_0)$$

with $u_3(x_0)$ being obtained from (8.88). For example if $x_0 = L/2$ then

$$\begin{aligned} Qu_3(L/2) &= Q\Big[u_3(0)\frac{L/2}{L^3}(L^2 + \frac{L^2}{2} - \frac{L^2}{2}) + u_3'(0)\frac{L/2}{L^2}(L^2 - L^2 + L^2/4) \\ &\quad + u_3(L)\frac{L/2}{L^3}(\frac{3L^2}{2} - \frac{L^2}{2}) + u_3'(L)\frac{L^2/4}{L^2}(-L/2)\Big] \\ &= Q\Big[u_3(0)\frac{1}{2} + u_3'(0)(\frac{L}{8}) + u_3(L)(\frac{L}{2}) + u_3'(L)(-\frac{L}{8})\Big] \end{aligned} \tag{8.91}$$

and so, on comparison with (8.86) the equivalent loading would be:

$$F_3^{(1)} = \frac{1}{2}Q \quad F_3^{(2)} = \frac{1}{2}Q \quad M_2^{(1)} = -\frac{1}{8}QL \quad M_2^{(2)} = \frac{1}{8}QL$$

Longitudinal pressure and point loadings may also be treated. In this case we would attempt to replace these forces by nodal longitudinal forces $F_1^{(1)}, F_1^{(2)}$ and use a displacement expression as in (8.5):

$$u_1(x_1) = u_1(0)[1 - \frac{x_1}{L}] + u_1(L)[\frac{x_1}{L}]$$

Constant pressure loads and point loads in *any* direction are analysed in terms of work-equivalent nodal forces by first resolving the given loading into transverse and longitudinal components and then treating each component separately.

All of this analysis has been carried through in local coordinates. Of course once equivalent nodal loadings have been obtained they can be transformed into appropriate expressions with respect to global coordinates through the use of an appropriate transformation matrix. The equivalent nodal forces, obtained in global coordinate form, can then be added into the applied loading matrix **G** and the stiffness matrix approach used as above to solve for the unknowns.

We can extend the stiffness matrix method to analyse structures built from more sophisticated structural elements. It is a relatively painless extension to develop stiffness matrices for plate elements. With elements like these, along with beam elements and pin-joint elements we can begin to think about real buildings which, to a large extent, are formed from combinations of these three basic elements.

Having developed a method for analysing structures we can begin to think about the *optimal* design of structures. I suppose it is fair to say that most structures we see around us are designed by architects to be aesthetically pleasing and structurally sound. Few could be described as optimal. For example, consider the three element truss designed to support a load P as shown in Figure 8.30.

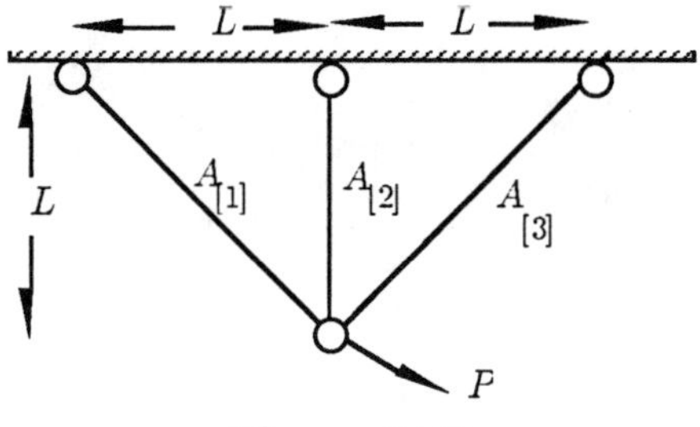

Figure 8.30

It is easy enough to build a structure adhering to certain safety considerations which support the load. But can we do better than this? If each element is to be constructed from structural steel then the only parameter to be 'designed' is the cross-sectional areas of each element. We can choose $A_{[1]}, A_{[2]}$ and $A_{[3]}$ so as to minimise the total amount of material used (called the *objective* function) subject to safety constraints on the longitudinal stresses. This problem can be solved analytically using standard techniques in optimisation. In this simple example of structural design the unknowns were the cross-sectional areas. In other examples the *shape* of the structure may be an unknown. In designing a column of minimum weight to support a given load subject to certain safety constraints on longitudinal stress and on buckling a design may change from an initial shape as in Figure 8.31a to a final shape as in 8.31c.

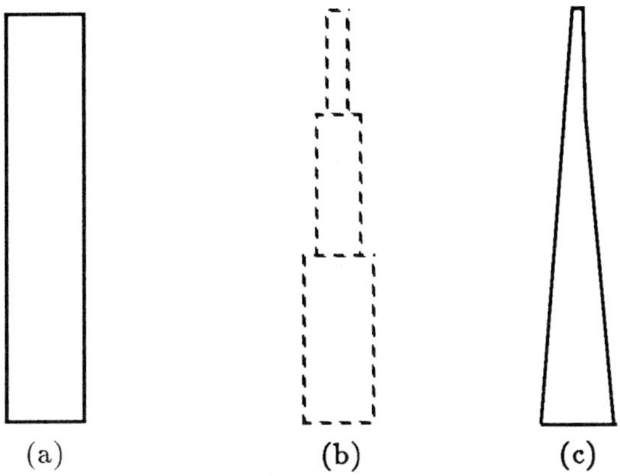

Figure 8.31

We will not pursue this line. The interested reader is referred to specific texts on structural analysis and on the optimal design of structures. Our aim here has been to give a basic introduction to the subject and to illustrate the way in which the subject is founded in linear elasticity theory.

EXERCISES

8.11 Repeat Exercise 8.10 by eliminating node 3 and regarding the structure as comprised of three elementds only with a single element connecting nodes 2 and 4. Obviously the point force P will need to be modelled in terms of equivalent joint forces.

8.12 A beam element is deformed by a transverse pressure varying linearly from $q^{(1)}$ at node 1 to $q^{(2)}$ at node 2. Determine equivalent joint forces for this loading.

8.13 (a) Determine the equivalent joint forces due to a transverse point force of strength Q acting at a distance ξ from one end.
(b) Is it possible to counteract the deforming effects of a uniform transverse pressure q with two transverse point forces placed at suitable points?

8.14 Determine equivalent joint forces for (a) uniform longitudinal pressure and (b) a point longitudinal force.

Chapter 9

TWO DIMENSIONAL ELASTOSTATICS

9.1 Plane Strain, Plane Stress and Generalised Plane Stress

It is always useful, for any mathematical model, to obtain exact solutions. One can then begin to appreciate how closely (or otherwise) the model conforms with reality. It is not as easy to draw conclusions on the validity of a mathematical model from a numerical solution since, by its nature, a numerical solution can only apply to one specific situation at a time. An exact solution is equivalent to obtaining a range of numerical solutions and it is worth expending a great deal of effort to obtain one.

Knowledge of exact solutions also helps with the analysis and appreciation of numerical procedures and it is fair to say that the proper understanding of a numerical analysis only follows after one has a good grasp of the analytical development of the subject under consideration.

Of course, with a system of equations as complicated as those of linear elasticity it is unlikely that exact solutions can be obtained without making further simplifying assumptions. One of the most obvious simplifications is to reduce (if possible) the dimensionality of the equations from three to two. From this desire to reduce the dimensionality of the full system of equations of linear elasticity (5.9) – (5.14) there has arisen the theory of *two dimensional elastostatics.*

As we shall see below, there are two basic ways that this reduction can be carried through. The first, obtained by simply assuming that the displacement vector has only two non-zero components (each of which is a function of two variables alone) is the subject of the theory of **plane strain**. The second approach, which removes one variable from the equations of linear elasticity by an averaging process, is the subject of the theory of **generalised plane stress**. Each of these theories have applications to important, physically realistic, elastic systems. It is an unexpected bonus that the equations which govern each of these separate theories are identical in form allowing both subjects to be treated in tandem.

In what follows, we shall always arrange matters so that it is the x_3-variable that is removed.

• Plane Strain

A state of plane strain is said to exist in an elastic body if the displacement components u_i take the *particular form*:

$$u_1 = f(x_1, x_2) \qquad u_2 = g(x_1, x_2) \qquad u_3 = 0 \tag{9.1}$$

That is, u_1 and u_2 are functions of x_1 and x_2 alone and the displacement component in the x_3-direction, u_3, is zero. This choice implies that the only non-zero strain components are $\varepsilon_{11}, \varepsilon_{12}$ and ε_{22} so that deformations only occur in the plane $x_3 =$ constant. Hence the terminology 'plane strain'.

Intuitively, we expect that surface stresses will need to be applied in order to maintain u_3 at zero. In a practical situation this may be difficult to arrange which to some extent devalues the usefulness of the theory that develops from (9.1). However, there are a number of practical situations in which a state of plane strain may be considered to exist. For example, within a thin cross-section of a dam or long pressure vessel, away from the ends, the assumption of plane strain conditions as defined in (9.1), would be a reasonably accurate approximation to the actual deformations on these sections. See Figure 9.1.

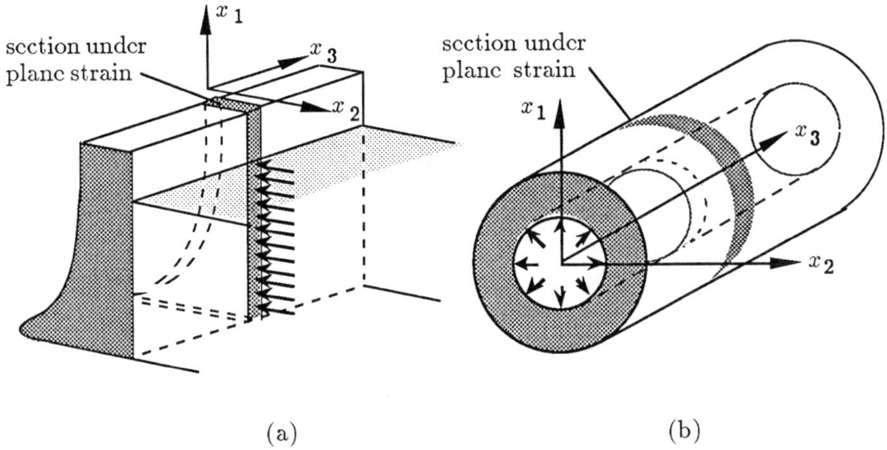

Figure 9.1

In the first part of this section therefore, we accept the assumption detailed in (9.1) as applying to a two dimensional region R (which is not necessarily simply connected) in the x_1x_2 plane bounded by a curve (or collection of curves) C. On the boundary, deforming planar forces may be applied. See Figure 9.2.

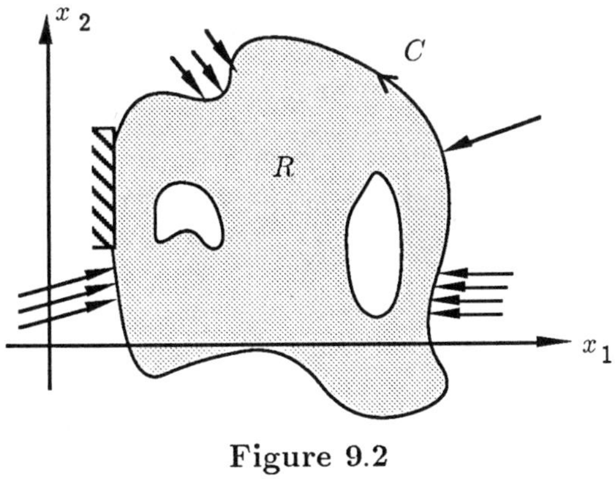

Figure 9.2

The elasticity equations to be solved when plane strain conditions hold are easily obtained from the full system (5.9) – (5.14).

From now on we shall use *lower case greek letters* to denote an index that ranges over the values 1 and 2 only. Thus, for example, $u_\alpha u_\alpha$ is shorthand for $u_1u_1 + u_2u_2$. This notation will greatly simplify expressions which arise in two dimensional elastostatics.

With this notation, and assuming (9.1), we confirm that the only non-zero components of strain are $\varepsilon_{\alpha\beta}$. Also, the only non-zero components of stress are $\sigma_{\alpha\beta}$ and σ_{33} with values:

$$\sigma_{\alpha\beta} = 2\mu\varepsilon_{\alpha\beta} + \lambda\delta_{\alpha\beta}\varepsilon_{\delta\delta} \tag{9.2a}$$

$$\sigma_{33} = \nu\sigma_{\alpha\alpha} \tag{9.2b}$$

In the absence of body forces (which we shall ignore throughout this chapter) the equilibrium equations reduce to:

$$\sigma_{\alpha\beta},_{\beta} = 0 \tag{9.3}$$

Also, the Cauchy relation (5.10) simplifies to:

$$T_\alpha^{(n)} = \sigma_{\alpha\beta}n_\beta \tag{9.4}$$

(the normal to C has only non-zero x_1 and x_2 components).

The compatibility conditions (if required) reduce from six independent relations (5.14) to just one:

$$\varepsilon_{11},_{22} + \varepsilon_{22},_{11} - 2\varepsilon_{21},_{21} = 0 \tag{9.5}$$

For a boundary value problem of the first kind, in which the stress vector $T_\alpha^{(n)}$ is specified at each point of the boundary, the satisfaction of the Beltrami-Mitchell equations throughout a simply-connected region R and of the boundary conditions (9.4) guarantee a unique single valued solution for the displacements. (Extra conditions may be required if R is not simply connected. See Chapter 7 on the torsion problem for an appreciation of this point).

In plane strain the Beltrami-Mitchell equations reduce to:

$$\nabla^2(\sigma_{\alpha\alpha}) = 0 \tag{9.6}$$

(To see this utilise the equilibrium equations (9.3) in (9.5) together with the 'inverse relations':

$$\varepsilon_{\alpha\beta} = \frac{1}{2\mu}(\sigma_{\alpha\beta} - \frac{\lambda\delta_{\alpha\beta}\sigma_{\delta\delta}}{2\mu + 2\lambda})$$

obtained from (9.2a) and the equilibrium equations (9.3)).

In the next section we show how to obtain exact solutions to the equations governing plane strain.

- **Plane Stress**

We have already defined this state of stress, in Section 3.6, as being one in which all the components of stress with respect to a given direction, say x_3, are zero:

$$\sigma_{13} = \sigma_{23} = \sigma_{33} = 0 \tag{9.7}$$

If an *exact* theory based upon the assumptions in (9.7) is to be useful then the displacement components (and hence *all* strain and stress components) should be dependent only on x_1, x_2. This would then be a two dimensional problem to which we could hope to find some exact solutions. However, it is easy to show that the theory which results from the assumptions in (9.7) cannot be two dimensional. For let us assume u_1 is independent of x_3, then $\frac{\partial u_1}{\partial x_3} = 0$ and from (5.12) with $i = j = 3$

$$u_3 = \int \varepsilon_{33} dx_3 = -\int \frac{\mu}{E}(\sigma_{11} + \sigma_{22}) dx_3 = F(x_1, x_2)[x_3 - const.]$$

where $F(x_1, x_2)$ is some function of x_1, x_2 alone. It follows that $\frac{\partial u_3}{\partial x_1} \neq 0$ in general and so $\varepsilon_{13} = \frac{1}{2}(u_{1,3} + u_{3,1}) = \frac{1}{2} u_{3,1} \neq 0$. As $\sigma_{13} = 2\mu\varepsilon_{13} \neq 0$, we arrive at a contradiction with (9.7). We conclude that in the *exact* analysis of an elastic body in a state of plane stress the theory *must* involve all three coordinates.

If we relax the statements in (9.7) so that these particular stress components are only approximately zero we obtain an essentially two dimensional state of stress. This approximation could apply to very thin plates, in the x_1x_2 plane, whose bounding faces are stress free, (see Section 3.6).

A useful two dimensional theory which can be applied to plates of moderate thickness is that obtained in generalised plane stress.

- **Generalised Plane Stress**

This state of stress relates to a plate, with centre plane in the x_1x_2 plane, of thickness t (see Figure 9.3).

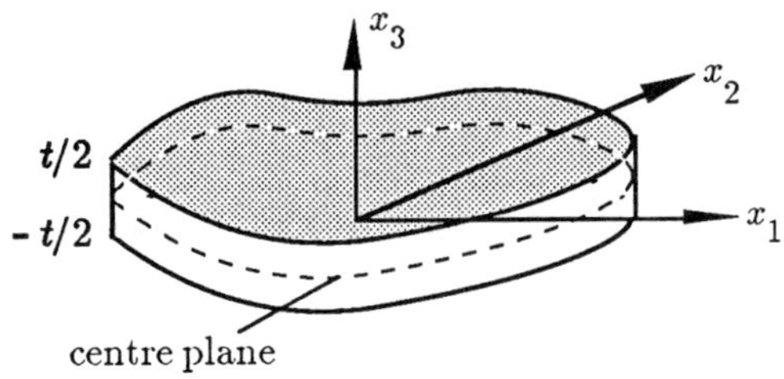

Figure 9.3

A state of generalised plane stress is said to exist if

$$\sigma_{13} = \sigma_{23} = 0 \quad \text{on} \quad x_3 = \pm\frac{t}{2} \qquad \sigma_{33} = 0 \quad \text{throughout the plate} \qquad \textbf{(9.8)}$$

As long as t is not too large such a state of stress will exist in a plate which is deformed by the action of in-plane edge forces alone. For, in such a physical system, the Cauchy relation implies

$$0 = \sigma_{ij} n_j \quad \text{on} \quad x_3 = \pm\frac{t}{2}$$

But, on the upper and lower faces of the plate, $n_j = (0, 0, \pm 1)$ and so

$$\sigma_{13} = \sigma_{23} = \sigma_{33} = 0 \quad \text{on} \quad x_3 = \pm\frac{t}{2} \qquad \textbf{(9.9)}$$

It follows that the rate of change of σ_{13} or σ_{23} *within* either surface $x_3 = \pm\frac{t}{2}$ is zero. In particular, $\sigma_{13,1} = 0$ and $\sigma_{23,2} = 0$ on $x_3 = \pm\frac{t}{2}$. We now employ the third equilibrium equation ($i = 3$ in (5.11)):

$$\sigma_{31,1} + \sigma_{32,2} + \sigma_{33,3} = 0$$

which is valid *throughout* the plate, to deduce that,

$$\sigma_{33,3} = 0 \quad \text{on} \quad x_3 = \pm\frac{t}{2} \tag{9.10}$$

Since, for this physical system, $\sigma_{33} = 0$ (by assumption) and $\sigma_{33,3} = 0$ (by deduction) on $x_3 = \pm\frac{t}{2}$ the contention that σ_{33} is small throughout the plate is supported and our *demand* in the theory of generalised plane stress that $\sigma_{33} = 0$ everywhere is a reasonable approximation as long as the plate thickness, t, is not too large.

In generalised plane stress we also demand that the edge forces are **symmetrical** with respect to the middle plane. See Figure 9.4.

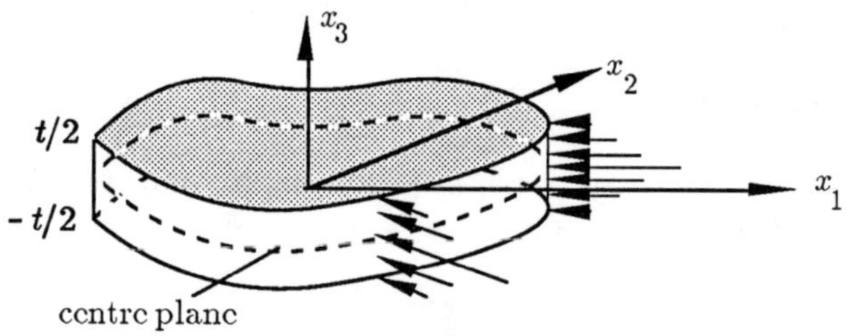

Figure 9.4

Mathematically, this loading implies that the displacement in the x_3-direction, u_3, is an **odd-function** of x_3 and so its **mean value** (denoted by an overbar) through the plate is zero. That is

$$\overline{u}_3 = \frac{1}{t}\int_{-t/2}^{t/2} u_3(x_1, x_2, x_3)\,dx_3 = 0 \tag{9.11}$$

Also using exactly the same argument u_1 and u_2 are both **even functions** of x_3. Hence the strain components $\varepsilon_{13} = \frac{1}{2}(u_{1,3} + u_{3,1})$ and $\varepsilon_{23} = \frac{1}{2}(u_{2,3} + u_{3,2})$ and the corresponding stress components σ_{13}, σ_{23} are each **odd functions** of x_3.

Of course it can, quite reasonably, be argued that for a thin plate the average values of the displacements, the strains and the stresses are just as useful as the actual point values of these quantities. Indeed in practical measurements of stress and strain it is an average value (albeit over a small region) that is generally obtained. Consider, for example, the measurement of stress using the photoelastic technique. The experimenter has to form a two dimensional model which is sufficiently thick for rigidity and ease of working but also to give adequate optical sensitivity. The sensitivity is directly proportional to thickness. So the optical effect in a photoelastic model is proportional to the *mean stress* in the birefringent specimen and it is mean stress values that are obtained – not point values.

We are therefore led to consider average values of strain and stress (average values will be denoted by an overbar):

$$\overline{\varepsilon}_{\alpha\beta} = \frac{1}{t}\int_{-t/2}^{t/2} \varepsilon_{\alpha\beta} dx_3 \qquad \overline{\sigma}_{\alpha\beta} = \frac{1}{t}\int_{-t/2}^{t/2} \sigma_{\alpha\beta} dx_3 \tag{9.12}$$

The equations satisfied by these averaged quantities can be obtained from averaged versions of the normal elasticity equations.

• Averaged Equilibrium Equations

Consider the first two equations of (5.11). Averaging through the thickness we obtain, when $i = 1$:

$$\frac{1}{t}\int_{-t/2}^{t/2} \sigma_{11},_1\, dx_3 + \frac{1}{t}\int_{-t/2}^{t/2} \sigma_{12},_2\, dx_3 + \frac{1}{t}\int_{-t/2}^{t/2} \sigma_{13},_3\, dx_3 = 0 \tag{9.13}$$

which, after carrying through the integration implied in the third term and interchanging integration and differentiation operators in the first two terms gives:

$$\overline{\sigma}_{11},_1 + \overline{\sigma}_{12},_2 + \frac{1}{t}\Big[\sigma_{13}\Big]_{-t/2}^{t/2} = 0$$

that is,

$$\overline{\sigma}_{11},_1 + \overline{\sigma}_{12},_2 = 0 \tag{9.14}$$

since, by (9.8), σ_{13} vanishes on the upper and lower faces of the plate. The second equilibrium equation ($i = 2$ in (5.11)) is treated similarly to give

$$\overline{\sigma}_{21},_1 + \overline{\sigma}_{22},_2 = 0 \tag{9.15}$$

Averaging the final equilibrium equation $i = 3$ of (5.11) implies:

$$\frac{1}{t}\int_{-t/2}^{t/2} \sigma_{31},_1\, dx_3 + \frac{1}{t}\int_{-t/2}^{t/2} \sigma_{32},_2\, dx_3 + \frac{1}{t}\int_{-t/2}^{t/2} \sigma_{33},_3\, dx_3 = 0$$

but, since σ_{31} and σ_{32} are both odd functions of x_3, the first two integrals vanish. The last integral is zero since $\sigma_{33} = 0$ throughout the plate. Hence this 'averaged' equilibrium equation is satisfied exactly.

Grouping (9.14), (9.15) together gives:

$$\overline{\sigma}_{\alpha\beta},_\beta = 0 \tag{9.16}$$

• Averaged Cauchy Relation

For the stress-free bounding faces of the plate $n_i = (0, 0, \pm 1)$ and so (5.10) implies

$$0 = \pm\sigma_{i3} \quad \text{at} \quad x_3 = \pm\frac{t}{2}$$

which, of course, is satisfied due to our basic assumption in (9.8). On the bounding edge of the plate $n_i = (n_1, n_2, 0)$ and so (5.10) implies

$$T_\alpha^{(n)} = \sigma_{\alpha\beta} n_\beta$$

Averaging this equation gives

$$\frac{1}{t}\int_{-t/2}^{t/2} T_\alpha^{(n)} dx_3 = \left[\frac{1}{t}\int_{-t/2}^{t/2} \sigma_{\alpha\beta} dx_3\right] n_\beta$$

which is,

$$\overline{T}_\alpha^{(n)} = \overline{\sigma}_{\alpha\beta} n_\beta \tag{9.17}$$

- **Averaged Constitutive Equations**

We can use either of the equivalent forms (5.8) or (5.13). We choose the former since we have made an assumption, in (9.8), concerning the stress components.

When i or j take the values 1,2 in (5.8) then on averaging we find

$$\overline{\varepsilon}_{\alpha\beta} = \frac{1}{2\mu}\left[\overline{\sigma}_{\alpha\beta} - \frac{\lambda}{2\mu + 3\lambda}\delta_{\alpha\beta}\overline{\sigma}_{\gamma\gamma}\right] \tag{9.18}$$

where we have used $\sigma_{33} = 0$. We can 'invert' (9.18) in the usual way by first contracting α and β:

$$\overline{\varepsilon}_{\alpha\alpha} = \frac{1}{2\mu}\left[\overline{\sigma}_{\alpha\alpha} - \frac{2\lambda}{2\mu + 3\lambda}\overline{\sigma}_{\alpha\alpha}\right] = \frac{(2\mu + \lambda)}{2\mu(2\mu + 3\lambda)}\overline{\sigma}_{\alpha\alpha}$$

and so, after a little algebra:

$$\overline{\sigma}_{\alpha\beta} = 2\mu\overline{\varepsilon}_{\alpha\beta} + \overline{\lambda}\delta_{\alpha\beta}\overline{\varepsilon}_{\gamma\gamma} \tag{9.19}$$

in which

$$\overline{\lambda} = \frac{2\mu\lambda}{2\mu + \lambda} \tag{9.20}$$

The remaining three constitutive equations produce, from the assumption (9.8)

$$\overline{\varepsilon}_{33} = -\frac{\lambda}{2\mu(2\mu + 3\lambda)}\overline{\sigma}_{\alpha\alpha} \qquad \overline{\varepsilon}_{13} = \overline{\varepsilon}_{23} = 0 \tag{9.21}$$

- **Averaged Compatibility Equations**

We must first realise that, in terms of averaged values there are only two unknown displacement components $\overline{u}_1, \overline{u}_2$ since ($\overline{u}_3 = 0$) and these are obtained by solving the equations, for *given* strain components $\overline{\varepsilon}_{\alpha\beta}$

$$\overline{\varepsilon}_{\alpha\beta} = \frac{1}{2}(\overline{u}_{\alpha,\beta} + \overline{u}_{\beta,\alpha}) \tag{9.22}$$

The compatibility equations corresponding to this system of partial differential equations are identical to (5.14) excepting that the indices are restricted to take the values 1 and 2:

$$\bar{\varepsilon}_{\alpha\beta},_{\gamma\gamma} + \bar{\varepsilon}_{\gamma\gamma},_{\alpha\beta} - \bar{\varepsilon}_{\gamma\beta},_{\gamma\alpha} - \bar{\varepsilon}_{\gamma\alpha},_{\gamma\beta} = 0 \tag{9.23}$$

It is an easy matter to check that there is only *one* independent relation here:

$$\bar{\varepsilon}_{11},_{\gamma\gamma} + \bar{\varepsilon}_{\gamma\gamma},_{11} - 2\bar{\varepsilon}_{\gamma1},_{\gamma1} = 0$$

or

$$\bar{\varepsilon}_{11},_{22} + \bar{\varepsilon}_{22},_{11} - 2\bar{\varepsilon}_{21},_{21} = 0 \tag{9.24}$$

The governing equations of generalised plane stress are (9.16), (9.17), (9.19) and (9.24) and are *identical* in form to the governing equations of plane strain (9.2a), (9.3), (9.4) and (9.5) excepting that the parameters (λ, μ) in plane strain are replaced by the parameters $(\bar{\lambda}, \bar{\mu}) \equiv (\frac{2\mu\lambda}{2\mu+\lambda}, \mu)$ in generalised plane stress (equivalently, in terms of the Poisson ratio, $\bar{\nu} \rightarrow \frac{\nu}{1+\nu}$).

From henceforth, since the notation is easier, we shall deal with the equations of plane strain though we should always remember that an exact solution to a problem in plane strain is easily amended (by replacing λ by $\frac{2\mu\lambda}{2\mu+\lambda}$) to a solution pertaining to a state of generalised plane stress.

9.2 Exact Solutions to Problems in Plane Strain

We now show that a problem in plane strain (or generalised plane stress) depends upon being able to solve the **bi-harmonic equation**.

The first equilibrium equation of (9.3):

$$\sigma_{11},_1 + \sigma_{12},_2 = 0$$

suggests the introduction of a function θ such that

$$\sigma_{12} = \frac{\partial\theta}{\partial x_1} \qquad \sigma_{11} = -\frac{\partial\theta}{\partial x_2} \tag{9.25}$$

whilst the second equilibrium equation of (9.3) suggests that we write the stress components σ_{21}, σ_{22} in terms of a function ϕ:

$$\sigma_{21} = -\frac{\partial\phi}{\partial x_2} \qquad \sigma_{22} = \frac{\partial\phi}{\partial x_1} \tag{9.26}$$

The existence of θ, ϕ is guaranteed since their mixed derivatives are equal:

$$0 = \sigma_{11},_1 + \sigma_{12},_2 \equiv -\frac{\partial^2\theta}{\partial x_1 \partial x_2} + \frac{\partial^2\theta}{\partial x_2 \partial x_1}$$

and

$$0 = \sigma_{21},_1 + \sigma_{22},_2 \equiv -\frac{\partial^2\phi}{\partial x_1 \partial x_2} + \frac{\partial^2\phi}{\partial x_2 \partial x_1}$$

However, since the stress tensor is symmetric we require, for consistency, $\sigma_{12} = \sigma_{21}$ and so

$$\frac{\partial \phi}{\partial x_2} = -\frac{\partial \theta}{\partial x_1} \tag{9.27}$$

This suggests the introduction of a third function Ψ:

$$\theta = -\frac{\partial \Psi}{\partial x_2} \qquad \phi = \frac{\partial \Psi}{\partial x_1} \tag{9.28}$$

The existence of Ψ being guaranteed by (9.27). It is possible to write all of the stress components in terms of the function Ψ alone:

$$\sigma_{11} = -\frac{\partial \theta}{\partial x_2} = \frac{\partial^2 \Psi}{\partial x_2^2} \qquad \sigma_{12} = \frac{\partial \theta}{\partial x_1} = -\frac{\partial^2 \Psi}{\partial x_1 \partial x_2} \qquad \sigma_{22} = \frac{\partial \phi}{\partial x_1} = \frac{\partial^2 \Psi}{\partial x_1^2}$$

The function Ψ is called the Airy stress function. The reduced Beltrami-Mitchell equation (9.6) implies that Ψ satisfies

$$\nabla^2 \left[\frac{\partial^2 \Psi}{\partial x_2^2} + \frac{\partial^2 \Psi}{\partial x_1^2} \right] \equiv \nabla^2(\nabla^2 \Psi) = 0 \tag{9.29}$$

or

$$\nabla^4 \Psi \equiv \frac{\partial^4 \Psi}{\partial x_1^4} + 2\frac{\partial^4 \Psi}{\partial x_1^2 \partial x_2^2} + \frac{\partial^4 \Psi}{\partial x_2^4} = 0 \tag{9.30}$$

This fourth order partial differential equation is called the **bi-harmonic equation**.

If we can solve this equation subject to the boundary conditions (9.4) then a solution to a plane strain problem will have been obtained.

Those readers who are familiar with the analytic treatment of the two dimensional Laplace equation $\nabla^2 G = 0$ will know that the use of a complex variable formulation can be used to great effect. It is not surprising, therefore, that the use of complex variables can also be used with great success in the analytical treatment of the bi-harmonic equation.

Although we shall employ a complex variable formulation of the equations of plane strain we shall not deal directly with the bi-harmonic equation. Rather, we begin with the displacement vector. As this is an introductory text, in what follows we shall ignore the mathematical intricacies which naturally arise in the use of complex variables in multi-connected regions.

We shall use the usual convention that $i = \sqrt{-1}$, and also introduce a complex variable z:

$$z = x_1 + i x_2 \qquad \text{with conjugate} \quad z^* = x_1 - i x_2$$

(all complex conjugates will be denoted by a star *). Equivalently, we can write

$$x_1 = \frac{1}{2}(z + z^*) \quad \text{and} \quad x_2 = \frac{1}{2i}(z - z^*)$$

and we can regard z, z^* as *independent* variables. Then, from the calculus:

$$\frac{\partial}{\partial z} \equiv \frac{\partial}{\partial x_1}\frac{\partial x_1}{\partial z} + \frac{\partial}{\partial x_2}\frac{\partial x_2}{\partial z} = \frac{1}{2}(\frac{\partial}{\partial x_1} - i\frac{\partial}{\partial x_2})$$
$$\frac{\partial}{\partial z^*} \equiv \frac{\partial}{\partial x_1}\frac{\partial x_1}{\partial z^*} + \frac{\partial}{\partial x_2}\frac{\partial x_2}{\partial z^*} = \frac{1}{2}(\frac{\partial}{\partial x_1} + i\frac{\partial}{\partial x_2}) \qquad \textbf{(9.31)}$$

Also

$$\frac{\partial^2}{\partial z^* \partial z} \equiv \frac{1}{4}(\frac{\partial^2}{\partial x_1^2} + \frac{\partial^2}{\partial x_2^2}) \equiv \frac{1}{4}\nabla^2 \qquad \textbf{(9.32)}$$

If we define the complex displacement: $D = u_1 + iu_2$ then it is an easy matter to confirm that the dilitation in plane strain, $\varepsilon_{\alpha\alpha}$, can be expressed in the form:

$$\varepsilon_{\alpha\alpha} = \frac{\partial u_1}{\partial x_1} + \frac{\partial u_2}{\partial x_2} \equiv \frac{\partial D}{\partial z} + \left[\frac{\partial D}{\partial z}\right]^* \qquad \textbf{(9.33)}$$

From (9.3) and (9.2a) we can write the two equilibrium equations in terms of the complex displacement (the reader should verify this construction)

$$\mu\nabla^2 u_1 + (\lambda + \mu)\frac{\partial}{\partial x_1}(\frac{\partial u_1}{\partial x_1} + \frac{\partial u_2}{\partial x_2}) = 0 \qquad \textbf{(9.34a)}$$

$$\mu\nabla^2 u_2 + (\lambda + \mu)\frac{\partial}{\partial x_2}(\frac{\partial u_1}{\partial x_1} + \frac{\partial u_2}{\partial x_2}) = 0 \qquad \textbf{(9.34b)}$$

Adding (9.34a) to i(9.34b) enables the two equilibrium equations to be expressed as a *single* complex equation:

$$\mu\nabla^2 D + 2(\lambda + \mu)\frac{\partial}{\partial z^*}\left\{\frac{\partial D}{\partial z} + \left[\frac{\partial D}{\partial z}\right]^*\right\} = 0$$

or, equivalently, using (9.32)

$$4\mu\frac{\partial^2 D}{\partial z^* \partial z} + 2(\lambda + \mu)\frac{\partial}{\partial z^*}\left\{\frac{\partial D}{\partial z} + \left[\frac{\partial D}{\partial z}\right]^*\right\} = 0$$

which is,

$$\frac{\partial}{\partial z^*}\left(4\mu\frac{\partial D}{\partial z} + 2(\lambda + \mu)\left\{\frac{\partial D}{\partial z} + \left[\frac{\partial D}{\partial z}\right]^*\right\}\right) = 0 \qquad \textbf{(9.35)}$$

This equation integrates immediately to

$$4\mu\frac{\partial D}{\partial z} + 2(\lambda + \mu)\left\{\frac{\partial D}{\partial z} + \left[\frac{\partial D}{\partial z}\right]^*\right\} = \theta' \qquad \textbf{(9.36a)}$$

in which $\theta(z)$ is an analytic function. (We write it as a derivative for later convenience). We can easily isolate $\dfrac{\partial D}{\partial z}$ from (9.36a) by first forming its complex conjugate:

$$4\mu\left[\frac{\partial D}{\partial z}\right]^* + 2(\lambda + \mu)\left\{\frac{\partial D}{\partial z} + \left[\frac{\partial D}{\partial z}\right]^*\right\} = (\theta')^* \qquad \textbf{(9.36b)}$$

implying

$$(6\mu + 2\lambda)\left[\frac{\partial D}{\partial z}\right]^* = (\theta')^* - 2(\lambda + \mu)\frac{\partial D}{\partial z}$$

and so, substituting into (9.36a), implies, after some algebra:

$$8\mu(2\mu + \lambda)\frac{\partial D}{\partial z} = (3\mu + \lambda)\theta' - (\lambda + \mu)(\theta')^* \tag{9.37}$$

Since $(\theta')^*$ is a function of z^* alone we can integrate this immediately to give

$$8\mu(2\mu + \lambda)D = (3\mu + \lambda)\theta - (\lambda + \mu)z(\theta')^* + \eta^* \tag{9.38}$$

where $\eta(z)$ is an arbitrary function. This expression can be written in the slightly more elegant form:

$$2\mu D = \kappa\Omega - z(\Omega')^* - \omega^* \tag{9.39}$$

in which

$$\Omega(z) \equiv \frac{(\lambda + \mu)}{4(\lambda + 2\mu)}\theta(z) \qquad \omega(z) \equiv -\frac{1}{4(\lambda + 2\mu)}\eta(z) \qquad \kappa = \frac{\lambda + 3\mu}{\lambda + \mu} \tag{9.40}$$

The two analytic functions $\Omega(z)$ and $\omega(z)$ govern any solution to a problem in plane strain.

The stresses can be obtained in terms of these two complex functions from (9.2a) and (9.33):

$$\sigma_{11} + \sigma_{22} = 2(\lambda + \mu)\left\{\frac{\partial D}{\partial z} + \left[\frac{\partial D}{\partial z}\right]^*\right\} = 2(\Omega' + (\Omega')^*) \tag{9.41}$$

whilst from (9.2a) and (9.31)

$$\sigma_{11} - \sigma_{22} + 2i\sigma_{12} = 4\mu\frac{\partial D}{\partial z^*} = -2(z(\Omega'')^* + (\omega')^*) \tag{9.42}$$

By making judicious choices for $\Omega(z)$ and $\omega(z)$ many interesting problems in plane strain can be solved exactly. This complex variable approach was developed by the Russian school of solid mechanics under the direction of the mathematician N. Muskhelishvili. A sophisticated knowledge of complex variable theory, particularly of the complex integral calculus and of the use of multi-valued functions is required if this approach is to be used in a consistent manner. To a large extent it is beyond the scope of a text at this level and so we shall not pursue this line of working too deeply. We shall content ourselves with examing some simple problems for which straightforward solutions may be proposed.

• Rigid Body Motion

Our first inquiry is to determine the functional forms of $\Omega(z)$ and $\omega(z)$ which give rise to zero stresses (and hence zero strains). Physically this will correspond to a rigid body motion. Clearly, as the two complex equations:

$$\sigma_{11} + \sigma_{22} = 0 \quad \text{and} \quad \sigma_{11} - \sigma_{22} + 2i\sigma_{12} = 0$$

imply $\sigma_{11} = \sigma_{22} = \sigma_{12} = 0$, we deduce that, in a zero stress situation $\Omega(z)$ and $\omega(z)$ satisfy:

$$\Omega' + (\Omega')^* = 0 \tag{9.43}$$

$$z(\Omega'')^* + (\omega')^* = 0 \tag{9.44}$$

To solve these equations we write $\Omega' = r(x_1, x_2) + is(x_1, x_2)$ in which (since Ω' is analytic) r and s are related through the Cauchy-Riemann equations:

$$\frac{\partial r}{\partial x_1} = \frac{\partial s}{\partial x_2} \qquad \frac{\partial r}{\partial x_2} = -\frac{\partial s}{\partial x_1} \tag{9.45}$$

Now (9.43) implies $r(x_1, x_2) = 0$ and so (9.45) demands that s must be a constant. Hence

$$\Omega'(z) = iA$$

where A is a real constant. Integrating we obtain

$$\Omega(z) = iAz + B$$

where B is a complex constant. Substituting this into (9.44) gives $(\omega')^* = 0$ and so

$$\omega(z) = C$$

where C is a complex constant. With these forms for $\Omega(z)$ and $\omega(z)$ we have, from (9.39),

$$\begin{aligned} 2\mu D &= \kappa(iAz + B) + izA - C^* \\ &= i(\kappa + 1)Az + (\kappa B - C^*) \end{aligned}$$

The first term describes a pure rotation and the second describes a pure translation. (See Example 4.1). The combination corresponds to a rigid body motion of the region in the x_1x_2 plane.

- **Uniform Tension**

As a first excursion into finding exact solutions to plane strain problems we shall make simple choices for $\Omega(z)$ and $\omega(z)$ and determine the resulting stress fields. To begin with if we *choose* $\Omega(z) = \frac{1}{2}Tz$ and $\omega(z) = 0$ in which T is a real constant, then, in (9.41) and (9.42):

$$\sigma_{11} + \sigma_{22} = 2(\frac{1}{2}T + \frac{1}{2}T) = 2T \quad \text{and} \quad \sigma_{11} - \sigma_{22} + 2i\sigma_{12} = 0$$

therefore

$$\sigma_{12} = 0 \qquad \sigma_{11} = \sigma_{22} = T$$

This stress field is called a state of uniform tension. To examine its physical characteristics consider an arbitrary line in the x_1x_2 plane on which there is assigned a unit normal $n_\alpha = (n_1, n_2)$. The stress vector on this line is, by the Cauchy relation (we consider a *unit thickness* of the plate)

$$T_\alpha^{(n)} = \sigma_{\alpha\beta} n_\beta = \sigma_{\alpha 1} n_1 + \sigma_{\alpha 2} n_2$$

therefore

$$T_1^{(n)} = T n_1 \qquad T_2^{(n)} = T n_2$$

and so the resultant force on an arbitrarily oriented length $\delta\ell$ is $\sqrt{T_\alpha^{(n)} T_\alpha^{(n)}} = T\delta\ell$.

If we wish to put a rectangular plate into a state of uniform tension we simply apply a normally directed force T per unit length to the edges all around the boundary (see Figure 9.5a). As a commonplace example each small section of the bounding 'skin' of an expanded balloon is, to a good approximation, in a state of uniform tension (see Figure 9.5b).

The reader should confirm that the *choice*

$$\Omega(z) = \frac{1}{4} T z \qquad \omega(z) = -\frac{1}{2} T e^{-2i\alpha} z \qquad (T \text{ real}) \tag{9.46}$$

describes a state of **uniaxial tension** at an angle α to the x_1-axis. (Check that for an elementary length $\delta\ell$ oriented at an angle $\pi/2 + \alpha$ that the resultant force is $T\delta\ell$, whilst for elementary lengths aligned *along* the line at angle α the resultant force is zero. See Figure 9.5c).

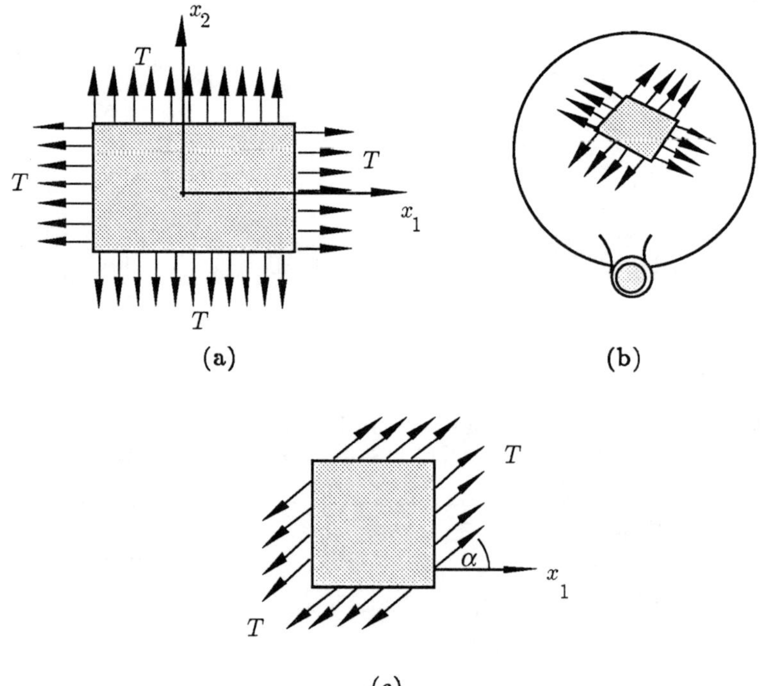

(c)

Figure 9.5

- **Unstressed Circular Hole in an Infinite Plate subjected to uniaxial Tension**

The plate with a circular hole, of radius a, is stretched by tensile forces along the x_1-direction as described in Figure 9.6. If the hole was absent the solution would be given by (9.46) with $\alpha = 0$. We shall see below, from an examination of the exact solution, that the existence of the hole, no matter how small, considerably weakens the ability of the plate to sustain the tensile stress T.

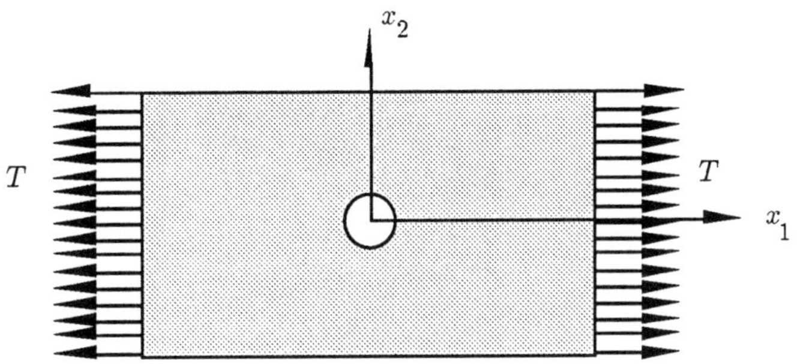

Figure 9.6

Although, theoretically, we have to deal with an infinite plate the solution we obtain will, to a good approximation, apply to the stress field of a large rectangular plate stretched as in the figure with a small hole at the centre.

The uniaxial tension along the x_1-axis is easy to describe (from (9.46) with $\alpha = 0$):

$$\Omega^0(z) = \frac{1}{4}Tz \qquad \omega^0(z) = -\frac{1}{2}Tz$$

We need to add terms to these expressions which ensure that the circular hole is stress free. It may be shown that to ensure finite stress levels 'at infinity' that the additive terms must only involve negative powers of z alone. Therefore we propose the solution:

$$\Omega(z) = \frac{1}{4}Tz + \sum_{k=1}^{\infty} \alpha_k z^{-k} \qquad \omega(z) = -\frac{1}{2}Tz + \sum_{k=1}^{\infty} \beta_k z^{-k} \tag{9.47}$$

in which α_k, β_k are constants to be determined.

We shall find it easier, in imposing the boundary conditions at the circular hole, to transform to new coordinates (x_1', x_2') obtained from (x_1, x_2) by a pure rotation through an angle θ so that the x_1'-axis is always directed radially. We have already worked out the stress transformation equations in Section 4.4 (for strain, though exactly the same relations will apply to stress components, see (4.16)). From those results it is easily verified that

$$\sigma_{11}' + \sigma_{22}' = \sigma_{11} + \sigma_{22} \tag{9.48}$$

and

$$\sigma_{11}' - \sigma_{22}' + 2i\sigma_{12}' = e^{-2i\theta}(\sigma_{11} - \sigma_{22} + 2i\sigma_{12}) \tag{9.49}$$

These two relations allow us to obtain the expression for the 'primed' stress components in terms of the complex functions $\Omega(z)$ and $\omega(z)$ introduced above (and using the polar form of a complex number $z = re^{i\theta}$ implying $e^{2i\theta} = z/z^*$):

$$\sigma'_{11} + \sigma'_{22} = 2[\Omega' + (\Omega')^*] \tag{9.50}$$

$$\sigma'_{11} - \sigma'_{22} + 2i\sigma'_{12} = -2[z^*(\Omega'')^* + \frac{z^*}{z}(\omega')^*] \tag{9.51}$$

Now we are in a position to impose the stress free boundary conditions on $|z| = a$. In the primed coordinates, at any point of the boundary with $n'_\alpha = (-1, 0)$ we demand:

$$0 = T'^{(n')}_\alpha = \sigma'_{\alpha\beta} n'_\beta = -\sigma'_{\alpha 1} \quad \text{on } |z| = a \tag{9.52}$$

that is,

$$\sigma'_{11} = \sigma'_{12} = 0 \quad \text{on } |z| = a \tag{9.53}$$

Adding (9.50) to (9.51) implies:

$$\sigma'_{11} + i\sigma'_{12} = \Omega' + (\Omega')^* - z^*(\Omega'')^* - \frac{z^*}{z}(\omega')^* \tag{9.54}$$

and our stress free boundary condition will be satisfied if

$$\Omega' + (\Omega')^* - z^*(\Omega'')^* - \frac{z^*}{z}(\omega')^* = 0 \quad \text{on } |z| = a \tag{9.55}$$

Now, using (9.47) we have, for the expansions of Ω', Ω'' and ω':

$$\begin{aligned}
\Omega' &= \frac{1}{4}T - \frac{\alpha_1}{z^2} - \frac{2\alpha_2}{z^3} + \ldots. \\
\Omega'' &= \frac{2\alpha_1}{z^3} + \frac{6\alpha_2}{z^4} + \ldots. \\
\omega' &= -\frac{1}{2}T - \frac{\beta_1}{z^2} - \frac{2\beta_2}{z^3} + \ldots.
\end{aligned}$$

and therefore, on the circle; $z = ae^{i\theta}$,

$$\begin{aligned}
\Omega' &= \frac{1}{4}T - \frac{\alpha_1}{a^2}e^{-2i\theta} + \ldots. \\
\Omega'' &= \frac{2\alpha_1}{a^3}e^{-3i\theta} + \frac{6\alpha_2}{a^4}e^{-4i\theta} + \ldots. \\
\omega' &= -\frac{1}{2}T - \frac{\beta_1}{a^2}e^{-2i\theta} - \frac{2\beta_2}{a^3}e^{-3i\theta} - \frac{3\beta_3}{a^4}e^{-4i\theta} + \ldots.
\end{aligned}$$

Substituting into (9.55) gives (after regrouping in powers of $e^{i\theta}$):

$$\begin{aligned}
e^{-2i\theta}[-\frac{\alpha_1}{a^2} + \frac{T}{2}] + [\frac{T}{2} + \frac{\beta_1^*}{a^2}] + e^{i\theta}[\frac{2\beta_2^*}{a^3}] \\
+ e^{2i\theta}[-\frac{\alpha_1^*}{a^2} - \frac{2\alpha_1^*}{a^2} + \frac{3\beta_3^*}{a^4}] + \ldots. = 0
\end{aligned}$$

Equating coefficients of $e^{i\theta}$ to zero implies:

$$\alpha_1 = \alpha_1^* = \frac{a^2T}{2} \quad \beta_1 = \beta_1^* = -\frac{a^2T}{2} \quad \beta_2^* = 0 \quad \beta_3^* = \frac{a^4T}{2} \tag{9.56}$$

with all the other α and β terms being zero. Hence *the* solution to this problem is

$$\Omega(z) = \frac{1}{4}Tz + \frac{a^2T}{2z} \qquad \omega(z) = -\frac{1}{2}Tz - \frac{a^2T}{2z} + \frac{a^4T}{2z^3} \tag{9.57}$$

The primed stress components are easily determined from (9.50), (9.51) and we find (after relabelling $\sigma'_{11}, \sigma'_{22}$ and σ'_{12} as $\sigma_{rr}, \sigma_{\theta\theta}$ and $\sigma_{r\theta}$ respectively, which we first did in Section 5.4)

$$\begin{aligned}
\sigma_{rr} &= \frac{T}{2}\left[(1-\frac{a^2}{r^2}) + (1 - \frac{4a^2}{r^2} + \frac{3a^4}{r^4})\cos 2\theta\right] \\
\sigma_{\theta\theta} &= \frac{T}{2}\left[(1+\frac{a^2}{r^2}) - (1 + \frac{3a^4}{r^4})\cos 2\theta\right] \\
\sigma_{r\theta} &= -\frac{T}{2}\left[1 + \frac{2a^2}{r^2} - \frac{3a^4}{r^4}\right]\sin 2\theta
\end{aligned} \tag{9.58}$$

On the boundary of the hole ($r = a$) the only non-vanishing stress component is the hoop stress $\sigma_{\theta\theta}$:

$$\sigma_{\theta\theta} = T(1 - 2\cos 2\theta) \tag{9.59}$$

which takes its maximum value of $3T$ at $\theta = \pi/2,\ 3\pi/2$. Thus the presence of the hole, no matter how small, leads to a three fold increase in stress levels at these particular points. Under large loading the plate will first begin to tear apart at these points.

Particular aspects of this exact solution can be compared with experiment. The photoelastic fringe pattern of a stretched plate with a small hole (small in comparison to the dimensions of the plate) is shown in Figure 9.7(a). (I am grateful to K Sharples of Sharples Engineering, Bamber Bridge, Preston, England for supplying this figure. The ratio of hole diameter to plate dimension is 1:10). The dark areas locate the lines along which the maximum shear stress takes constant values. These are the isochromatic curves:

$$(\sigma_{rr} - \sigma_{\theta\theta})^2 + 4\sigma_{r\theta}^2 = \text{constant} \tag{9.60}$$

If the isochromatic curves are plotted from the exact solution we obtain the contour plot shown in Figure 9.7(b). The gross features of the two sets of curves correspond closely. A detailed calculation using the photoelastic results confirm the maximum hoop stress levels predicted from the exact solution.

photoelastic results (a) exact solution (b) finite element solution (c)

Figure 9.7

We could proceed in this manner, outlined above, to develop further exact solutions although we would prefer to use a more methodical approach than the 'guesswork' technique that we have employed. Muskhelishvili has developed a more formal approach using general expressions for the functions $\Omega(z)$ and $\omega(z)$ in the form of infinite power series. He has shown how the power series coefficients can be determined using various integral theorems in complex analysis. (This method is well illustrated in the text 'Complex Variable Methods in Elasticity' by A H England published by Wiley (1971)).

However, no matter how ingenious the approach there is a constraint to almost all of the problems treated by this or indeed any other approach used to obtain analytical solutions. The region R and the boundary conditions display such a high degree of 'regularity' that they can be described as applying only to *ideal* physical situations. In realistic applications of two dimensional elastostatics an exact solution of a similar *ideal problem* can throw a good deal of light, in a qualatitive sense, as to the kind of behaviour to expect but only a numerical analysis or an experimental model can hope to provide quantative results on the actual problem under consideration.

In the next section we look in detail at one particular numerical treatment – the Finite Element Method – which has gained considerable prominence in recent years (in the main due to the advent of high-speed computing facilities). In many respects it is a generalisation of the techniques used in the matrix analysis of structures to the continuum. Mathematically, it is founded on the principle of minimum potential energy.

EXERCISES

9.1 Show that the complex potentials: $\Omega(z) = iCz^2 \quad \omega(z) = -iCz^2 \quad C$ real, represents a state of pure bending in plane strain (check that $\sigma_{22} = \sigma_{12} = 0$ everywhere). Calculate u_1, u_2 and the bending moment $(M = \int_{-b}^{b} x_2\sigma_{11}dx_2)$ across a section $-b < x_2 < b$. Show that the line of centroids $(x_2 = 0)$ is deformed into a circle (approximately) with radius $\mu/C(\kappa + 1)$.

9.2 If $\Omega(z) = Az^3$ make an appropriate choice for $\omega(z)$ so the lines $x_2 = \pm b$ are stress free, that is: $\sigma_{22} + i\sigma_{12} = 0$ on $z - z* = \pm 2ib$. Show that a 'block' of material $-b \le y \le b \quad -d \le\le 0$ is in static equilibrium i.e. that the total force and the total couple on the body are zero.

9.3 A large elastic plate in plain strain contains a hole of radius a under pressure p. Verify that the complex potentials:

$$\Omega(z) = 0 \qquad \omega(z) = -\frac{pa}{z}$$

adequately describes this situation. Show that the hoop stress is $\sigma_{\theta\theta} = p$ and the radial displacement at the hole is $u_r = \dfrac{pa}{2\mu}$.

9.4 Use the complex potential of 9.3 to show that a radial stress $\sigma_{rr} = -\dfrac{2\mu r_0 a^2}{r^2}$ is required to increase the radius of the hole by ar_0.

9.5 Verify that complex potentials: $\Omega(z) = -\frac{1}{2}pz \qquad \omega(z) = 0$ can be used to describe a disc, in plain strain conditions, subjected to uniform pressure on its perimeter. Show that for a disc of radius a the radial displacement of the perimeter is $u_r = -\dfrac{p}{E_d}(1 - \nu_d)a$. where E_d and ν_d are the Young's modulus and Poisson ratio.
If an oversize **elastic** disc of radius $a(1 + r_0)$ is forced to occupy a hole of undeformed radius a in a large elastic plate use the results of 9.3 to show that the pressure at the common boundary is:

$$p = \frac{r_0 E_d E_p}{E_d(1 + \nu_p) + E_p(1 - \nu_d)}$$

where E_p, ν_p are the elastic constants of the plate.

9.6 Show that the complex potentials: $\Omega(z) = Ax \qquad \omega(z) = B/z$ where A and B are real can be used to describe the deformation of a cylinder of internal and external radii a and b under pressures p_1 and p_2 respectively. Check your conclusions with results obtained in Section 5.4

9.3 Approximations in Two Dimensional Elastostatics

- **Potential Energy**

We can, from the general expression in (6.24), form a potential energy per unit thickness appropriate to the analysis of plane strain or of generalised plane stress.

Consider a two dimensional region R bounded by a curve C. Let $(T_\alpha^{(n)}, B_\alpha, u_\alpha, \varepsilon_{\alpha\beta}, \sigma_{\alpha\beta})$ be a two dimensional elastic force system associated with the body and related through the equations (9.2a), (9.3) and (9.4). The system satisfies the boundary conditions:

$$\sigma_{\alpha\beta} n_\beta = T_\alpha^{(n)} \quad \text{on } C_0 \qquad u_\alpha = h_\alpha \quad \text{on } C - C_0 \tag{9.61}$$

in which $T_\alpha^{(n)}$ and h_α are *given* functions on the boundary. (C_0 may be a single section of the boundary or it may be a collection of discrete parts). The two dimensional potential energy U_2 is *defined* to be:

$$U_2 = \frac{1}{2} \iint_R \theta_{\alpha\beta}(2\mu\theta_{\alpha\beta} + \lambda\delta_{\alpha\beta}\theta_{\tau\tau})dR - \int_{C_0} T_\alpha^{(n)}\gamma_\alpha d\ell - \iint_R \rho B_\alpha \gamma_\alpha dR \tag{9.62}$$

As with its three dimensional counterpart, γ_α is an *arbitrary* vector field except that it satisfies the *same* boundary conditions as u_α (namely, $\gamma_\alpha = h_\alpha$ on $C - C_0$) and that $\theta_{\alpha\beta}$ is related to γ_α through:

$$\theta_{\alpha\beta} = \frac{1}{2}(\gamma_{\alpha},_{\beta} + \gamma_{\beta},_{\alpha}) \tag{9.63}$$

that is, $\theta_{\alpha\beta}$ is the 'strain tensor' corresponding to γ_α.

Using the same procedure as outlined in Section 6.5 we may show that of all the values of U_2 obtained by using different vector fields γ_α in (9.62), the minimum value is attained when γ_α is the *true* displacement field u_α (and hence $\theta_{\alpha\beta}$ is the *true* strain tensor). This extremum principle (and its three-dimensional form in Section 6.5) is the basis of an important numerical procedure, the Finite Element Method, which has been developed in the last thirty years, and which we shall now (briefly) outline.

- **Finite Elements**

In Chapter 6 we determined the approximate solution of a number of problems of practical interest by first *proposing* ('intelligent guesswork') the form taken by the displacement vector. The 'form' depended on a number of parameters (see Example 6.2, in which the parameters were the components of the displacement at node B namely (γ_1, γ_2)). We then obtained *specific* values for these parameters by assigning values to them which minimised the potential energy. (We were careful to choose a proposed displacement vector which was concordant with the displacement boundary conditions). To a large extent the practical implementation of the principle of minimum potential energy is dependent upon making *simplifying* assumptions as to the deformation in the elastic body under consideration. For the two member structure in Example 6.2 the simplifying assumption is that each member can only sustain stress along its axis, which is not exactly true, though for long slender beams is a good approximation. Indeed in the analysis of structures (see Chapter 8) similar assumptions are used throughout and one can view the whole of that theory as a methodical implementation of the principle of minimum potential energy.

The first stage in a finite element analysis is to discretize the region R into a number of elements. The elements may be of varying shapes and sizes, though invariably they are triangular or quadrilateral in shape. The position of the elements within R and the shape of their edges are defined by the element nodes. See Figure 9.8. Although not shown here, each of the nodes and each of the elements will be assigned an identifying number. The numbers associated with the nodes are called **global node numbers**.

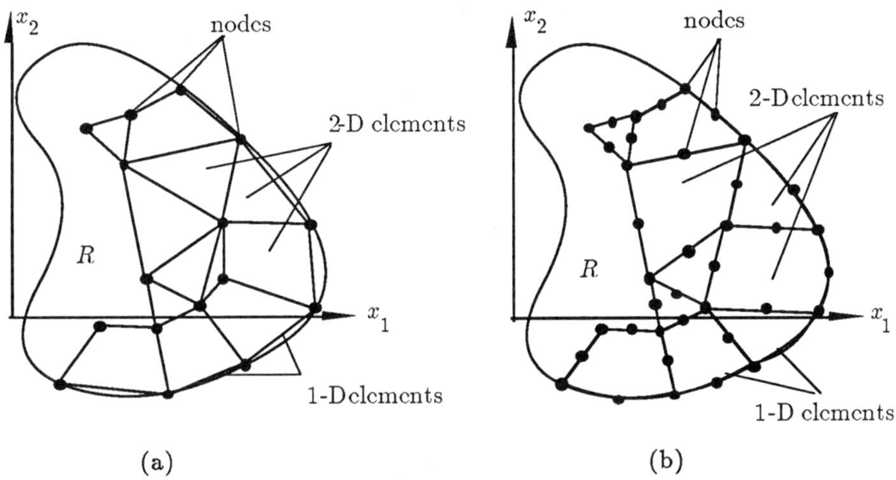

Figure 9.8

Three-noded triangular and four-noded quadrilateral elements have straight edges. If the boundary of R is curved, then no matter how small the elements, the boundary cannot be modelled exactly. A better approach in such cases is to use six-noded triangular or eight-noded quadrilateral elements which, by choosing the mid-side nodes appropriately, can have (parabolic) curved edges and so approximate curved boundaries more accurately (see Figure 9.8(b)). As long as the element preserves its basic triangular or quadrilateral shape the positions of the nodes are at our disposal and so, for a given problem, there are many possible 'meshes' which can be used.

Having discretised the domain R we note that this *naturally* induces a discretisation on the boundary C. The use of three-noded triangular or four-noded quadrilateral elements discretises the boundary into two-noded one dimensional elements. Correspondingly, if we use six-noded triangular or eight-noded quadrilateral elements then the boundary of the region is split into (possibly curved) three-noded one-dimensional elements. In the remaining discussion on finite elements we will exclusively consider eight-noded quadrilateral elements.

Having 'meshed' the region R, we now make certain simplifying assumptions as to the behaviour or variation of the proposed displacement γ_α throughout each element. The simplification will be dependent upon the type of element we choose to use in the meshing. Due to the variable nature of the position and size of the 'finite elements' it will prove advantageous to firstly transform each element into a so-called standard element and then to impose the simplification on the displacement with respect to this standard element. A separate transformation will be needed for each element. (This process is rather similar to the use of local and global coordinates in the matrix analysis of structures. There, we first of all derived a relation between forces and displacements with respect to local coordinates – because it was relatively easy to do that – then we used a transformation matrix to deduce the 'more complicated' relation with respect to global coordinates).

The standard eight-noded quadrilateral element is defined with respect to (ξ, η) coordinates (to be transformed into (x_1, x_2) coordinates shortly) as in Figure 9.9. The square extends from $-1 \leq \xi, \eta \leq 1$ and the nodes are numbered as shown. The numbers $1, ..., 8$ attached to the nodes of the standard element are called **local node numbers**. Each of the eight-noded quadrilateral elements used in the mesh will be transformed, using a coordinate transformation to be specified later, into the standard element. There is therefore a relation between the global node numbers associated with an element $[e]$ and the local node numbers attached to the standard element.

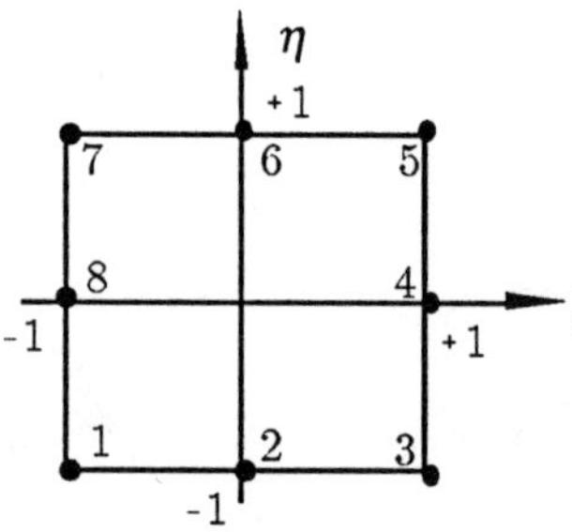

Figure 9.9

It is the practice, in finite elements, to approximate the variation in each of the proposed displacement components γ_α within the standard element by a multinomial in (ξ, η) and then to uniquely 'tag' this multinomial to the values of γ_α at the nodes.

With an eight-noded element it seems sensible to use a multinomial with just eight coefficients. We therefore propose that the variation in the displacement components, within, element $[e]$, be:

$$\gamma_1^{[e]} = a_0 + a_1\xi + a_2\eta + a_3\xi^2 + a_4\xi\eta + a_5\eta^2 + a_6\xi^2\eta + a_7\xi\eta^2 \qquad \textbf{(9.64a)}$$

$$\gamma_2^{[e]} = b_0 + b_1\xi + b_2\eta + b_3\xi^2 + b_4\xi\eta + b_5\eta^2 + b_6\xi^2\eta + b_7\xi\eta^2 \qquad \textbf{(9.64b)}$$

In each expansion in (9.64) the choice of the first six multinomial terms $1, \xi, \eta, \xi^2, \xi\eta, \eta^2$ was obvious. The choice of the remaining two cubic terms $\xi^2\eta$, $\eta^2\xi$, less so. Of the four cubic terms possible, we choose these two so that, on the edges of the element ($\xi = \pm 1$ or $\eta = \pm 1$) the multinomial variation in γ_1 or γ_2 is a quadratic. This quadratic is uniquely 'tagged' to the values of the displacement at the three edge nodes alone. This implies that the displacement components vary continuously from one element to another across common edges in R. This is an important mathematical requirement.

Obviously, the more multinomial terms we use the more 'accurate' will be the element though greater complication in the theoretical development will be involved. Greater accuracy can be achieved by using smaller elements (within R). In practice eight-noded elements have been found to work well and their use is a reasonable compromise between accuracy and intracy.

There are many possible ways a given multinomial may be expressed. In finite elements it is the practice to use parameters with a direct physical interpretation. An obvious choice is to use the values of the displacement components γ_1, γ_2 at the nodes as the eight parameters in each of (9.64a,b). Define

$$\gamma_{1(k)}^{[e]} = \gamma_1^{[e]}(\xi_k, \eta_k) \qquad \gamma_{2(k)}^{[e]} = \gamma_2^{[e]}(\xi_k, \eta_k) \quad \text{each } k = 1, ..., 8 \tag{9.65}$$

then

$$\begin{aligned}
\gamma_{1(1)}^{[e]} &= \gamma_1^{[e]}(-1,-1) = a_0 - a_1 - a_2 + a_3 + a_4 + a_5 - a_6 - a_7 \\
\gamma_{1(2)}^{[e]} &= \gamma_1^{[e]}(0,-1) = a_0 - a_2 + a_5 \\
\gamma_{1(3)}^{[e]} &= \gamma_1^{[e]}(+1,-1) = a_0 + a_1 - a_2 + a_3 - a_4 + a_5 - a_6 + a_7 \\
\gamma_{1(4)}^{[e]} &= \gamma_1^{[e]}(+1,0) = a_0 + a_1 + a_3 \\
\gamma_{1(5)}^{[e]} &= \gamma_1^{[e]}(+1,+1) = a_0 + a_1 + a_2 + a_3 + a_4 + a_5 + a_6 + a_7 \\
\gamma_{1(6)}^{[e]} &= \gamma_1^{[e]}(0,+1) = a_0 + a_2 + a_5 \\
\gamma_{1(7)}^{[e]} &= \gamma_1^{[e]}(-1,+1) = a_0 - a_1 + a_2 + a_3 - a_4 + a_5 + a_6 - a_7 \\
\gamma_{1(8)}^{[e]} &= \gamma_1^{[e]}(-1,0) = a_0 - a_1 + a_3
\end{aligned}$$

When these eight equations are solved for $a_i \quad i = 0, ...7$ and substituted back into (9.64a) we find, after a rearrangement,

$$\gamma_1^{[e]} = \sum_{i=1}^{8} \gamma_{1(i)}^{[e]} \Phi_{(i)}^{[e]}(\xi, \eta) \tag{9.66a}$$

and for $\gamma_2^{[e]}$

$$\gamma_2^{[e]} = \sum_{i=1}^{8} \gamma_{2(i)}^{[e]} \Phi_{(i)}^{[e]}(\xi, \eta) \tag{9.66b}$$

in which:

$$\begin{aligned}
\Phi_{(1)}^{[e]} &= -\frac{1}{4}(1-\xi)(1-\eta)(1+\eta+\xi) & \Phi_{(2)}^{[e]} &= \frac{1}{2}(1-\xi^2)(1-\eta) \\
\Phi_{(3)}^{[e]} &= -\frac{1}{4}(1+\xi)(1-\eta)(1+\eta-\xi) & \Phi_{(4)}^{[e]} &= \frac{1}{2}(1-\eta^2)(1+\xi) \\
\Phi_{(5)}^{[e]} &= -\frac{1}{4}(1+\xi)(1+\eta)(1-\eta-\xi) & \Phi_{(6)}^{[e]} &= \frac{1}{2}(1-\xi^2)(1+\eta) \\
\Phi_{(7)}^{[e]} &= -\frac{1}{4}(1-\xi)(1+\eta)(1-\eta+\xi) & \Phi_{(8)}^{[e]} &= \frac{1}{2}(1-\eta^2)(1-\xi)
\end{aligned} \tag{9.67}$$

The eight functions $\Phi^{[e]}_{(i)}(\xi,\eta)$ $i = 1,...8$ are called **shape functions** appropriate to a standard eight-noded quadrilateral element.

We emphasize that the actual behaviour of u_α over this element is *approximated* by the multinomial variation detailed in (9.66). At this stage the nodal values $(\gamma^{[e]}_{1(k)}, \gamma^{[e]}_{2(k)})$ $k = 1,...8$ of the displacement components (γ_1, γ_2) are unknowns, and are to be determined. Once these values are known then the other physical quantities of interest; the stresses and the strains, within the element $[e]$, may be determined in the usual way.

We have now dealt with the variation of the displacement components with respect to the standard element. The variation with respect to the actual element (in (x_1, x_2) coordinates) is achieved by applying a coordinate transformation from (ξ,η) to (x_1, x_2) variables. Although there are infinitely many possible coordinate transformations that could be employed, in finite elements we generally make further use of the shape functions when specifying the coordinate transformation.

We focus attention on a particular element $[e]$. If $(x^{[e]}_{1(k)}, x^{[e]}_{2(k)})$ $k = 1,...8$ are the coordinate positions of the eight nodes then we *define* the coordinate transformation between (ξ,η) and (x_1, x_2) as:

$$x_1 = \sum_{k=1}^{8} x^{[e]}_{1(k)} \Phi^{[e]}_{(k)}(\xi,\eta) \qquad x_2 = \sum_{k=1}^{8} x^{[e]}_{2(k)} \Phi^{[e]}_{(k)}(\xi,\eta) \tag{9.68}$$

Note that the 'form' of each of these equations is exactly that assumed for the variation in the displacement components in (9.66). The consequence of this definition is that each edge of the standard element is transformed into a parabola (which can degenerate into a straight line) which passes through the three edge nodes. For example, if we consider the edge $\eta = -1$, $-1 \le \xi \le 1$ then using (9.68) and (9.67) we find

$$\begin{aligned} x_1 &= x^{[e]}_{1(1)}\left[\frac{1}{2}\xi(\xi-1)\right] + x^{[e]}_{1(2)}\left[1-\xi^2\right] + x^{[e]}_{1(3)}\left[\frac{1}{2}\xi(\xi+1)\right] \\ x_2 &= x^{[e]}_{2(1)}\left[\frac{1}{2}\xi(\xi-1)\right] + x^{[e]}_{2(2)}\left[1-\xi^2\right] + x^{[e]}_{2(3)}\left[\frac{1}{2}\xi(\xi+1)\right] \end{aligned} \tag{9.69}$$

Since x_1, x_2 are expressed *quadratically* in terms of the parameter ξ this curve is a parabola in the x_1x_2 plane passing through the points $(x^{[e]}_{1(1)}, x^{[e]}_{2(1)})$, $(x^{[e]}_{1(2)}, x^{[e]}_{2(2)})$ and $(x^{[e]}_{1(3)}, x^{[e]}_{2(3)})$.

Thus, by a judicious choice in the position of the nodes and the size of the element any curved boundary can be modelled as accurately as is desired. The mesh is purely at the discretion of the finite element modeller. A coarse mesh will provide a crude numerical approximation. A finer mesh will provide a better numerical approximation though, generally, at a higher cost in computing time. In numerical work there is always a balance to be drawn between accuracy and cost.

• Finite Element Formulation of Two Dimensional Elasticity

We begin with the expression for the potential energy in the absence of body forces:

$$U_2 = \frac{1}{2}\iint_R \theta_{\alpha\beta}(2\mu\theta_{\alpha\beta} + \lambda\delta_{\alpha\beta}\theta_{\tau\tau})dR - \int_{C_0} T_\alpha^{(n)}\gamma_\alpha d\ell \tag{9.70}$$

in which $\theta_{\alpha\beta} = \frac{1}{2}(\gamma_{\alpha},_{\beta} + \gamma_{\beta},_{\alpha})$. When this is expanded we find:

$$\begin{aligned} U_2 = \frac{1}{2}\iint_R \Big\{(2\mu+\lambda)\Big[(\frac{\partial\gamma_1}{\partial x_1})^2 + (\frac{\partial\gamma_2}{\partial x_2})^2\Big] + 2\lambda\frac{\partial\gamma_1}{\partial x_1}\frac{\partial\gamma_2}{\partial x_2} \\ + \mu\Big[(\frac{\partial\gamma_1}{\partial x_1})^2 + (\frac{\partial\gamma_2}{\partial x_2})^2 + 2\frac{\partial\gamma_1}{\partial x_2}\frac{\partial\gamma_2}{\partial x_1}\Big]\Big\}dR \\ - \int_{C_0}(T_1^{(n)}\gamma_1 + T_2^{(n)}\gamma_2)d\ell \end{aligned} \tag{9.71}$$

We split R into a number of eight-noded elements in which the boundary of each element is a parabola (or a straight line). This defines a total number of nodes nde. In what follows roman capital letters used as indices will be assumed to range from 1 to nde and the summation convention will be amended accordingly. We now express (γ_1, γ_2) in terms of the nodal values of this proposed displacement through the relations:

$$\gamma_1 = u_K R_K \qquad \gamma_2 = v_K R_K \tag{9.72}$$

in which (u_K, v_K) $K = 1, ...nde$ are the nodal values of (γ_1, γ_2) and the functions R_K $K = 1, ...nde$ are called global interpolation functions (or 'roof' functions) and are defined over the complete domain in terms of the local shape functions through the relation

$$R_K = \begin{cases} \Phi_{(r)}^{[e]} & \text{in element } [e] \\ \Phi_{(m)}^{[f]} & \text{in element } [f] \\ . & \\ . & \\ . & \\ \Phi_{(p)}^{[g]} & \text{in element } [g] \\ 0 & \text{elsewhere} \end{cases} \tag{9.73}$$

Here node K is a node attached to elements $[e], [f], ...[g]$ and $r, m, ...p$ are (possibly different) local node numbers corresponding to global node number K. See Figure 9.10. In that figure node K is 'attached' to elements $[f]$ and $[g]$ whilst node J is 'attached' to elements $[e], [f], [g]$ and $[h]$. Substituting (9.72) into (9.71) implies:

$$\begin{aligned} U_2 = \frac{1}{2}\iint_R \Big\{(2\mu+\lambda)\Big[u_K u_J\frac{\partial R_K}{\partial x_1}\frac{\partial R_J}{\partial x_1} + v_K v_J\frac{\partial R_K}{\partial x_2}\frac{\partial R_J}{\partial x_2}\Big] + 2\lambda u_K v_J\frac{\partial R_K}{\partial x_1}\frac{\partial R_J}{\partial x_2} \\ + \mu\Big[u_K u_J\frac{\partial R_K}{\partial x_2}\frac{\partial R_J}{\partial x_2} + v_K v_J\frac{\partial R_K}{\partial x_1}\frac{\partial R_J}{\partial x_1} + 2u_K v_J\frac{\partial R_K}{\partial x_2}\frac{\partial R_J}{\partial x_1}\Big]\Big\}dR \\ - \int_{C_0}(T_1^{(n)}u_K + T_2^{(n)}v_K)R_K d\ell \end{aligned} \tag{9.74}$$

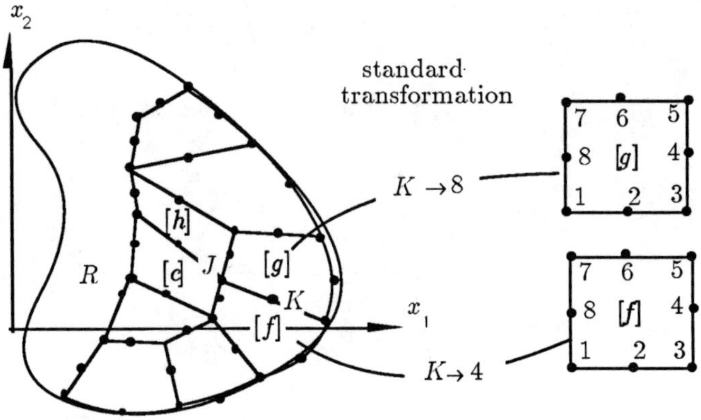

Figure 9.10

Now minimising the potential energy U_2 with respect to each of the nodal parameters (u_K, v_K) $K = 1, ...nde$ implies

$$\frac{\partial U_2}{\partial u_K} = 0 \qquad \frac{\partial U_2}{\partial v_K} = 0 \quad K = 1, ...nde \tag{9.75}$$

Carrying through the operations indicated in (9.75) shows that the parameters are the solutions to the matrix equations:

$$\begin{aligned} A_{KJ}u_J + B_{KJ}v_J &= F_{1K} \\ C_{KJ}u_J + D_{KJ}v_J &= F_{2K} \end{aligned} \tag{9.76}$$

in which

$$A_{KJ} = \iint_R \left[(2\mu + \lambda)\frac{\partial R_K}{\partial x_1}\frac{\partial R_J}{\partial x_1} + \mu\frac{\partial R_K}{\partial x_2}\frac{\partial R_J}{\partial x_2}\right] dR \qquad A_{KJ} = A_{JK} \tag{9.77a}$$

$$B_{KJ} = \iint_R \left[\lambda\frac{\partial R_K}{\partial x_1}\frac{\partial R_J}{\partial x_2} + \mu\frac{\partial R_K}{\partial x_2}\frac{\partial R_J}{\partial x_1}\right] dR \qquad C_{KJ} = B_{JK} \tag{9.77b}$$

$$D_{KJ} = \iint_R \left[(2\mu + \lambda)\frac{\partial R_K}{\partial x_2}\frac{\partial R_J}{\partial x_2} + \mu\frac{\partial R_K}{\partial x_1}\frac{\partial R_J}{\partial x_1}\right] dR \qquad D_{KJ} = D_{JK} \tag{9.77c}$$

and

$$F_{1K} = \int_{C_0} T_1^{(n)} R_K d\ell \qquad F_{2K} = \int_{C_0} T_2^{(n)} R_K d\ell \tag{9.77d}$$

Now, if $\underline{U}$ is a $(2nde \times 1)$ column vector of the form:

$$\underline{U} = [u_1, v_1, u_2, v_2, ..., u_{nde}, v_{nde}]^T \tag{9.78}$$

then the equations in (9.76) can be written as a single matrix equation:

$$\underline{G}\underline{U} = \underline{H} \tag{9.79}$$

in which $\underline{G}$ is a symmetric matrix:

$$\underline{G} = \begin{bmatrix} A_{11} & B_{11} & A_{12} & B_{12} & . & . & . \\ C_{11} & D_{11} & C_{12} & D_{12} & . & . & . \\ A_{21} & B_{21} & A_{22} & B_{22} & . & . & . \\ C_{21} & D_{21} & C_{22} & D_{22} & . & . & . \\ . & . & . & . & . & . & . \\ . & . & . & . & . & . & . \\ . & . & . & . & . & . & . \end{bmatrix} \quad \text{and} \quad \underline{H} = \begin{bmatrix} F_{11} \\ F_{21} \\ F_{12} \\ F_{22} \\ . \\ . \\ . \end{bmatrix} \tag{9.80}$$

Since, theoretically, the functions R_K are known through (9.73), each of the members of the matrices $\underline{G}$ and $\underline{H}$ may be determined. The boundary conditions on u_α are imposed on those nodes on $C - C_0$ in a manner similar to that used in the matrix analysis of structures. The system (9.79) is then solved using any standard procedure for solving systems of linear equations (utilising the symmetric nature of $\underline{G}$) and the finite element analysis is all but complete. What remains is the post-processing:- using the obtained solution to determine, within each element, the strains and the stresses and to display these results in a convenient and informative graphical form.

I have, thus far, skimmed over what could be argued as the most important aspect (and to the beginner the most difficult and confusing) of finite element 'detail' – the evaluation of the integrals in (9.77a,b,c,d).

Consider one of the double integrals, say A_{KJ} from the matrix $\underline{A}$ of (9.77a). To evaluate this integral we utilise the discretised region and re-express the integral over the complete domain as a sum of integrals taken over each element:

$$A_{KJ} = \sum_{e=1}^{nel} \iint_{[e]} \left[(2\mu + \lambda) \frac{\partial R_K}{\partial x_1} \frac{\partial R_J}{\partial x_1} + \mu \frac{\partial R_K}{\partial x_2} \frac{\partial R_J}{\partial x_2} \right] dR \tag{9.81}$$

in which there are nel elements in total. For a particular element $[e]$, a contribution to A_{KJ} will only be made if K, J are 'attached' to that element (this follows from (9.73): $R_K \equiv 0$ unles K is attached to the element $[e]$). In fact, for an eight-noded element there will be 64 possible non-zero contributions to the matrix $\underline{A}$. For example, if under a standard coordinate transformation (as in (9.68)), element $[e]$ is transformed as shown in Figure 9.11, then the global/local node correspondence is:

$$(K, N, L, J, P, Q, S, T) \to (1, 2, 3, 4, 5, 6, 7, 8) \tag{9.82}$$

then

$$\begin{aligned} &\iint_{[e]} \left[(2\mu + \lambda) \frac{\partial R_K}{\partial x_1} \frac{\partial R_J}{\partial x_1} + \mu \frac{\partial R_K}{\partial x_2} \frac{\partial R_J}{\partial x_2} \right] dR \\ &\qquad \equiv \iint_{[e]} \left[(2\mu + \lambda) \frac{\partial \Phi_{(1)}^{[e]}}{\partial x_1} \frac{\partial \Phi_{(4)}^{[e]}}{\partial x_1} + \mu \frac{\partial \Phi_{(1)}^{[e]}}{\partial x_2} \frac{\partial \Phi_{(4)}^{[e]}}{\partial x_2} \right] dR \end{aligned} \tag{9.83}$$

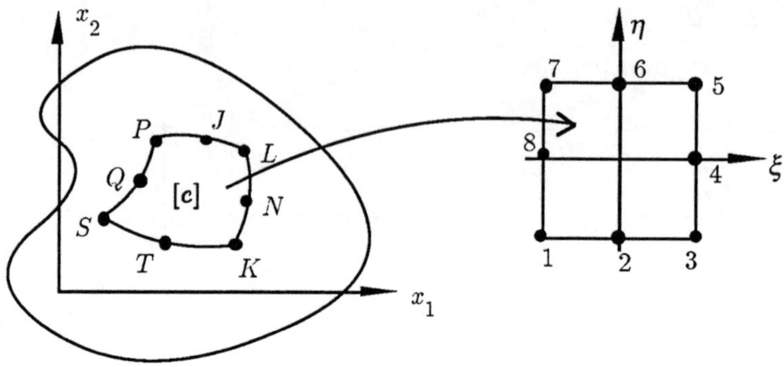

Figure 9.11

In finite element practice it is more convenient to work 'backwards'. Define the 'local matrix' $\underline{a}^{[e]}$ with components $a_{ij}^{[e]}$ $i,j = 1,...8$ through the relations:

$$a_{ij}^{[e]} \equiv \iint_{[e]} \left[(2\mu+\lambda) \frac{\partial \Phi_{(i)}^{[e]}}{\partial x_1} \frac{\partial \Phi_{(j)}^{[e]}}{\partial x_1} + \mu \frac{\partial \Phi_{(i)}^{[e]}}{\partial x_2} \frac{\partial \Phi_{(j)}^{[e]}}{\partial x_2} \right] dR \qquad \textbf{(9.84)}$$

Each member of this 'local' matrix $\underline{a}^{[e]}$ is superimposed into apropriate positions (governed by the node number correspondence (9.82)) in the 'global' matrix $\underline{A}$ as in the matrix theory of structural analysis. For example, $a_{11}^{[e]}$ would be added to A_{KK} whilst $a_{26}^{[e]}$ is added to A_{NQ} and so on.

The superposition procedure is repeated for every element in R.

In the evaluation of $a_{ij}^{[e]}$, of course, due attention must be taken of the transformation equations in (9.68) so that, for example,

$$\frac{\partial \Phi_{(i)}^{[e]}}{\partial x_1} = \frac{\partial \Phi_{(i)}^{[e]}}{\partial \xi} \frac{\partial \xi}{\partial x_1} + \frac{\partial \Phi_{(i)}^{[e]}}{\partial \eta} \frac{\partial \eta}{\partial x_1} \qquad \textbf{(9.85)}$$

The partial derivatives, $\dfrac{\partial \xi}{\partial x_1}, \dfrac{\partial \eta}{\partial x_1}$ are obtained by solving the relations (obtained directly from (9.68)):

$$1 = \sum_{k=1}^{8} x_{1(k)}^{[e]} \left[\frac{\partial \Phi_{(k)}^{[e]}}{\partial \xi} \frac{\partial \xi}{\partial x_1} + \frac{\partial \Phi_{(k)}^{[e]}}{\partial \eta} \frac{\partial \eta}{\partial x_1} \right] \qquad \textbf{(9.86a)}$$

$$0 = \sum_{k=1}^{8} x_{2(k)}^{[e]} \left[\frac{\partial \Phi_{(k)}^{[e]}}{\partial \xi} \frac{\partial \xi}{\partial x_1} + \frac{\partial \Phi_{(k)}^{[e]}}{\partial \eta} \frac{\partial \eta}{\partial x_1} \right] \qquad \textbf{(9.86b)}$$

A similar treatment is used to determine $\dfrac{\partial \Phi_{(i)}^{[e]}}{\partial x_2}$. Then, following the implementation of the coordinate transformation (9.68) we find:

$$a_{ij}^{[e]} = \int_{-1}^{1}\int_{-1}^{1}\left[(2\mu+\lambda)\frac{\partial\Phi_{(i)}^{[e]}}{\partial x_1}\frac{\partial\Phi_{(j)}^{[e]}}{\partial x_1} + \mu\frac{\partial\Phi_{(i)}^{[e]}}{\partial x_2}\frac{\partial\Phi_{(j)}^{[e]}}{\partial x_2}\right]|J|d\xi d\eta \qquad \textbf{(9.87)}$$

in which J is the Jacobian of the transformation:

$$J = \begin{vmatrix} \dfrac{\partial x_1}{\partial \xi} & \dfrac{\partial x_2}{\partial \xi} \\ \dfrac{\partial x_1}{\partial \eta} & \dfrac{\partial x_2}{\partial \eta} \end{vmatrix} \qquad \textbf{(9.88)}$$

(each member of this determinant can be obtained from (9.68)).

Generally, the integrals in (9.87) will need to be evaluated numerically. Since the integrations are, always, (due to the coordinate transformation) over the standard region $-1 \leq \xi, \eta \leq 1$ this presents no problem. Working (invariably on a computer) through each element in this way leads to a methodical construction of the matrix $\underline{A}$. The other matrices $\underline{C}, \underline{D}$ and $\underline{F}$ are treated in a similar manner.

The more detailed examination of the finite element method is beyond the scope of this text. The interested reader should consult more specialised texts though there are few which the author could recommend as being 'easy to read'. The authors own text (*The Finite Element Method – Principles and Applications* Addison-Wesley 1991) is slanted towards the soution of boundary value problems though the level of mathematical knowledge required for its understanding is about the same as is required for this text.

A typical application (omitting the detail) is the problem of a plate, with an unstressed circular hole, subject to uniaxial tension which we discussed earlier in this chapter. In a numerical treatment we will need to exchange the word 'infinite' for the phrase 'very large'. Due to the inherent symmetry of R and of the load only a quarter of the plate need be analysed. Obviously along the 'cut-edges' appropriate boundary conditions, consistent with the original configuration, need to be imposed. These can be deduced using symmetry arguments. The diameter of the hole is taken to be one tenth of the dimension of the square.

Using 64 eight-noded elements graded in size (small elements near the hole, increasing in size as we proceed to 'infinity') we obtain a numerical solution to the problem. From this solution for the displacement components the stresses and strains can then be determined. If we plot the isochromatic curves

$$(\sigma_{11} - \sigma_{22})^2 + 4\sigma_{12}^2 = \text{constant}$$

we obtain the contour plot shown in Figure 9.7(c). In this plot we have only displayed a selection of the elements near the hole. The gross features of the curves are in excellent agreement with the exact solution to this problem.

The exact solution proved to be relatively easy to obtain due to the high degree of symmetry, not only in the region R, but also in the loading conditions. If no such symmetry is present an exact solution would be very difficult, if not impossible, to obtain. In such cases a finite element analysis can prove to be an excellent alternative. In most practical situations, it and the photoelastic technique are the only realistic ways of analysing two dimensional elastostatic problems which provide an acceptable degree of accuracy.

Appendix 1

THE VARIATIONAL CALCULUS

A1.1 The Fundamental Lemma

The variational calculus is essentially concerned with the optimisation of objects called functionals (which are introduced below) and is based, to a large extent, on the application of a single result from mathematical analysis known as the fundamental lemma. This lemma is loosely based on the following simple result: 'If k and f are two numbers and it is demanded that the product kf should vanish for **all** values of k, that is,

$$kf = 0 \qquad \text{all } k$$

then we can conclude that $f \equiv 0$'.

All we need do to verify this statement is to choose (say) $k = 1$ and then $1.f = 0$ implies $f = 0$.

This rather trivial result may be generalized firstly to sums of pair-products of numbers and then to integrals of pair-products of functions.

If k_i and f_i, $i = 1, 2, \ldots, n$ are numbers and it is demanded that

$$\sum_{i=1}^{n} k_i f_i = 0 \qquad \text{all } k_i,\ i = 1, 2, \ldots, n$$

then it is easy to deduce that this statement can only be valid if $f_i \equiv 0$, $i = 1, 2, \ldots, n$. To show this we choose firstly

$$k_1 = 1, \quad k_j = 0 \qquad j = 2, 3, \ldots, n$$

to give

$$1.f_1 + 0.f_2 + 0.f_3 + \ldots + 0.f_n = 0$$

that is

$$f_1 = 0$$

Then by choosing $k_1 = 0,\ \ k_2 = 1,\ \ k_j = 0,\ \ j = 3, 4, \ldots, n$ we obtain $f_2 = 0$ and so on.

The generalization of these basic results to integrals of pairs of functions requires considerably more care. As the result we obtain is fundamental to variational methods we shall deduce the result, known as the **fundamental lemma of the variational calculus**, rigorously.

Lemma

Let $f(x)$ be a continuous function in the interval $a \leq x \leq b$. Suppose $k(x)$ is a continuously differentiable function, i.e. a function with a continuous derivative. If it is demanded that

$$\int_a^b f(x)k(x)\,dx = 0 \tag{A1.1a}$$

for **every** such function $k(x)$ then it follows that

$$f(x) \equiv 0 \qquad a \leq x \leq b \tag{A1.1b}$$

Proof

The proof is by contradiction. We suppose that (A1.1b) does not hold.

If $f(x)$ is not identically zero inside $a \leq x \leq b$ then there must be a point c, satisfying $a \leq c \leq b$, such that $f(c) \neq 0$. To be definite we suppose $f(c) > 0$. Now, since $f(x)$ is continuous there must be an interval (no matter how small) throughout which $f(x) > 0$. Let this interval be $a' < c < b'$

We now choose a particular function $k(x)$ satisfying the above conditions:

$$k(x) = \begin{cases} 0 & a \leq x \leq a' \\ (x-a')^2(x-b')^2 & a' \leq x \leq b' \\ 0 & b' \leq x \leq b \end{cases}$$

Here $k(x)$ is continuously differentiable throughout $a \leq x \leq b$.

Now for this particular $k(x)$, the integral (A1.1a) becomes

$$\int_{a'}^{b'} f(x)(x-a')^2(x-b')^2\,dx$$

However, since $f(x) > 0$ everywhere inside $a' \leq x \leq b'$ the integrand is positive everywhere inside this interval and so the integral must also be positive which contradicts (A1.1a). A similar contradiction can be obtained if it is assumed that $f(c) < 0$. The result (A1.1b) is therefore proved.

A1.2 Functionals and the Variational Calculus

As is well-known, a **function** is essentially a rule which transforms a number into another number. For example the function $\sin x$ transforms any number x into a number y, where $y = \sin x$. The set of values of y (here $-1 \leq y \leq +1$) is called the range of the function. In many applications it is important to know the maximum and minimum values of y and the values of x at which they occur. To determine local maxima and minima the techniques of differential calculus are often used and the equation

$$\frac{dy}{dx} = 0$$

is solved to determine possible extreme values of y.

A **functional** is a generalization of a function in that it is a rule which transforms a function into a number. Perhaps the most common example is a definite integral:

$$J[\phi] = \int_a^b \phi(x)\, dx$$

where the limits a and b are given. (The notation $J[\phi]$ is common for a functional.) Clearly $\phi(x)$ is any function which we can integrate between a and b.

If, for example, $a = 0$ and $b = 1$ so that the functional is

$$J[\phi] = \int_0^1 \phi(x)\, dx$$

then $\phi(x) = \sin x$ gives $J[\phi] = 1 - \cos 1 = 0.4597$ whilst choosing $\phi(x) = x^2$ gives $J[\phi] = 1/3$ and so on. The transformation from a function to a number is clear. The question of particular interest to us is which particular function $\phi(x)$ (for which $J[\phi]$ is defined) gives the minimum (or maximum) value of $J[\phi]$.

An important type of functional that we need to study is

$$J[\phi] = \int_a^b F(x, \phi, \phi')\, dx \tag{A1.2}$$

where F is a **given** function of its arguments $x, \phi(x)$ and $\phi'(x) \equiv d\phi/dx$. The precise set of functions $\phi(x)$ for which $J[\phi]$ in (A1.2) is defined will depend on the given function F but will normally include only functions which are differentiable between a and b. We shall refer to such functions as **admissible**.

A number of functionals of the form (A1.2) can be quoted:

(i) $\quad J[\phi] = \displaystyle\int_a^b \sqrt{1 + (\phi')^2}\, dx$

which is the length of the curve $y = \phi(x)$ between $x = a$ and $x = b$

(ii) $\quad J[\phi] = 2\pi \displaystyle\int_a^b \phi\sqrt{1 + (\phi')^2}\, dx$

The interpretation of $J[\phi]$ here is that it is the area, between the planes $x = a$ and $x = b$, of the surface obtained when the curve $y = \phi(x)$ is rotated about the x-axis.

(iii) $\quad J[\phi] = \dfrac{1}{\sqrt{2g}} \displaystyle\int_a^b \frac{\sqrt{1 + (\phi')^2}}{\sqrt{\phi}}\, dx$

The interpretation of $J[\phi]$ in this case comes from a problem in elementary mechanics: given a frictionless wire whose equation is $y = \phi(x)$, calculate the time taken for a particle to slide between the points $(x = a, y = A)$ and $(x = b, y = B)$. This time clearly depends on the particular path $y = \phi(x)$ involved and the basic problem is to determine which path minimizes the sliding time. (A child's slide made using this curve would give the fastest ride.) This problem was one of the first studied in the subject of variational calculus which is concerned with minimizing (or maximizing) functionals.

The reason for studying this topic in a text on solid mechanics is that, as we have seen, particularly in Chapter 6, the application of the theorem of minimum potential energy naturally gives rise to problems in functional optimisation. Indeed in Chapter 6 we claimed that optimising the potential energy functional in (6.38) was equivalent to finding the solution to a boundary value problem. We now examine problems of this type in more detail.

Example A1.1

Show that minimizing the functional

$$J[\phi] = \int_0^1 (\phi^2 + (\phi')^2)\, dx$$

with $\phi(0) = u_0$ and $\phi(1) = u_1$ is equivalent to solving a boundary value problem.

Solution

The problem is to find the specific function say $u(x)$ (from the set of admissible functions for which $J[\phi]$ is defined) which minimizes the value of J. That is, u must be such that

$$J[u] < J[\phi], \quad \text{all} \quad \phi \neq u$$

We proceed by writing

$$\phi(x) = u(x) + \varepsilon\, \eta\,(x) \tag{A1.3}$$

where ε is a parameter independent of x. The function $\eta(x)$ is an almost arbitrary function but it must satisfy

$$\eta(0) = 0 \quad \text{and} \quad \eta(1) = 0 \tag{A1.4}$$

since $\phi(x) = u(x)$ (for all ϕ) at $x = 0$ and $x = 1$. Geometrically, we are considering a range of curves $\phi(x)$ in the 'close neighbourhood' of the function we seek. Each 'comparison curve' must, of course, pass through the points $(0, u_0); (1, u_1)$ as this is a constraint imposed on the functional (see Figure A1.1).

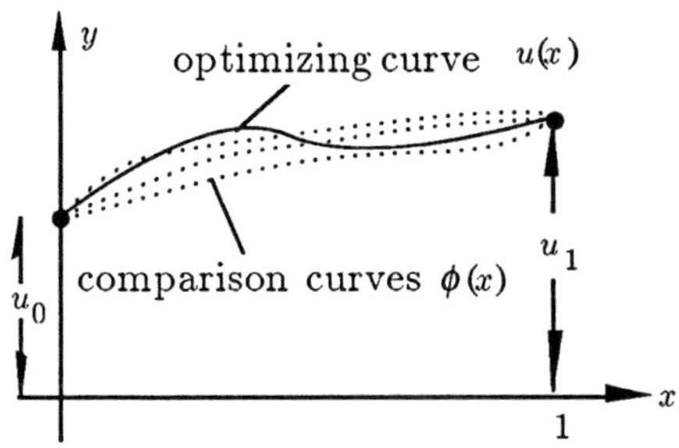

Figure A1.1

We substitute the assumed form (A1.3) into the given functional producing

$$J[u + \varepsilon\,\eta] = \int_0^1 \{(u + \varepsilon\eta)^2 + (u' + \varepsilon\eta')^2\}\, dx \qquad \textbf{(A1.5)}$$

which, when we carry out the integration, is a function only of the parameter ε and will be denoted by $\Phi(\varepsilon)$.

We find

$$\Phi(\varepsilon) = \int_0^1 \{u^2 + (u')^2 + (\varepsilon\eta)^2 + (\varepsilon\eta')^2 + 2\varepsilon(u\eta + u'\eta')\}\ dx$$

$$= J[u] + \varepsilon^2 J[\eta] + 2\varepsilon \int_0^1 (u\eta + u'\eta')\ dx$$

so that

$$\frac{d\Phi}{d\varepsilon} = 2\varepsilon J[\eta] + 2\int_0^1 (u\eta + u'\eta')\ dx$$

If this is to be a minimum when $\varepsilon = 0$ we have from elementary calculus

$$\int_0^1 (u\eta + u'\eta')\ dx = 0$$

Integrating the second term by parts and using (A1.4) we obtain

$$\int_0^1 (u\eta - u''\eta) dx = 0$$

or

$$\int_0^1 \eta(u - u'') dx = 0$$

from which using the fundamental lemma of the variational calculus proved in A1.1 we deduce:

$$u'' - u = 0 \qquad 0 \le x \le 1$$

This differential equation, together with the given conditions $u(0) = u_0$ and $u(1) = u_1$, constitutes a standard boundary value problem which is readily solved.
We now do another, but more difficult, example showing the equivalence of minimizing a functional to solving a boundary value problem.

Example A1.2

Deduce the boundary value problem equivalent to minimizing the functional

$$J[\phi] = \int_1^3 \{-(\phi')^2 - 4\phi\}dx + 12\,\phi(3)$$

where

$$\phi(1) = 1$$

Solution

Clearly the admissible functions $\phi(x)$ must be differentiable and must each satisfy $\phi(1) = 1$.

Proceeding in a similar manner to the previous example, the function $u(x)$ that we seek is obtained by writing any other admissible function $\phi(x)$ as

$$\phi(x) = u(x) + \varepsilon\,\eta\,(x)$$

where $\eta(1) = 0$ but otherwise $\eta(x)$ is arbitrary.

Then

$$\begin{aligned}\Phi(\varepsilon) &= J[u + \varepsilon\,\eta] \\ &= \int_1^3 \{-(u' + \varepsilon\,\eta')^2 - 4(u + \varepsilon\,\eta)\}\,dx + 12(u(3) + \varepsilon\,\eta(3)) \\ &= \int_1^3 \{-(u')^2 - 4u - 2\varepsilon u'\eta' - (\varepsilon\eta')^2 - 4\varepsilon\eta\}\ dx \\ &\qquad\qquad + 12(u(3) + \varepsilon\eta(3)) \\ &= J[u] - \varepsilon\left\{\int_1^3 \{2u'\eta' + 4\eta\}\ dx - 12\eta(3)\right\} - \varepsilon^2 \int_1^3 (\eta')^2\ dx\end{aligned}$$

Hence

$$\frac{d\Phi}{d\varepsilon} = -\left\{\int_1^3 \{2u'\eta' + 4\eta\}\ dx - 12\eta(3)\right\} - 2\varepsilon \int_1^3 (\eta')^2\ dx$$

so to obtain $\dfrac{d\Phi}{d\varepsilon} = 0$ when $\varepsilon = 0$ we must have

$$\int_1^3 \{-2u'\eta' - 4\eta\}dx + 12\,\eta\,(3) = 0 \qquad \textbf{(A1.6)}$$

The first term in (A1.6) can be transformed using integration by parts:

$$\begin{aligned}\int_1^3 u'\eta' dx &= [u'\eta]_1^3 - \int_1^3 u''\eta\,dx \\ &= u'(3)\eta(3) - \int_1^3 u''\eta\,dx\end{aligned}$$

Hence (A1.6) can be written

$$\int_1^3 (2u'' - 4)\eta\, dx + (12 - 2u'(3))\,\eta\,(3) = 0 \tag{A1.7}$$

We now seek to extract $u(x)$ from (A1.7). It is vital to realize that this is not a single equation but must be satisfied for **all** functions $\eta(x)$ such that $\eta(1) = 0$ (and there are infinitely many such functions). Suppose that from the set of all such functions we choose a subset satisfying both $\eta(1) = 0$ **and** $\eta(3) = 0$ (see Figure A1.2).

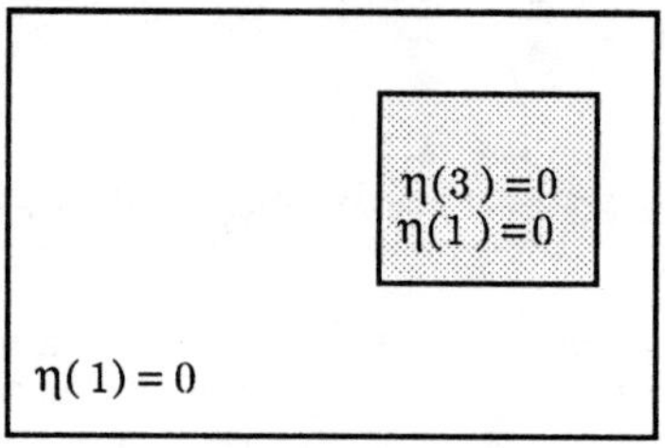

Figure A1.2

For this 'smaller' set of functions, which still contains infinitely many members, (A1.7) becomes

$$\int_1^3 (2u'' - 4)\eta\, dx = 0$$

Since $\eta(x)$ here is arbitrary apart from satisfying $\eta(1) = 0$ and $\eta(3) = 0$ and we can conclude, again from the fundamental lemma of the variational calculus that

$$2u'' - 4 = 0 \qquad 1 \le x \le 3 \tag{A1.8}$$

We now revert to (A1.7) and to functions for which only $\eta(1) = 0$ is demanded. Using (A1.8) this becomes

$$(12 - 2u'(3))\,\eta\,(3) = 0$$

This equation must hold for **all** functions $\eta(x)$ in the set of functions satisfying $\eta(1) = 0$ and not just for those in the smaller set satisfying $\eta(3) = 0$. Consequently

$$u'(3) = 6 \tag{A1.9}$$

In other words we have shown that the particular function $u(x)$ (taken from the general class of functions satisfying $\phi(1) = 1$) which minimizes the given functional is the solution to the boundary value problem

$$u'' = 2 \qquad 1 \le x \le 3 \tag{A1.10a}$$

with (from the original constraint $\phi(1) = 1$)

$$u(1) = 1 \tag{A1.10b}$$

and

$$u'(3) = 6 \tag{A1.10c}$$

Clearly the exact solution to this problem is

$$u(x) = x^2$$

which yields a minimum value for the given functional of

$$J[u] = \int_1^3 (-4x^2 - 4x^2)dx + (12)(9)$$

$$= \frac{116}{3}$$

The reader should experiment with a number of other admissible functions to show that a value smaller than 116/3 is never obtained.

A1.3 Construction of functionals

In Examples A1.1 and A1.2 we derived the boundary value problems equivalent to two particular variational problems. A number of techniques are available to enable us to do this for more general problems.

Perhaps the most straightforward case is when

$$J[\phi] = \int_a^b F(x, \phi, \phi')dx \tag{A1.11}$$

and ϕ is the set of functions each satisfying $\phi(a) = A$ and $\phi(b) = B$ where A and B are given constants.

For this case, the particular function $u(x)$ which makes $J(\phi)$ an extremum (maximum or minimum) is a solution of the **Euler-Lagrange equation**

$$\frac{\partial F}{\partial u} - \frac{d}{dx}\left(\frac{\partial F}{\partial u'}\right) = 0 \qquad a \leq x \leq b \tag{A1.12}$$

with

$$u(a) = A \quad \text{and} \quad u(b) = B \tag{A1.13}$$

The proof of this result may be found in any book on variational calculus. For our purposes it is sufficient to note that (A1.12) and (A1.13) constitute the boundary value problem equivalent to the given variational problem. We note that (A1.12) is a **second**-order differential equation. The reader is invited to re-work Example A1.1 using (A1.12) and (A1.13); a straightforward exercise.

The Euler-Lagrange equation can be generalized beyond (A1.12). If the integrand in (A1.11) is $F(x, \phi, \phi', \phi'', \phi''', \ldots)$ then the corresponding Euler-Lagrange equation can be shown to be

$$\frac{\partial F}{\partial u} - \frac{d}{dx}\left(\frac{\partial F}{\partial u'}\right) + \frac{d^2}{dx^2}\left(\frac{\partial F}{\partial u''}\right) - \ \ldots \ = 0 \quad a \leq x \leq b \tag{A1.14}$$

In general if the highest derivative of u appearing as an argument in F is of nth order then (A1.14) is an ordinary differential equation of order $2n$. This is typical of a differential equation equivalent to a variational process – such an equation is always of **even** order.

In the context of this text a second order boundary value problem may be interpreted as the model of a beam deformed by longitudinal forces (see Section

8.1). In such a model boundary conditions which contain a derivative term apply directly to the imposition of forces at a boundary (see, for example (8.3) or (8.3a)). Of course the absence of a derivative term in a boundary condition implies that the constraint is a displacement boundary condition.

The following theorem displays the precise relation between a second order boundary value problem and its corresponding variational problem.

Theorem A1.1

The second-order boundary value problem:

$$-\frac{d}{dx}\left(p(x)\frac{du}{dx}\right) + q(x)\, u = r(x) \qquad a \le x \le b \tag{A1.14a}$$

$$\begin{aligned} \alpha_0\, u(a) + \alpha_1\, u'(a) &= \gamma_1 \\ \beta_0\, u(b) + \beta_1\, u'(b) &= \gamma_2 \end{aligned} \tag{A1.15b}$$

may be written as the necessary conditions to be satisfied by the solution of the variational problem:

$$J[\phi] = \text{extremum}$$

for each of the following cases.

- case 1 $\alpha_1 \neq 0,\ \beta_1 \neq 0$

that is, with derivative (or force) boundary conditions at $x = a$ and $x = b$. In this case the corresponding functional is

$$J[\phi] = \int_a^b \{p(\phi')^2 + q\phi^2 - 2r\phi\}\, dx + B_a + B_b \tag{A1.16a}$$

where

$$\begin{aligned} B_a &= \frac{p(a)}{\alpha_1}\left(-\alpha_0\phi^2(a) + 2\,\gamma_1\,\phi(a)\right) \\ B_b &= \frac{p(b)}{\beta_1}\left(\beta_0\phi^2(b) - 2\,\gamma_2\,\phi(b)\right) \end{aligned} \tag{A1.16b}$$

and $\phi(x)$ need not satisfy either boundary condition.

- case 2 $\alpha_1 = 0,\ \beta_1 \neq 0$

that is, with a displacement boundary condition at $x = a$ and a force boundary condition at $x = b$. Here $J[\phi]$ is identical to (A1.16a) except that B_a is omitted and the functions $\phi(x)$ **must** satisfy the displacement boundary condition

$$\alpha_0\, \phi(a) = \gamma_1$$

at $x = a$.

- case 3 $\alpha_1 \neq 0,\ \beta_1 = 0$

There is a force boundary condition at $x = a$ and an displacement boundary condition at $x = b$. Here $J[\phi]$ is identical to (A1.16a) except that B_b is omitted and the functions $\phi(x)$ **must** satisfy the displacement boundary condition

$$\beta_0\, \phi(b) = \gamma_2$$

at $x = b$.

- case 4 $\alpha_1 = 0,\ \beta_1 = 0$

Here both boundary conditions are displacement type. In this case $J[\phi]$ is identical to (A1.16a) except that both B_a and B_b are omitted and the functions $\phi(x)$ **must** satisfy both the displacement boundary conditions

$$\alpha_0\, \phi(a) = \gamma_1 \qquad \beta_0\, \phi(b) = \gamma_2$$

It is easy to confirm, using the above theorem and the identifications:

$$p \to \frac{1}{2}EA \quad q \to 0 \quad r \to 0 \quad \phi \to u_3 \quad a \to 0 \quad b \to L$$

$$\alpha_0 \to 0 \quad \beta_0 \to 0 \quad \alpha_1 \to -EA \quad \beta_1 \to EA \quad \gamma_1 \to P_0 \quad \gamma_2 \to P_2$$

that the boundary value problem resulting from the optimisation of the potential energy functional (up to an irrelevant constant factor) of (8.1) is that given in (8.2) and (8.3).

It is perhaps not immediately obvious that all second-order, linear ordinary differential equations can be expressed in the form (A1.15a) which is called the **Sturm-Liouville** form.

To show that this is the case, we consider any second-order linear ordinary differential equation of the form

$$a_2(x)u'' + a_1(x)u' + a_0(x)u = c(x) \tag{A1.17}$$

We note firstly that the form of (A1.15a) is arbitrary up to a multiplicative function, say $f(x)$. Thus we can write:

$$-f(x)\, (p(x)u')' + f(x)\, q(x)u = f(x)\, r(x)$$

or, on expanding,

$$-f(x)\, p(x)u'' - f(x)\, p'(x)\, u' + f(x)\, q(x)\, u = f(x)r(x) \tag{A1.18}$$

For (A1.18) to be equivalent to (A1.17) we must have

$$a_2(x) = -f(x)\,p(x)$$
$$a_1(x) = -f(x)\,p'(x)$$
$$a_0(x) = f(x)\,q(x)$$
$$c(x) = f(x)\,r(x)$$

from which we deduce that

$$\frac{p'(x)}{p(x)} = \frac{a_1(x)}{a_2(x)}$$

Hence

$$\ell n\; p(x) = \int \frac{a_1(x)}{a_2(x)}\,dx$$

or

$$p(x) = \exp\left\{\int \frac{a_1(x)}{a_2(x)}\,dx\right\} \qquad \textbf{(A1.19)}$$

giving

$$f(x) = -\frac{a_2(x)}{p(x)} \quad q(x) = \frac{a_0(x)}{f(x)} \quad \text{and} \quad r(x) = \frac{c(x)}{f(x)} \qquad \textbf{(A1.20)}$$

Example A1.3

Determine the variational problem corresponding to the boundary value problem:

$$x^2\,u'' + x\,u' + (x^2-1)u = 0 \qquad 0 \le x \le 1$$
$$u(0) = 1 \qquad u'(1) = 6$$

Solution

We first re-write the given differential equation in Sturm-Liouville form. The coefficients of u and its derivatives are, by comparison with (A1.17),

$$a_2(x) = x^2 \qquad a_1(x) = x \qquad a_0(x) = x^2 - 1$$

Hence from (A1.19) and (A1.20) we find

$$p(x) = \exp\left\{\int \frac{1}{x}\,dx\right\} = \exp(\ell n\; x) = x$$

$$f(x) = -\frac{x^2}{x} = -x \qquad q(x) = -\frac{(x^2-1)}{x} \qquad r(x) = 0$$

Thus the given differential equation may be expressed in the form (A1.15a) as

$$-\frac{d}{dx}\left(x\,\frac{du}{dx}\right)-\left(x-\frac{1}{x}\right)u=0$$

which is the form required for the application of Theorem A1.1. We find using (A1.16) (case 2)

$$J[\phi]=\int_0^1\left\{x(\phi')^2-\left(x-\frac{1}{x}\right)\phi^2\right\}dx-12\,\phi\,(1)$$

and the functions $\phi(x)$ must all satisfy $\phi(0)=1$.

The reader is invited to work Examples A1.1 and A1.2 in reverse i.e. to use Theorem A1.1 (case 4 and case 2 respectively) to re-obtain $J[\phi]$ in each case from the derived boundary value problem.

A1.4 One Dimensional Fourth-order Problems

One important example of a fourth-order boundary value problem arises in the theory of the transverse deformation of beams.

Consider a beam of length ℓ as shown in Figure A1.3. As indicated in the figure the cross section of the beam may be variable and the x-axis is chosen to lie along the line of centroids of each cross section. The beam is supported at its extremities $x=0$ and $x=\ell$ so as to prevent it moving as a rigid body. These supports can take many forms, simple and clamped supports being the simplest. The original configuration of the beam is altered by the action of transverse deforming forces which may be point forces placed at various positions along the length and/or distributed forces acting along a portion of the surface.

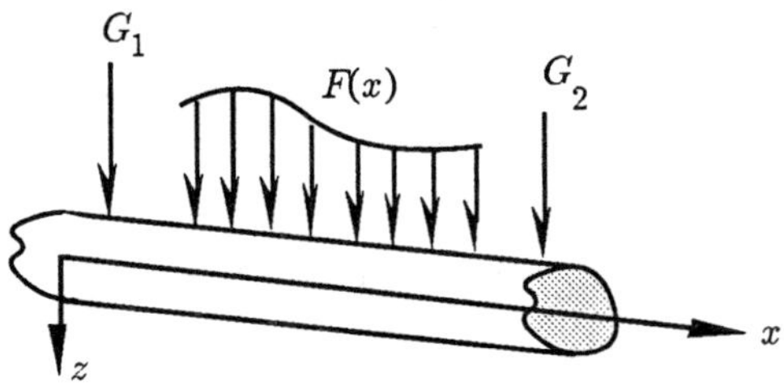

Figure A1.3

For long slender beams the Euler–Bernoulli theory of beam bending in which attention is focused on particles along the centre-line of the beam (the x-axis) is generally regarded as an adequate model (see Section 6.5). The theory predicts that the curve $z=u(x)$ assumed by the centre-line as a result of the deforming forces is a solution of the **fourth-order differential equation**:

$$\frac{d^2}{dx^2}\left(EI(x)\frac{d^2u}{dx^2}\right)=F(x) \qquad 0\le x\le\ell \tag{A1.21}$$

in which E is the Young's Modulus characterising the material from which the beam is formed, and $I(x)$ is the (possibly variable) second moment of area of the cross section. We have assumed in (A1.21) that the deforming forces $F(x)$ are distributed. (Point forces can be regarded as a special case of distributed forces.)

Equation (A1.21) is a fourth-order linear equation and consequently four boundary conditions, two at each end of the beam, must be specified before a unique solution can be obtained. For example, if the beam is clamped rigidly at both ends as in Figure A1.4(a) the boundary conditions are

$$u(0)=0, \quad \frac{du}{dx}(0)=0, \quad u(\ell)=0, \quad \frac{du}{dx}(\ell)=0 \qquad \textbf{(A1.22)}$$

On the other hand if the beam is clamped at $x=0$ and unconstrained (free) at $x=\ell$ as in Figure A1.4(b) then the boundary conditions are

$$u(0)=0, \ \frac{du}{dx}(0)=0, \quad \frac{d^2u}{dx^2}(\ell)=0, \quad \frac{d}{dx}\left(EI\frac{d^2u}{dx^2}\right)(\ell)=0 \qquad \textbf{(A1.23)}$$

The boundary conditions at $x=\ell$ imply the vanishing at this point of the bending moment $M(\equiv EI\,d^2u/dx^2)$ and of the shear force $S\,(\equiv -d/dx(EI\,d^2u/dx^2))$ respectively.

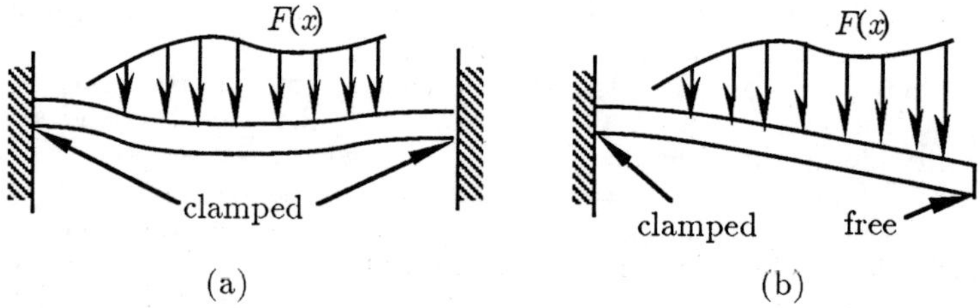

Figure A1.4

Another common method of beam support involves the use of hinges, a hinge being called a simple support. In Figure A1.5 there is a simple support at the left-hand end, whilst a clamp is placed at the right-hand end.

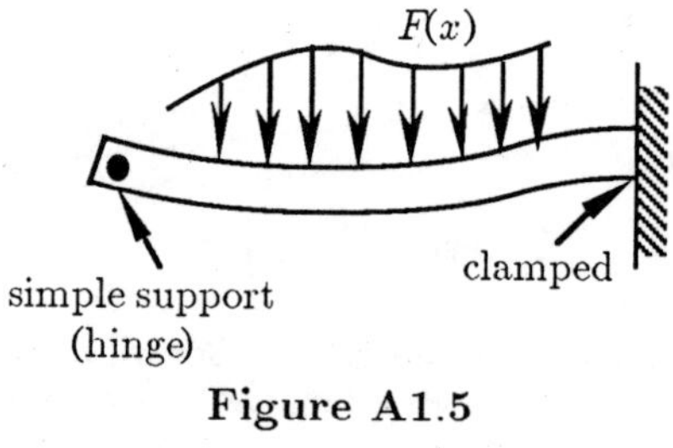

Figure A1.5

The boundary conditions in this case are

$$u(0)=0, \quad \frac{d^2u}{dx^2}(0)=0, \quad u(\ell)=0, \quad \frac{du}{dx}(\ell)=0 \qquad \textbf{(A1.24)}$$

A hinge cannot support or exert any bending moment. Hence the bending moment must vanish, which leads to the second boundary condition in (A1.24).

The differential equation (A1.21) together with four boundary conditions (such as (A1.22), (A1.23) or (A1.24)) form a **one dimensional fourth-order boundary value problem.** We see, as with second order boundary value problems, that the boundary conditions naturally split into two distinct types:– those involving second order or higher derivatives, which apply to the specification of forces (bending moments or shear forces) and those not involving second order or higher derivatives which apply to the specification of displacements (either linear or rotational) in the system.

The following example illustrates the method of solution.

Example A1.4

Determine the deflection of a uniform cantilever beam (clamped at its left hand end) deformed by a tip force P as shown in Figure A1.6.

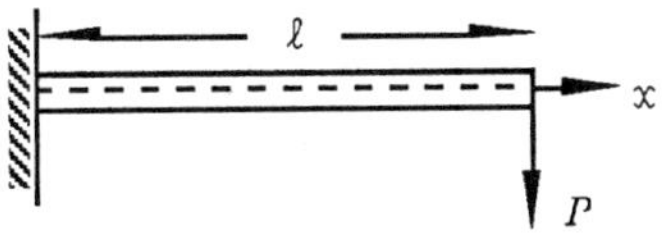

Figure A1.6

Solution

A point force P can be modelled by a Dirac delta-function. However, since in this case P is applied at the end-point $x = \ell$, it is easier to absorb it into the boundary conditions. Thus the boundary value problem describing this physical situation is

$$\frac{d^2}{dx^2}\left(EI\frac{d^2u}{dx^2}\right) = 0 \quad 0 \le x \le \ell \qquad \textbf{(A1.25a)}$$

$$u(0) = 0, \quad \frac{du}{dx}(0) = 0 \qquad \textbf{(A1.25b)}$$

$$\frac{d^2u}{dx^2}(\ell) = 0, \quad -\frac{d}{dx}\left(EI\frac{d^2u}{dx^2}\right)(\ell) = P \qquad \textbf{(A1.25c)}$$

The solution is easily obtained by successively integrating (A1.25a):

$$\frac{d}{dx}\left(EI\frac{d^2u}{dx^2}\right) = C_1 \qquad 0 \le x \le \ell$$

where $C_1 = -P$ due to the second equation of (A1.25c).

Integrating again

$$EI\frac{d^2u}{dx^2} = -Px + C_2$$

where $C_2 = P\ell$ due to the first equation of (A1.25c).

Continuing, we find (since I is constant for a uniform beam)

$$EI\frac{du}{dx} = -P\frac{x^2}{2} + P\ell x + C_3$$

where $C_3 = 0$ due to the second equation of (A1.25b).

Finally

$$EI\,u = -P\frac{x^3}{6} + P\ell\frac{x^2}{2} + C_4 \qquad \textbf{(A1.26)}$$

But $C_4 = 0$ due to the first equation of (A1.26) so that

$$u(x) = \frac{P}{EI}\left(-\frac{x^3}{6} + \frac{x^2}{2}\ell\right) \qquad \textbf{(A1.27)}$$

In realistic problems the deflection (A1.27) would be hardly noticeable (unless P is very large) because of the large value of Young's modulus E for most structural materials.

More complicated fourth-order boundary value problems than (A1.25) can easily be constructed. For example a beam elastically supported along its entire length can be modelled by the differential equation

$$\frac{d^2}{dx^2}\left(E_b\,I(x)\frac{d^2u}{dx^2}\right) + E_s\,u = F(x) \qquad \textbf{(A1.28)}$$

where E_b, E_s are the Young's modulus of the beam and support materials respectively.

Both (A1.21) and (A1.28) are special cases of the general linear fourth-order equation in Sturm-Liouville form:

$$\frac{d^2}{dx^2}\left(p(x)\frac{d^2u}{dx^2}\right) - \frac{d}{dx}\left(q(x)\frac{du}{dx}\right) + r(x)u = s(x) \quad a \le x \le b \qquad \textbf{(A1.29)}$$

where $p(x)$, $q(x)$, $r(x)$ and $s(x)$ are given functions. Many (but not all) linear fourth-order differential equations can be written in this form. This is because many fourth order differential equations in problems of physical interest can be derived using a variational principle from which the Sturm-Liouville form naturally arises. This is particularly true in linear elasticity theory which is the source of many fourth-order boundary value problems both in one and two dimensions.

A brief inspection of (A1.29), and indeed of the special cases (A1.21) and (A1.28), suggests that finding exact solutions to fourth-order boundary value problems is not possible unless, as in Example 10.1, the coefficient functions $p(x), q(x), r(x)$ and $s(x)$ have very simple forms. Thus, as in the treatment of second-order boundary value problems, we would generally need to consider numerical approximations many of which are based on an appropriate variational principle (essentially the potential energy).

A1.5 Variational formulation of Fourth-order Problems

In this section we again consider the special case $q(x) \equiv 0$ of (A1.29). In a similar manner to Theorem A1.1 it is possible to recast linear fourth-order boundary value problems in variational form. The prescription for the procedure is as follows.

Theorem A1.2

The fourth-order linear differential equation:

$$\frac{d^2}{dx^2}\left(p(x)\frac{d^2u}{dx^2}\right) + r(x)u = s(x) \quad a \le x \le b \tag{A1.30}$$

can be written as the necessary conditions to be satisfied by the solution of the variational problem

$$J[\phi] = \text{extremum}$$

where

$$J[\phi] = \int_a^b \left\{ p(x)\left(\frac{d^2\phi}{dx^2}\right)^2 + r(x)\phi^2 - 2\, s(x)\phi \right\} dx + B_a + B_b \tag{A1.31}$$

The following cases should be noted:

- Case 1. If all the boundary conditions are displacement-type, that is

$$u(a) = \alpha, \quad \frac{du}{dx}(a) = \beta, \quad u(b) = \gamma, \quad \frac{du}{dx}(b) = \delta$$

where $\alpha, \beta, \gamma, \delta$ are given constants then we can take $B_a \equiv B_b \equiv 0$ but the admissible functions $\phi(x)$ **must** satisfy all the boundary conditions.

- Case 2. If the boundary conditions at one end, say $x = a$, are force-type, say

$$\frac{d^2u}{dx^2}(a) = k_1, \quad \frac{d^3u}{dx^3}(a) = k_2$$

where k_1, k_2 are given constants then we must take $B_b \equiv 0$ and

$$B_a = -2(k_2\, p(a) + k_1\, p'(a))\, \phi(a) + 2\, k_1\, p(a)\, \phi'(a)$$

Similarly if the force-type conditions apply at $x = b$, say

$$\frac{d^2u}{dx^2}(b) = h_1 \qquad \frac{d^3u}{dx^3}(b) = h_2$$

then we would take $B_a \equiv 0$ and

$$B_b = 2\,(h_2\, p(b) + h_1\, p'(b))\, \phi(b) - 2\, h_1\, p(b)\, \phi'(b)$$

In both these situations $\phi(x)$ need only satisfy the displacement boundary conditions.

- Case 3. If the boundary conditions at both ends are force-type, of the form given in case 2 above, then B_a and B_b must both be taken as in case 2 but in this case the admissible functions $\phi(x)$ need not satisfy any boundary conditions.

Using this theorem it is now easily checked that the boundary value problem given in (6.39) (applying to a beam with uniform cross section) is equivalent, up to an irrelevant constant factor, the variational problem (extremum of the potential energy functional) given in 6.38.

REFERENCES

Bland D.R. *Lecture Notes*
Cranfield Institute of Technology

Sokolnikoff I.S. (1956) *Mathematical Theory of Elasticity.*
McGraw-Hill Publishing Company

Timoshenko S.P. and Goodier J.N. *Theory of Elasticity*
McGraw-Hill Publishing Company.

Pearson C.E. (1959). *Theoretical Elasticity.*
Harvard Monographs in Applied Science.

Landau L.D. and Lifshitz E.M. (1959). *Theory of Elasticity.*
Pergamon Press.

England A.H. (1971) *Complex Variable Methods in Elasticity*
Wiley-Interscience.

Godfrey D.E.R (1949). *Theoretical Elasticity and Plasticity for Engineers* Thames and Hudson

Index

Mechanics

SOLID MECHANICS AND ITS APPLICATIONS

Series Editor: G.M.L. Gladwell

Aims and Scope of the Series

The fundamental questions arising in mechanics are: *Why?*, *How?*, and *How much?* The aim of this series is to provide lucid accounts written by authoritative researchers giving vision and insight in answering these questions on the subject of mechanics as it relates to solids. The scope of the series covers the entire spectrum of solid mechanics. Thus it includes the foundation of mechanics; variational formulations; computational mechanics; statics, kinematics and dynamics of rigid and elastic bodies; vibrations of solids and structures; dynamical systems and chaos; the theories of elasticity, plasticity and viscoelasticity; composite materials; rods, beams, shells and membranes; structural control and stability; soils, rocks and geomechanics; fracture; tribology; experimental mechanics; biomechanics and machine design.

1. R.T. Haftka, Z. Gürdal and M.P. Kamat: *Elements of Structural Optimization*. 2nd rev.ed., 1990 ISBN 0-7923-0608-2
2. J.J. Kalker: *Three-Dimensional Elastic Bodies in Rolling Contact*. 1990 ISBN 0-7923-0712-7
3. P. Karasudhi: *Foundations of Solid Mechanics*. 1991 ISBN 0-7923-0772-0
4. N. Kikuchi: *Computational Methods in Contact Mechanics*. (forthcoming) ISBN 0-7923-0773-9
5. Y.K. Cheung and A.Y.T. Leung: *Finite Element Methods in Dynamics*. 1991 ISBN 0-7923-1313-5
6. J.F. Doyle: *Static and Dynamic Analysis of Structures*. With an Emphasis on Mechanics and Computer Matrix Methods. 1991 ISBN 0-7923-1124-8; Pb 0-7923-1208-2
7. O.O. Ochoa and J.N. Reddy: *Finite Element Modelling of Composite Structures*. (forthcoming) ISBN 0-7923-1125-6
8. M.H. Aliabadi and D.P. Rooke: *Numerical Fracture Mechanics*. ISBN 0-7923-1175-2
9. J. Angeles and C.S. López-Cajún: *Optimization of Cam Mechanisms*. 1991 ISBN 0-7923-1355-0
10. D.E. Grierson, A. Franchi and P. Riva: *Progress in Structural Engineering*. 1991 ISBN 0-7923-1396-8
11. R.T. Haftka and Z. Gürdal: *Elements of Structural Optimization*. 3rd rev. and exp. ed. 1992 ISBN 0-7923-1504-9; Pb 0-7923-1505-7
12. J.R. Barber: *Elasticity*. 1992 ISBN 0-7923-1609-6; Pb 0-7923-1610-X
13. H.S. Tzou and G.L. Anderson (eds.): *Intelligent Structural Systems*. 1992 ISBN 0-7923-1920-6
14. E.E. Gdoutos: *Fracture Mechanics*. An Introduction ISBN 0-7923-1932-X
15. J.P. Ward: *Solid Mechanics*. An Introduction. 1992 ISBN 0-7923-1949-4

Kluwer Academic Publishers – Dordrecht / Boston / London

Mechanics

FLUID MECHANICS AND ITS APPLICATIONS

Series Editor: R. Moreau

Aims and Scope of the Series

The purpose of this series is to focus on subjects in which fluid mechanics plays a fundamental role. As well as the more traditional applications of aeronautics, hydraulics, heat and mass transfer etc., books will be published dealing with topics which are currently in a state of rapid development, such as turbulence, suspensions and multiphase fluids, super and hypersonic flows and numerical modelling techniques. It is a widely held view that it is the interdisciplinary subjects that will receive intense scientific attention, bringing them to the forefront of technological advancement. Fluids have the ability to transport matter and its properties as well as transmit force, therefore fluid mechanics is a subject that is particularly open to cross fertilisation with other sciences and disciplines of engineering. The subject of fluid mechanics will be highly relevant in domains such as chemical, metallurgical, biological and ecological engineering. This series is particularly open to such new multidisciplinary domains.

1. M. Lesieur: *Turbulence in Fluids.* 2nd rev. ed., 1990 ISBN 0-7923-0645-7
2. O. Métais and M. Lesieur (eds.): *Turbulence and Coherent Structures.* 1991 ISBN 0-7923-0646-5
3. R. Moreau: *Magnetohydrodynamics.* 1990 ISBN 0-7923-0937-5
4. E. Coustols (ed.): *Turbulence Control by Passive Means.* 1990 ISBN 0-7923-1020-9
5. A. A. Borissov (ed.): *Dynamic Structure of Detonation in Gaseous and Dispersed Media.* 1991 ISBN 0-7923-1340-2
6. K.-S. Choi (ed.): *Recent Developments in Turbulence Management.* 1991 ISBN 0-7923-1477-8
7. E.P. Evans and B. Coulbeck (eds.): *Pipeline Systems.* 1992 ISBN 0-7923-1668-1
8. B. Nau (ed.): *Fluid Sealing.* 1992 ISBN 0-7923-1669-X
9. T.K.S. Murthy (ed.): *Computational Methods in Hypersonic Aerodynamics.* 1992 ISBN 0-7923-1673-8
10. R. King (ed.): *Fluid Mechanics of Mixing.* Modelling, Operations and Experimental Techniques. 1992 ISBN 0-7923-1720-3
11. Z. Han & X. Yin: *Shock Dynamics.* 1992 ISBN 0-7923-1746-7

Kluwer Academic Publishers – Dordrecht / Boston / London